MINERALS OF MEXICO

MINERALS OF MEXICO

William D. Panczner

VNR VAN NOSTRAND REINHOLD COMPANY
New York

Manufactured in the United States of America.

Published by Van Nostrand Reinhold Company Inc.
115 Fifth Avenue
New York, New York 10003

Van Nostrand Reinhold Company Limited
Molly Millars Lane
Wokingham, Berkshire RG11 2PY, England

Van Nostrand Reinhold
480 La Trobe Street
Melbourne, Victoria 3000, Australia

Macmillan of Canada
Division of Canada Publishing Corporation
164 Commander Boulevard
Agincourt, Ontario MIS 3C7, Canada

15 14 13 12 11 10 9 8 7 6 5 4 3 2 1

The endsheet maps are reprinted from *The Mexican Mining Industry
1890-1950* by Marvin D. Bernstein, by permission of the State University
of New York Press; copyright © 1965 by the State University of New York.

Library of Congress Cataloging-in-Publication Data
Panczner, William D. (William David), 1938-
  Minerals of Mexico.
  Bibliography: p.
  1. Mineralogy—Collectors and collecting—Mexico.
  2. Mines and mineral resources—Mexico. I. Title.
QE377.M6P36 1987   549.972   86-13325
ISBN 0-442-27285-5

This book is dedicated to the people of México
and to the three people who directly affected me and this project,
but unfortunately did not live long enough to see it completed,

my father and mother,
WILLIAM J. and KATHERINE H. PANCZNER,
and my dear friend,
DR. EDUARDO SCHMITTER

CONTENTS

FOREWORD

After many years of geographical and bibliographical journeys, William Panczner has completed a project that many of us would have loved to initiate, but did not undertake because of its magnitude and intrinsic complexity. Not since L. Salazar Salinas, who is credited with authoring *Boletín números 40* and *41* (Instituto Geológico de México, 1922, 1923), has an author been able to provide readers with a comprehensive volume containing information that is both authentic and reliable on Mexican mineralogy, mineral species, and localities.

This volume is the most complete synthesis about Mexican minerals and their occurrences to date. It is richly illustrated with photographs and drawings, is well documented, and is organized into four sections, making it easy to use and enjoyable to read.

The introduction contains an interesting summary of the mining history and the development of mineralogy. It also describes, in a condensed but accurate and stimulating manner, the geography and the mineralogy of the country, dividing it into eleven mineral provinces. The author discusses eight of the more important mining districts in México, which produce fine mineral specimens. There is also a chronology of historical, geological, and mineralogical events in México. This is followed by a bibliography with over 500 references on the subject.

The second section is a catalog describing more than 600 mineral species from more than 4,500 localities, including 55 mineral species first found in México. Included in this section are over 75 drawings and photographs and mineralogical and historical facts about many of the minerals and their locations.

In the appendix there is a complete updated listing of all the Mexican states and their counties [*municipios*].

Anyone interested in Mexican minerals should own a copy of *Minerals of México*. This work is highly recommended not only for geologists, mineralogists, and mining engineers, but also for mineral and gem collectors. It is the hope and wish of the Sociedad Mexicana de Mineralogía, which promotes the study of Mexican minerals, that readers will be stimulated by this book to further investigate the subject.

Miguel Romero PhD
President
Sociedad Mexicana de Mineralogía
Tehuácan, Puebla

PREFACE

This book is the first detailed and comprehensive modern guide to the minerals of México and their occurrences. It will be helpful to both the professional and the hobbyist interested in minerals. It is not a book on mineralogy, but rather a current catalog of minerals and locations throughout México.

The statistics of over 450 years of active mining with México are impressive. Over $6 billion worth of silver and gold alone have been mined. During 300 years of Spanish rule The Royal treasury received an average of $12.00 a minute from the mines of México. Today, México is still a world leader in silver production. It is also first in the production of fluorite and arsenic; second for celestite and sodium sulfate; third for bismuth and antimony minerals; fourth for graphite, diatomite, mercury, selenium, and sodium sulfate minerals; and fifth for sulfur, lead, zinc, feldspar, and cadmium minerals. Needless to say, a large number of mineral specimens were also produced during the past several centuries. Most, however have been lost or destroyed and few specimens now survive that can be traced much past the last century. Serious interest in Mexican minerals for collection or scientific study arose only after the establishment of the Colegio de Mineria (College of Mining) in México City on January 1, 1792. Before then minerals were important, for the most part, only for the metals that could be extracted from them. Fausto de Elhuyar (1755-1833), the first director of Colegio de Mineria appointed the renowned Spanish scientist Andrés del Río (1764-1849) to teach the first course in Mineralogy in 1795. Lacking textbooks, he first translated a book by his former professor, Abraham Gottlieb Werner (1750-1817) of Freiburg, Saxony, but shortly thereafter he wrote his own, *Elementos de Orictognosia* [*Elements of Mineralogy*]. The volume was published in México City in 1795 and was the first mineral book written in the new world.

Although several of del Río's students continued his interest in Mexican minerals, it was not until the middle 1800s that the next treatise on Mexican minerals appeared. They included Dr. Antonio del Castillo's *Cuadro de Especies Mineralógicas;* Carlos F. de Landero's *Sinopsis Mineralógica* and *Catálogo de*

Especies Minerales de Jalisco; D. Santiago Ramírez's *La Riqueza Minera de México,* and his translation of Dana's *Tratado de Mineralógia.* In 1898, Jose Guadalupe Aguilera authored *Boletín número 11* of the Instituto Geológico de México was published, which contained the *Systematic and Geographical Catalog of the Mineral Species of the Republic of México* based on previous works. In 1922, Dr. L. Salazar Salinas wrote for the Instituto Geológico de México *Boletín número 40,* Catálogo Sistemático de Especies Minerales de México y Sus Aplicaciónes Industriales and the following year, *Boletín número 41,* Catalogo Geográfico de las Especies Minerales de México. These works were actually authored by José Guadalupe Aguilera, but because of Dr. L. Salazar Salinas being the director of the Instituto Geológico de México, credit was given to him as being the author of the volumes. Since that time, many bulletins have been published by both the Instituto Geológico de México and the United States Geological Survey that have described the geology and economics of broad geologic regions, mining districts, and specific deposits. This book is a result of gathering relevant information from various sources and carefully checking the compiled data for accuracy.

PLAN OF THIS WORK

To avoid needless repetition of general mineralogical data that is readily available in many textbooks, I have not included this in the catalog on mineral occurrences. I have, however, included chemical composition and a brief description of the geological setting for each mineral, as well as detailed information on each locality and the unique features and associations of the minerals that occur in each locality.

Species are listed in alphabetical order and apportioned among states, municipios (the equivalent of counties), cities or geographical features. Mineral names are taken from the latest edition of Michael Fleischer's, *Glossary of Mineral Species* (1983), Mineralogical Record, and *The Encyclopedia of Mineralogy* (1981), edited by Keith Frye.

ACKNOWLEDGMENTS

The writing of a volume of this nature would have been almost impossible to accomplish without the assistance and cooperation of many individuals, corporations, museums, universities, and governmental bodies.

My deepest thanks must go to John Sinkankas; Miguel Romero; Eduardo Schmitter; and Jeff Kurtzeman, who originally came to me with the idea of writing this book.

I would like to thank the director of the Arizona-Sonora Desert Museum who allowed me to travel extensively throughout México to research this book while I was employed as Curator of Earth Sciences and Director of Mexican Programs.

A very special thanks to the staffs of Romero Museo de Mineralógia; Smithsonian Institution; Royal Ontario Museum; National Museum of Canada; University of Paris; National Museum of France; French School of Mines; British Museum of Natural History; Pinch Mineralogical Museum; Universidad

Nacional Autonoma de México; Instituto Geológico de México; Instituto Nacional de Historia y Anthropologia; Consejo de Recursos Minerales; The United States Geological Survey; The Universidad de Guanajuato; Universidad de Sonora; and the University of Arizona, for providing me with access to their libraries, catalogs, collections, and permitting me to photograph whatever was needed for this book.

I extend my thanks to the personnel at Industrial Minera de México (IMMSA), ASARCO, AMAX, Comission de Fomento Minero, Peñoles, Compañia Minera Fresnillo, and Compañia Santa Rosalía, who allowed me to photograph, gather samples, information, and permitted access to records of their mining operations in México.

Special thanks also to Don Pedro Mahieux, Miguel Romero, Eduardo and Rebecca Schmitter, Isidoro Bonilla, Benny Fenn, George Griffith, Peter Megaw, John Whitmire, Paul Desautels, Jose Luis Garza-Nieto, A. P. Salas-Guillero, Eugene Schlepp, Joseph Urban and Enrique and Jose Angle Toricillas for sharing their extensive knowledge of the minerals of México and their occurrences. I would also like to thank Sister Barbara Ann Gamboa for her patient review of the diacritical marks that appear throughout this volume. Many other people helped on this project in countless ways, but space does not permit listing of their names.

My close friends Emanuel Hecht, Arthur Holton, John Sinkankas, Arthur Rowe, and James Doyle reviewed much of the text and made numerous helpful suggestions, Robert Jones, Harold and Erica Van Pelt, Mike Havstad, Wendell Wilson, William Pinch provided several photographs, and Julian Blakely gave technical advice on all photographic work associated with this book.

A special thanks must go to Mr. and Mrs. Emanuel Hecht who provided financial support for this project.

Lastly, this book could not have been completed without the cooperation of my family who had to put up with all the problems with the project over many years and thousands of hours I spent as its author. My wife, Sharon, encouraged me throughout its completion, and my daughter, Shawna, did the line drawings; my son, Chris, traveled with me throughout México and made numerous suggestions on the project.

William D. Panczner

MINERALS OF MEXICO

IMPORTANT MINING DISTRICTS

México has been referred to as a land of vast extremes. This statement is not only true of its land, people, and history, but also in its mineral wealth. To the mineral collector and mineralogist, the country has been, and will continue to be, a prolific producer of outstanding specimens from what appears to be a cornucopia of mineral wealth. The great African empire-builder, Cecil Rhodes, once said, "México will one day furnish the gold, silver, and copper of the world; that from her hidden vaults, her subterranean treasure houses will come the gold, silver, copper and precious stones that will build the empires of tomorrow."

México includes 1,969,344 square kilometers (760,365 square miles) of land area of which slightly more than 60 percent is mountainous. Its coastline ranges from 7,338 kilometers (4,560 miles) bordering the Pacific Ocean and the Gulf of California (Sea of Cortés) in the west to 2,805 kilometers (1,743 miles) bordering the Gulf of México and the Caribbean Sea in the east (see front map). The topography varies from high mountains to broad plateaus and highlands that give way to vast tropical jungles toward the south. Altitudes range from 6 meters (20 feet) below sea level near Mexicali to heights of 5,700 meters (18,700 feet) at Pico de Orizaba volcano, and includes 22 mountain peaks with greater than 3,050 meters (10,000 feet). In the north the climate is temperate and toward Central America it becomes tropical. A wide variety of fauna and flora have developed because of this wide climatic range. There are thousands of different species and varieties of animals and plants. For example, there are more than 1,000 different species of birds alone. There are more cacti species in México than any other country in the entire world.

Describing México physiographically, two major mountain chains, the Sierra Madre Oriental to the east, and the Sierra Madre Occidental to the west, run north to south and are separated by a large central plateau. A third, smaller mountain range, Sierra del Sur, runs east to west in the southern portion of the country. Because of the rugged mountains and deep canyons throughout México, transportation and communication have always been difficult. Even today, after more than twelve thousand years of habitation and over four hundred years of European occupation and influence, large areas of the Country remain isolated and undeveloped. There are only 324,350 kilometers (201,547 miles) of roads throughout the country and only 58,958 kilometers (36,636 miles) are paved.

PROVINCES

Based on regional geography, geology, and mineralogy, México is conveniently divided into eleven regions as shown on the map at the end of the book. The provinces of Tehuantepec, the Coastal Plain, and Baja California (Fig. 1) produced specimens from such places as the La Fé Mine in the state of Chiapas and Santa Rosalía in the state of Baja California Sur. There are excellent opportunities for discovering new mineral locations in the future, however, extensive field exploration in remote areas will be required because large areas

Figure 1. The mineral provinces of Tehuantepec, the Gulf Coastal Plain, and Baja California. (Modified from Berstein, 1964, front cover; copyright 1964 by State University of New York, Albany)

are extremely mountainous and travel in the remote areas must be on foot, or on mules and horses.

The provinces of the Sierra Madre Oriental and the Mesa Central (Fig. 2) are closely related geologically. The Sierra Madre Oriental is a continuation of the Rocky Mountains with some elevations over 3,350 meters (10,991 feet). Mesa Central is very similar to the Great Basin of the United States and is considered to be the mineralogical heartland of all the Mexican provinces. In both provinces, outstanding mineral specimens have been, and probably will continue to be, produced from such world famous places as Mapimí in the state of Durango; Naica and Santa Eulalia in the state of Chihuahua; and Zacatecas in the state of Zacatecas. The potential for new mineral locations is good and field exploration is being done that will uncover new deposits in these provinces.

The Sierra Madre Occidental (Fig. 3) is a lava capped plateau located west of the Mesa Central. The western side of the high Sierra Madre is cut by rugged canyons over 1,599 meters (5,246 feet) deep. Many mountains are over 3,050 meters (10,000 feet) high. Mineralogically, this region is of great interest to the collector and includes such productive locations as Cerro de Mercado in the state of Durango and Batopilas in the state of Chihuahua. The future for finds within the vast volcanic areas is very promising. But because of the remoteness and ruggedness of the terrain, travel is mostly confined to narrow trails and must be on foot, mule, or horseback.

Figure 2. The mineral provinces of the Sierra Madre Oriental and Mesa Central. (Modified from Berstein, 1964, front cover, copyright 1964 by State University of New York, Albany)

Figure 3. The mineral province of Sierra Madre Occidental. (Modified from Berstein, 1964, back cover, copyright 1964 by State University of New York, Albany)

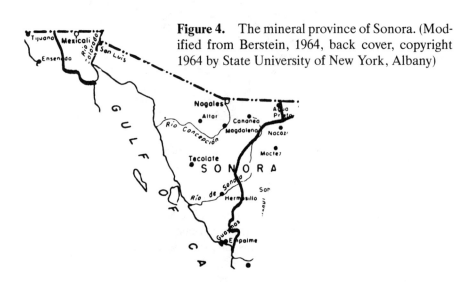

Figure 4. The mineral province of Sonora. (Modified from Berstein, 1964, back cover, copyright 1964 by State University of New York, Albany)

The province of Sonora, located in the northwest corner of México (Fig. 4) is an extension of both the basin and range province, and the copper porphyry belt of the western United States. There are many areas of interest to collectors, such as Moctezuma, Cucurpe, and Arispe in the state of Sonora. Prospects for future productive locations in this richly oxidized and heavily volcanic region is excellent, especially at the currently developing copper mines in the province.

The belt of active volcanism in south central México is the home of the province of Sierra de Los Volcanes (Fig. 5). Fine mineral specimens came from Guanajuato in the state of Guanajuato and Zimapán in the state of Hidalgo. The area has good potential for future production of mineral specimens if more field exploration is conducted.

South of Sierra de los Volcanes is the extremely rugged province of Sierra del Sur (Fig. 6), a structurally complex region composed of old, eroded mountains

Figure 5. The mineral province of Sierra de Los Volcanes. (Modified from Berstein, 1964, back cover, copyright 1964 by State University of New York, Albany)

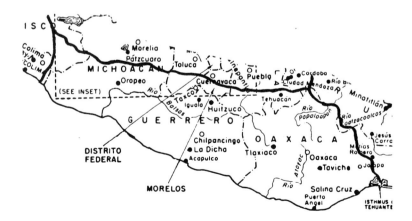

Figure 6. The mineral province of Sierra del Sur. (Modified from Bernstein, 1964, back cover, Copyright by State University of New York, Albany)

covered with jungle vegetation. Several areas, such as Taxco de Alcarón and Amatitlán in the state of Guerrero, have produced fine specimens. The potential is extremely good for future production from this province. As a whole, however, the region is very remote and inhospitable and again travel is either on foot or on the backs of mules or horses.

The provinces of Yucatán and the Pacific Coastal Plain (Fig. 7) are not mineralogically important areas. Their records of mineral specimen production are poor, and their futures are not promising. Only occasionally do drill cores of the many oil wells in the Yucatan region produce any specimens.

The development of México over its 12,000 year history has been for the most part oriented around its minerals. With the arrival of the Spaniards in 1517, the needs for minerals increased. Spain needed the gold and silver to finance its territorial aggrandizement, defend Catholicism, and to help meet the needs of Europe's expanding commerce. But, it was not until after the conquest in 1521, that the Spanish discovered the great mines that supplied the Royal treasury with its great wealth.

With over 450 years of Mining in México, countless mineral specimens have been mined from hundreds of mining districts and thousands of mines. Many of these districts and mines have been closed for many years. Unfortunately, most of the specimens produced have been lost, with only the written records of their discovery remaining. Less than two dozen of the older mineral producing districts remain active today and only occasionally do they produce outstanding mineral specimens.

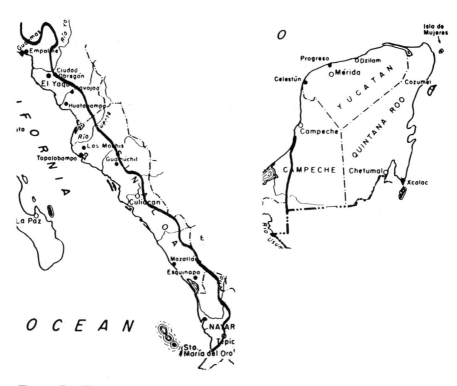

Figure 7. The mineral provinces of the Yucatan and the Pacific Coastal Plain. (Modified from Berstein, 1964, inside back cover, copyright by State University of New York, Albany)

Eight of these older districts will be discussed fully, because of their importance in Mexican history and their production of quality specimens cherished by the collectors: Santa Eulalia, Chihuahua; Naica, Chihuahua; Mapimí, Durango; Sombrerete, Zacatecas; Zacatecas, Zacatecas; Guanajuato, Guanajuato; Taxco de Alcarón, Guerrero; Zimapán, Hidalgo (Fig. 8). The dollar values of ores used in this section are based on dollar and metal values of the time period.

SANTA EULALIA

The Santa Eulalia mining district is located 19 kilometers (12 miles) southeast of the city of Chihuahua, the capital of the state of Chihuahua. To the collector, it is one of México's better known sources of mineral specimens. Several years ago, the Mexican government changed the names of the municipio and several of the district's mining camps. The district is now located in the municipio of Aquiles Serdán, while the town of Santa Eulalia itself is now Aquiles Serdán. The mining camp of Santo Domingo was renamed Francisco Portillo. The name of the mining district and all else within the area remain unchanged.

History

The actual date of mineral discovery in the Santa Eulalia area is uncertain. Records in Spain cite that silver was discovered in 1591 in the mountains, in what is now the Santa Eulalia mining district. State records, however, give the date as 1703. Because of the remoteness and the hostility of the Apache Indians who lived in the region, this discovery was not systematically developed for many years. It was not until the establishment of a Spanish garrison at Chihuahua that mining could really begin in the area. In the early 1700s a band of bandits took refuge in a remote side canyon of the Sierra de la Santa Eulalia where they accidentally rediscovered silver. While sitting around a camp fire, they noticed silver running out of the red-hot rocks. The bandits offered the information of their silver discovery to the parish priest in Chihuahua, hoping to receive absolution for their evil ways. They offered the padre enough silver to build the greatest cathedral in all of the Americas.

Once word of the discovery was out, mining soon began in the Sierra's Mina Vieja, San Antonio, and Santo Domingo mines. These mines became the first in the district. From 1705 to 1737, the mines produced $2 million per year in silver. All of the mining at that time was done by manual labor. By 1791, the mines produced $150 million, but only $112 million was reported and taxed. These were considered to be the bonanza days of the Real de la Santa Eulalia.

A small mining camp, located below the mines in the arroyo de Delores had grown to six thousand inhabitants. The district now had sixty-three rather primitive reduction plants, 188 smelting furnaces, and sixty-five cupelling furnaces to process the ore. The city of Chihuahua also had an additional twenty smelting furnaces processing the district's ore.

A cathedral was begun in Chihuahua in 1738 and was completed in 1750. It was built from the taxes levied on the silver produced by the district's mines; twelve cents for every 250 grams (8 troy ounces) of silver. The final cost was between eight-hundred thousand and one million dollars.

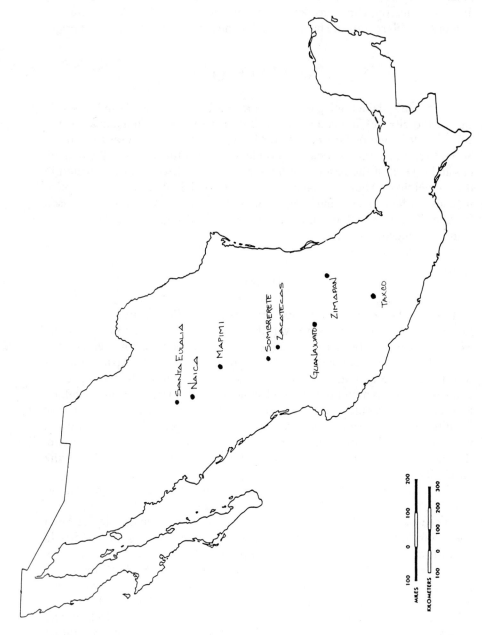

Figure 8. The locations of the mining districts of Santa Eulalia, Naica, Mapimi, Sombrerete, Zacatecas, Guanajuato, Zimapan, and Taxco.

The earliest production of silver came primarily from the Parcionera, Bustillos, Mina Vieja, San Juan, and Santo Domingo mines. The San Jose, Santa Rita, Rosario, San Antonio, Galdeana, Dolores, and Las Animas mines also produced silver ore during this period, but in lesser amounts. Although several of the mines extended downwards to 120 meters (393.7 feet) below the surface, most of the silver ore during this period was found near the surface in large caves. Some of these caves and mine stopes were so large that it was said that the cathedral in Chihuahua could be placed within them. It was not uncommon for ores to assay up to 3.1 kilograms (99.7 troy ounces) of silver per metric ton, 1-2 percent lead, and over 20 percent silica. Because mining districts to the south in the state of Durango needed silica rich silver ore as flux to mix with their silver ores, much of the ores from Santa Eulalia was sent on the backs of burros southward.

During the early 1800s, because of the political unrest in southcentral México, Spain withdrew most of its military forces from Chihuahua to strengthen its troops in the south. Once again, the Apache Indians began to raid the region. The Spanish mine owners tried to continue operations, but with México winning its independence from Spain in 1821, the Spaniards left the district and the mines closed. Even though the district was in a state of depression, the mint at Chihuahua continued to coin almost sixty-three million dollars in silver from 1811 to 1895. The revenue was generated mostly from silver mined by gambocinos (individual miners) from upper, older levels of the mines.

Mining began again in 1880, when American interests entered the district and formed the Santa Eulalia Silver Mining Company. In 1882, The Mexican Central Railroad was completed as far north as Chihuahua. Later that year, a narrow gauge railroad was built connecting the district with the Mexican Central Railroad. The Santa Eulalia Silver Mining Company was sold in 1891, after removing considerable ore from the Santo Domingo and Galdeana mines. The Chihuahua Mining Company now took over operations within the district after receiving land concessions from the Mexican government that covered a strip of land 8 kilometers (5 miles) wide and 24 kilometers (14.9 miles) long, covering the entire district. The new federal laws that taxed all mines and mining claims caused the Chihuahua Mining Company to relinquish much of its holdings within the district in 1884. As a result, the district opened up to other mining companies, ASARCO, formally the American Smelting and Refining Company, and other American interests began to develop and operate mines in the district.

The railroad enabled the import of new equipment to modernize the mines and mills. Locally generated electricity was installed along with improved and faster ore transportation systems. The mines were deepened and increased their ore production into the sulfide ore zone. Cheaper hydroelectric power was not introduced into the district until 1921. Problems caused by the Mexican Revolution of 1910 briefly closed the mines. Pancho Villa raided the district several times and held the staff from the El Potosí Mine and the ASARCO operations for ransom.

Mining properties are concentrated in the east and west camps as shown in Figure 9. East camp mines include San Antonio, Santa Juliana, Las Tres Mercedes, Josefina and Dolores, while the Santo Domingo, El Potosí, Buena Tierra, Gasolina, Bustillos, Galdeana, Juárez, Reina de Plata, Coronel (group), Democracia, Inglaterra, San Antonio Chico, Mina Vieja, Las Animas, Velardeña, Esmeralda, Santander, Santa Rita, and Parcionera are west camp mines.

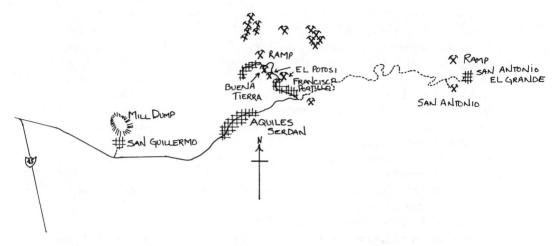

Figure 9. The mines and mining camps of the Santa Eulalia mining district.

In the early 1900s the mines were operated by ASARCO, Chihuahua Mining Company, El Potosí Mining Company, Santa Eulalia Exploration Company, and the San Toy Mining Company (Fig. 10). During the 1920s Peñoles entered the district by buying out San Toy Mining Company and in the next thirty years companies bought out each other until finally only two major operations, ASARCO and El Potosí Mining Company, remained in the district. Eventually, these two enterprises became owned and operated by two Mexican companies, Industrial Minera de México S. A. (IMMSA) and Minerales Nacionales de México S. A. (MINAMEX).

By the 1920s the El Potosí Mine was ranked as one of the four largest producers of lead in the world. Between 1930 and 1940 the San Antonio Mine became a major world producer of tin and vanadium. Current production comes principally from MINAMEX's El Potosí Mine and IMMSA's Buena Tierra Mine in the west camp and IMMSA's San Antonio Mine in the east camp (Fig. 11 and 12). IMMSA is expanding and modernizing its San Antonio Mine operation and building a new mine shaft to develop new ore bodies in the west camp of the District.

The production of the district, in silver and lead, until 1886 was almost $400 million. The district has produced, to date, over 12,442 metric tons (400 million troy ounces) of silver; 2.7 million metric tons of lead; 1.8 million metric tons of zinc; 5,000 metric tons of copper; 4,000 metric tons of tin; 700 metric tons of vanadium; and 1 metric ton of gold since its beginning in 1705. Using 1980 values, these totals place the Santa Eulalia Mining district as one of México's leading producers with slightly less than $3 billion in metals extracted from the mines of Sierra de Santa Eulalia.

Geology and Mineralogy

The Santa Eulalia Mining district is a limestone replacement deposit typical of México's famous manto and chimney type deposits (Fig. 13). The district is underlain by deep seated igneous rocks which are intruded into a series of

Figure 10. The mining camp of Santo Domingo and the El Potosi Mine-Santo Domingo #5 Shaft, Santa Eulalia mining district, circa 1910 (Courtesy of the University of Texas, El Paso)

Figure 11. The El Potosi Mine-Santo Domingo #5 shaft, Santa Eulalia mining district, 1979.

Figure 12. The San Antonio Mine, Santa Eulalia mining district 1979.

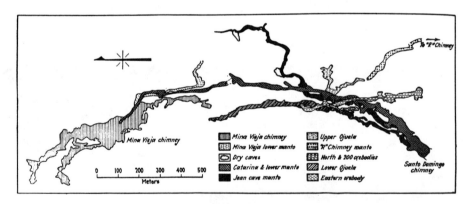

Figure 13. A plan view of the mantos and chimneys of the West Camp of the Santa Eulalia mining district. (Prescott, 1926, p. 248, copyright 1926 by Engineering and Mining Journal/courtesy of ASARCO)

impure and shaley limestone and evaporites with thick pure limestone beds over these lower units. The sedimentary rocks were faulted, gently folded, uplifted, and eroded before being capped by a thick sequence of volcanic rocks.

The ore bodies form continuous chimneys and mantos of great length (Fig. 14). For example, chimneys have been found that extend, unbroken, vertically for almost 489 meters (1,604 feet) and reach diameters of 18 meters to 40 meters (59.1 feet-131.2 feet). Some mantos are almost 3.2 kilometers (2 miles) long. Several of these enormous chimneys have produced over 1 million metric tons of ore.

A rather common feature in the upper ore zone within the district are caves,

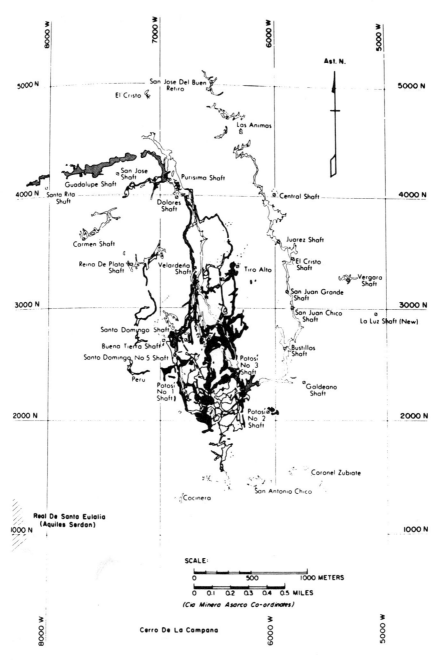

Figure 14. The mines and ore bodies of the West Camp of the Santa Eulalia mining district. (Hewitt, 1968, p. 231, copyright 1968 by Society of Mining Engineers Transactions/ASARCO)

classed as either ore caves or dry caves. The former are caused by oxidation of sulfide ore bodies and are found above the oxide mantos, along the edge of a chimney, or within the chimney itself. Some have been found as deep as the transition zone between the oxide and sulfide ore. These caves are commonly lined with calcite and aragonite speleothems (cave growths) of many types.

The dry caves are so called because of the lack of ore mineralization and are usually filled with water. They are found in both the oxide and sulfide ore zones throughout the district. Dry caves are usually lined with large crystals of calcite which are often color zoned. These caves are mining hazards because of the large amount of water they contain that can quickly flood portions of the mine.

Near the surface, the caves are often lined with crystals of gypsum variety selenite of enormous size. In 1978, for example, a most remarkable selenite cave was encountered along an exploration ramp in the Industrial Minera de México S. A. mine in the west camp. Mining officials of the company had the direction of the ramp changed in order to save its crystal contents. Caverna de Santo Domingo, as it was officially named by the author, consists of three cave rooms. The main gallery is over 100 meters (328.1 feet) long, 10 meters (32.8 feet) wide, and 15 meters (49.2 feet) high (Fig. 15). A few of the water-clear, selenite crystals reach just under 4 meters (13.1 feet) in length (Fig. 16). While the floor of the cave was covered with selenite crystals, the walls and ceilings were coated with ramshorn gypsum and extremely small fluorite crystals. Even though the district's water table is very deep, the zone of oxidation varies considerably. The distribution of surface cap rock in the immediate vicinity controls downward percolating water and oxidation.

A large area of barren rock separates the major ore body areas between the east and west sides of the district. The geology of the east camp varies slightly from that of the west camp. Both are characterized by manto and chimney ore body replacements in limestone. But, in the east side, a major contact-metamorphic skarn zone is also found. Caves are numerous on the east side, some are 305 meters (1,001 feet) long and 31 meters (101.7 feet) in diameter. They occur especially in the vanadium bearing horizon and in a few cases

Figure 15. A view of one end of the main room of the Caverna de Santo Domingo, Santa Eulalia mining district.

contained calcite stalactites coated with vanadinite crystals. These brownish-yellow stalactites have been found up to almost 20 centimeters (7.9 inches) in length and over 2.5 centimeter (1 inch) in diameter.

The mineralization of the west ore bodies changes from north to south. Mines on the northern end contain more silver and fewer lead minerals, whereas the reverse is true on the southern end. Iron minerals are also more prevalent in the northern portion. The mineralogy of the east side of the district is somewhat more complex due to the presence of a calc-silicate, zinc-rich chimney, which is zoned from skarn to massive sulfide ores; smaller lead-rich mantos connect to this chimney. Two smaller chimneys, the Tin Chimney and the Cock's Ore body are zoned from top to bottom, however, relationships to the tin ore bodies and other ore bodies are not presently understood. Both chimneys are above the water table and occur within the unzoned oxide ore body. In the San Antonio Mine, the principle mine of the east camp and presently the major producer of the entire district, the eighth level marks the point where the oxide ores change to sulfide ores. Over the past few years the San Antonio Mine has produced some of the most remarkable crystallized mineral specimens found in México.

NAICA

Naica is a major mineral specimen producing location as well as being one of the few major silver mining areas in México not directly discovered by the

Figure 16. The floor of the Caverna de Santo Domingo are covered with sprays of gypsum (selenite) crystals up to 4 meters in length.

Spanish. Naica is in the state of Chihuahua, approximately 130 kilometers (80.8 miles) southeast of capital city of Chihuahua, and 25 kilometers (15.5 miles) west of the small rail center of Conchos.

History

By the late 1500s the Spanish conquistadores were exploring for precious metals in the area near Naica. Prospecting parties, however, quickly passed by the Sierra de Naica because of the lack of the usual surface indications and the bands of hostile Indians that were protecting the area. Silver was discovered on the Sierra de Naica in 1794. Large scale mining did not begin until 1830 because of the hostile Indians and the outbreak of the Mexican War of Independence from Spain (1810-1821).

In the ensuing years the mines, mining camps, and local ranchos were continually raided by Indian war parties. In 1861 Indians raided Naica and captured three miners. They were taken to the top of the Sierra de Naica where they were tortured, killed, and partially eaten. The problem with the Indians became so extreme that the mines had to be abandoned for several years until protection could be obtained from the Mexican military. The mines and the mining camps were destroyed during the time they were left unguarded. Eventually, military protection was provided and mines were opened and developed on the eastern side of the Sierra de Naica.

In 1894, rich silver ore was discovered. The Maravillas Mine was legally claimed by Don Santiago Stoppelli, who with his friend, Don Saturmino González, formed an association in 1896 to develop the new mine. That same year, Stoppelli questionably sold his share of the Maravillas Mine to the newly formed Compañía Minera De Naica for 25 percent interest in the company. Forty percent of the company was owned by Compañía Minera de Peñoles which in turn was owned by the American Metal Company, a U.S. company controlled by interests in Frankfurt, Germany.

The silver ore from the Maravillas Mine was extremely rich, which caused heirs of both Stoppelli and González to take action to recover ownership of the mine. The families regretted the error of their fathers in losing control of the mine and their loss of the great fortune being made from the Maravillas Mine. Major legal problems arose from their heirs that made the mining district more famous among the legal profession than among mining engineers. In 1905, the Compañía Minera de Naica settled with both estates, but this did not satisfy all the heirs, who continued litigation until the Mexican courts ordered the return of the mine to them in the early 1920s. Although the Compañía Minera de Naica returned ownership of the Maravillas Mine to the heirs, they leased it back from the estate and continued mining. During the 1910 Mexican Revolution, the mines were closed briefly by strikes. In 1915, Pancho Villa took control of the mines for several months. After the revolution, the American owned Eagle-Picher Mining Company began mining in the Gibraltar Mine, which was destined to become the second most important operation within the district. Compañía Minera de Peñoles, now free of its German control, took over the Compañía Minera de Naica and combined the Maravillas, Lepanto, Ramon

Corona, and several other smaller mines into a single operation called the Naica Mine.

Until the 1920s, all mining had been done above the water table, in the rich upper oxide ore zone. From 1924 to 1928, Peñoles began to modernize the mines. They installed pumps capable of removing 45,425 liters (12,000 gallons) per minute of hot water (52°C), enabling development of the sulfide ore zone below the water table. However, between 1932 and 1937, a reemergence of legal problems that were thought to be settled caused Peñoles to begin to lose interest in the district.

In 1951, the Compañía Minera Fresnillo acquired the mining rights in the district. In 1956, they bought out the Eagle-Picher's mining operations at the Gibraltar Mine. Finally all the mines were under one ownership and now Compañía Minera Fresnillo referred to all the mines collectively as the Naica Mine (Fig. 17). It is interesting to note that Compañía Minera Fresnillo is presently owned by Peñoles (60%) and by the U.S. mining company of AMAX (40%).

Presently, Compañía Minera Fresnillo is mining a swarm of 11 mantos and more than 50 chimneys that produce silver, gold, lead, zinc, copper, tungsten, and molybdenum. The Naica Mine has been developed on eighteen levels with the major production coming from stopes between the 290 meter and 430 meter levels. Considerable ore is also being mined from stopes between the 150 meter and 290 meter levels. The Naica shaft and ramp are being used on the west side of the Sierra de Naica to develop ore along the Naica Fault, Gibraltar, and the large Fifth mantos.

Figure 17. A view of the mining camp of Naica, the Naica Mine, and the Sierra de Naica 1979.

Two major problems faced the miners underground at Naica: water and heat. The water within the mine is extremely hot (52°C). An enormous quantity of water 2,725,488 liters (720 thousand gallons) per hour must be pumped out of the mine. Large fans, blowing 37,160 cubic meters (1.3 million cubic feet) per minute into the mine via four ventilation shafts, are required to maintain satisfactory working temperatures.

Geology and Mineralogy

The ore deposits of Naica are typical limestone replacement deposits, occurring within a simple dome of the Sierra de Naica. The area has been intensely faulted before and after the mineralization. The deepest rocks are sandstones, overlain by shales and capped with limestones. The mineralization occurred in association with acid volcanic dikes and sills and may have been controlled by the early premineral faults. The fifty mineralized chimneys and eleven mantos of the district are found within a one square kilometer area on the northeast flank of the Sierra de Naica.

Ore chimneys at Naica have been found ranging from 3 meters to 80 meters (9.8 feet–262.5 feet) in diameter. Total lengths have not been determined, but are known to exceed 780 meters (2,559.1 feet). Mantos range from 1 meter to 10 meters (3.3 feet–32.8 feet) thick and reach a length of 700 meters (2,296.6 feet), most are only 45 meters to 50 meters (147.6 feet–164 feet) long (Fig. 18).

The ore minerals of Naica can be divided into three general classes: (1) oxidized, silver-lead-zinc carbonates; (2) sulfides; and (3) skarns. At the present time more than 80 minerals have been found with sulfides of lead, zinc, and copper predominating in the chimneys and silicates, carbonates, and some sulfides being more abundant in the mantos. Most crystallized mineral specimens of interest to the collector are found in the mantos within the district.

Mining at Naica uncovered many caves beneath the Sierra de Naica. Most of the caves are lined with crystals of gypsum variety selenite, but calcite-lined caves, and even one lined with anhydrite crystals occur. Most caves are small, but in the 1940s, in the Gibraltar Mine, two remarkable caverns on the 120 meter level were found. They were named the Xochitl Cave and the Cave of the Swords. The Xochitl Cave is smaller and contains selenite crystals up to about 40 centimeters (15.8 inches) long, most etched, but showing fine phantoms and moveable gas-liquid bubbles inclusions. Many years ago, miners tell of using small selenite crystals that contained moveable bubbles as levels in surveying and in the mines' carpenter shops.

The Cave of the Swords is much larger, over 100 meters (328.1 feet) long and 10 meters (32.8 feet) high, and is lined with selenite crystals up to 2.5 meters (8.2 feet) long. Crystals growing from the walls or ceilings are much smaller in size, with rounded terminations resulting from dissolution when the water level changed during formation. In 1950, the cave was flooded temporarily when lightning struck the company's power station. Before the power station could be repaired, the mine flooded. To preserve the cave and make it available for special tours, Compañía Fresnillo installed railed cat-walks and ladders made of wood, and some lighting. Today the area where the cave is located is not being mined. Ventilation is not available, and the caves are hot (37°C) and

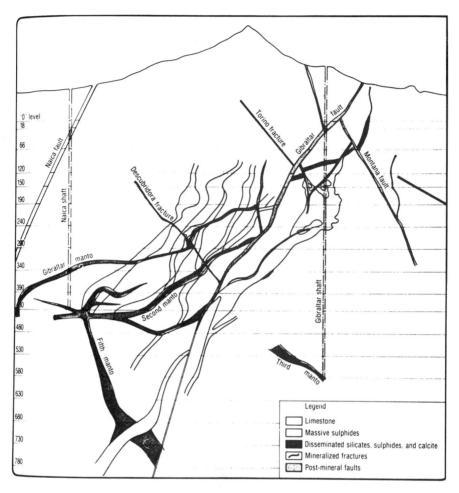

Figure 18.   A cross-sectional view of the ore bodies of the Naica Mine, Sierra de Naica. (White, 1980, p. 149, copyright 1980 by Engineering and Mining Journal)

humid. Tours of the Naica Mine always make their last stop at the Cave of the Swords.

MAPIMI

Mapimí, to the collector, is the best known of all Mexican mineral locations. Even today it produces excellent mineral specimens. The mining district is not located in the town of Mapimí, but rather some 9.7 kilometers (6 miles) southeast in the Bufa de Mapimí in the northeast portion of the state of Durango (Fig. 19). Mapimí is a quiet, dusty desert town of slightly more than 2 thousand inhabitants and is the seat of government for the Municipo de Mapimí. In the mining heyday of the late 1890s to early 1900s the population reached almost twenty thousand.

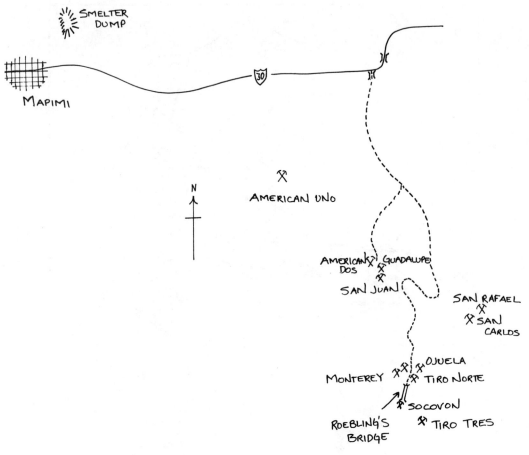

Figure 19. The Mapimi mining district.

History

In 1598, the Spanish discovered an outcrop of high grade silver ore in the side of one of the many canyons far up the mountainside from what is now the town of Mapimí. Eventually, this outcrop was exploited to become the famous Ojuela Mine.

Because of a fancied resemblance to leaves (hojas) noted in the ores, the Spanish named the mine leaf-like or Hojuela. Sometime in the next 200 years, the "H" was dropped and the mine was simply spelled Ojuela.

Traces of the original discovery site can still be seen on the canyon wall at what is now known as the Boca de Mina (Fig. 20). In order to reach the new silver strike, the Spanish carved steps into the solid limestone from the canyon's edge down 60 meters (196.7 feet) to the outcrop. For many years, these steps were part of a precarious pathway that provided not only access to the workings, but also the route that ore extracted from the mine was carried out on the backs of miners. Later a shaft was dug near the edge of the canyon and the pathway was abandoned.

The ore bodies are irregular chimneys and mantos. Because of the way the

Figure 20. The cavelike opening in the side of the canyon is the original Spanish Mine entrance known at Mapimi as the Boca de Mina.

ore chimneys and mantos would change direction abruptly, they were exploited by a mining method called gophering or rat hole mining (Fig. 21). Also, the ore bodies were so small that the miners dug their drifts no wider than necessary to remove the ore, often using cow horns and scrapers instead of shovels. Even as mining technology improved, older miners still refused to use shovels and preferred instead to use long handled hoes held between their legs to push ore onto pans that were then loaded into ore cars. For deeper portions of the mine, notched tree trunks called "chicken ladders" were used. As the years passed, these ladders were modified by using wood planks with 10 centimeters (3.9 inch) wood cleats nailed to them.

The Ojuela and other mines within the district were operated very successfully up to the beginning of the Mexican Revolution in 1810. With the expulsion of Spain and the declaration of México's independence in 1821, the new government took over all the former Spanish mining industry. Mapimí, now owned and operated by Mexicans, continued to produce ore from 1821 to 1867, but on a reduced scale and with only moderate success. From 1867 to 1884, the mines were abandoned for the most part, and it was not until 1890, when the Mexican government relaxed restrictions on foreign companies operating within the country that mining resumed. Two U.S. companies began to mine in the district, but were not successful. In 1893, the newly formed Compañía Minera de Peñoles, a Mexican subsidiary of American Metal Company purchased the Ojuela Mine and at first encountered problems in mining the irregular ore bodies before discovering profitable ores. Peñoles introduced new mining techniques and technologies, electricity was now used in the mines. Three smelters were built to handle the complex arsenical silver-lead ore with furnace

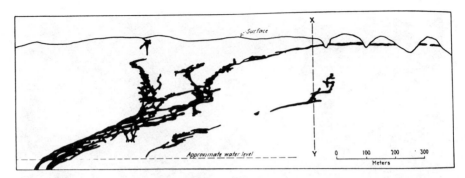

Figure 21. A cross-section view of one of the irregular chimneys and mantos (Cumbres ore body) of the Mapimi mining district. (Prescott, 1926, p. 292, copyright 1926 by Engineering and Mining Journal/Penoles)

draft provided by two unique 104 meter (341.2 foot) wood chimneys, which later had to be replaced because they caught fire from the hot furnace gasses (Fig. 22).

In the late 1890s, the old burro trail up to the Ojuela Mine was replaced by a 76.2 centimeter (30 inch) narrow gauge railroad. The railroad was conventional from the smelters at Mapimí to the foot of the Bufa de Mapimí, but changed to a cog-railroad for the steep ascent to the mine. The famous "Great Bridge" in the Mapimí mining district was built to connect the Socavón Mine across the steep and narrow canyon from the Ojuela Mine so as to increase ore production and to lower costs. The world-famous builders of the Brooklyn Bridge, the John A. Roebling Company of Trenton, New Jersey, were called in to construct the bridge using a combination of Roebling steel cables for suspension and wood for the body. Finished in a little over a year (1899) with a length of 325.3 meters (1,066.9 feet), 1.8 meters (5.9 feet) wide, and 80 meters (262 feet) above the canyon floor at the center of the span. It was the longest wooden suspension bridge built at that time, and is still one of the longest suspension bridges in México. Four ore cars were run back and forth across the bridge by a specially built steam engine (Fig. 23).

Peñoles replaced the homes that had been built around the Ojeula Mine with new homes with electricity. These homes were provided free to the miners and their families. The company also provided: wash houses and toilets; schools; a clinic and hospital; stores; billiards, dance, and gambling halls; and a large theater. The Ojuela mining camp became not only the finest in México, but also North America. Several tragedies hit the mining camp in the early part of the 1900s. The cog-railroad train while returning up the hill to the camp derailed just before entering the camp; catapulting four train cars down on the camp's baseball field, killing 40 people including the camp's baseball team. In 1907, a spectacular fire, starting in the camp's billiard and game hall by a spark from the carbon arc lamp, almost burned down the entire mining camp.

Several shafts compose the Ojuela Mine. The main one, Tiro Norte reached a depth of over 793 meters (2,601.7 feet). Above the eleventh level (450 meters) workings are at irregular intervals, but below the eleventh, the levels are regularly separated by 50 meters (164 feet). By 1910, the mine was down to the eighteenth level and eventually reached the twenty-third level before closing thirty years later. Water had to be pumped from a special tunnel and adit on the

Figure 22. The last of the wooden smelter smokestacks at the Mapimi smelter, circa 1910. (Rice, 1908, p. 374, copyright 1908 by The Engineering and Mining Journal)

213 meter level at a rate of 22,712.4 liters (6,000 gallons) per minute to keep the lower mine workings dry. The ore was hand broken and sorted into oxidized or sulfide ores for shipment to the smelters at Mapimí.

In 1899, production was over $4 million from as many as 216 mines within the Mapimí mining district. The German investors received up to $100 thousand in monthly dividends between the late 1890s and early 1900s. By 1913, just under 6 percent of all of México's total mineral production was from the district. Each day, 272,160 kilograms to 453,600 kilograms (300 to 500 tons) of ore were produced. From 1893 to 1931, Peñoles produced over 3.4 billion kilograms (3.8 million tons) of ore. It was only during the Mexican Revolution of 1910, that mining and smelting were briefly interrupted.

During most of World War I, Peñoles sold its strategic metals to both the United States and Germany, making great profits. Eventually the U. S. govern-

Figure 23. The Roebling bridge, the Ojuela Mine, and the mining camp of Ojuela, circa. 1910.

ment forced the German company that had controlling interest in American Metal Company to sell its holdings. Shortly thereafter, the Mexican government did the same with the German holdings in Peñoles.

Prior to 1920, the silicious ore from the mines began to run out, creating problems in smelting. For the first time silica flux had to be imported. Other major problems arose at the same time; ores became lower in metal content, mining costs increased with depth, and transportation costs from Ojuela to the smelters at Mapimí also increased. A decision was made to sink a new shaft nearer to Mapimí at the base of the mountain. Tiro Americados was completed and replaced the shafts at the Ojuela mining camp. By the late 1930s, the old mining camp was abandoned and fell into ruins.

In September 1921, in order to keep the Mapimí operations profitable, the smelters were dismantled and the Mapimí ores shipped directly via Peñoles narrow gauge railroad to Bermejillo, where it was loaded on cars of the Mexican Central Railroad and sent to the company's new smelter at Torreón. A few years later the ore was shipped to the more modern smelter at Monterrey. By 1924, the ore production within the district no longer was able to maintain a profit. In 1926, Peñoles completed a flotation plant to concentrate the lead sulfide (galena) for shipment to the smelter at Torreón. Despite these steps, the mines and all operations were closed by the later part of the 1930s.

Peñoles resumed mining in the early 1940s because of the increased demand for metals created by World War II. After the war, the 386 year old mine was leased to a local cooperative to clean out whatever ore remained in the more than 402.3 kilometers (250 miles) of underground workings. In the early 1980s the co-op, operating from the Americados Shaft, produced 16,330 kilograms (18 tons) per day, using 63 miners and 3 mules. By 1985, production had dropped to just over 900 kilograms (1 ton) of ore per day. Because Peñoles removed the pumps in the mine in the late 1940s, water has flooded the deeper levels of the mine. If the Cooperative could find the financing for new pumps and their installations, the life of the mine can be extended for several years. If not, the end of mining is nearing for the Mapimí mining district.

The district's mines from 1893 to 1931 produced for Peñoles over 3.8 million tons of ore, that had an average assay of 3.7 grams of gold, 462 grams of silver, and 14.9 percent lead. It is thought that of the tonnage figure, only 5 percent was from the deeper sulfide ore zone. The total amount of ore production by Peñoles, and later by the co-op yielded over 1,866,000 kilograms (59.9 million troy ounces) of silver; 14,306 kilograms (459,938 troy ounces) gold; 271,944 kilograms (271.9 metric tons) of lead; and 158,757 kilograms (158.8 metric tons) of arsenic. Combining estimates from the early Spanish period of the district with the previously mentioned values, yields an approximate combined total of 6 million kilograms (16 million troy pounds) of silver and 49,000 kilograms (131 thousand troy pounds) of gold for the district. Based on 1980 metal values, these totals would place the Mapimí mining district as one of México's more important metal producers, with just slightly more than $1.8 billion in metal being produced in the district's 387 years of operation.

Over the years, 216 mines and shafts have been developed within the district. The Ojuela Mine as it interconnected underground with the Americauno, Americados, Guadalupe, San Juan, Monterrey, Tiro Tres, and Socavón mines, it became the district's largest mine (Figs. 24 and 25). Six major shafts and hundreds of small shafts and adits allowed access to the more than 403 kilometers (250 miles) of underground workings within the six Ojuela ore bodies. All the mine levels, ore bodies, stopes, and watercourses have been named and numbered, thus allowing areas producing ore to be identified on the mine maps. Other mines of the district that remained separate from the Ojuela mine included: Talpa, San Rafael, San Carlos, and El Padre (La Esperanza).

In the late 1970s, the state of Durango replaced the old cog-railroad bed with cobblestones and built pulloffs so vehicles could drive up to the old mining camp of Ojuela. The new road is a single lane and provides thrilling views for most of its 3.2 kilometer (2 mile) climb. The Ojuela camp itself now lies in ruins. Some years ago the Roebling bridge had its wood flooring partially burned out, but this has been restored and is passable to foot traffic.

Geology and Mineralogy

The geology of the Mapimí mining district is complex and its geologic structure has been upset and distorted. The Mapimí area exhibits typical basin

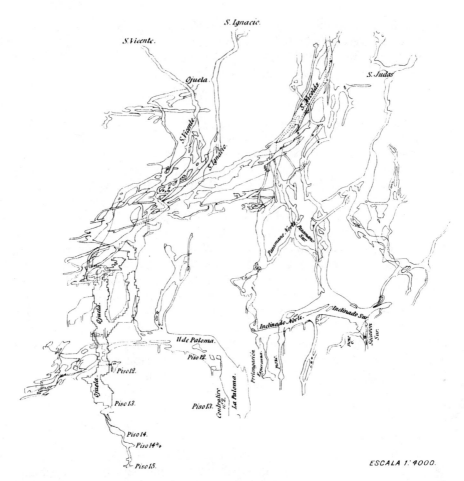

Figure 24. A cross-sectional view of the Ojuela Mine, circa 1906. (Villarello, 1906, back page, copyright 1906 by Litog y Tipog. S. A.)

and range topography with its uplifted and subsequent faulted mountains. The district's basement rocks are granites overlain by a moderately thick layer of sandstone. Above the sandstone is a transition zone of shale grading into a shaley limestone, then to a pure limestone of considerable thickness. This limestone consists of the Aurora formation (lower Cretaceous) on the bottom and the Indidura formation (upper Cretaceous) of heavily folded limestone and shales on top. Intruded into these sedimentary rocks are irregular bodies of igneous and metamorphic rocks.

The ore bodies within the district are considered a typical limestone replacement deposit with mantos and chimneys. There is not one chimney, but seven; each differing not only in metal content, but also mineral assemblages. The Ojuela-Paloma, San Jorgé-San Juan, Cumbres, Santa Rita, Carmen, San Carlos, and San Diego make up the district's chimney ore bodies. Most of the chimneys within the district reach depths greater than 900 meters (2,952.8 feet) and apparently continue to much greater depths; most of the mantos are found above the district's water table. The zone of oxidization in the district is very porous and ends 518 meters (1,699.5 feet) below the surface at the water table

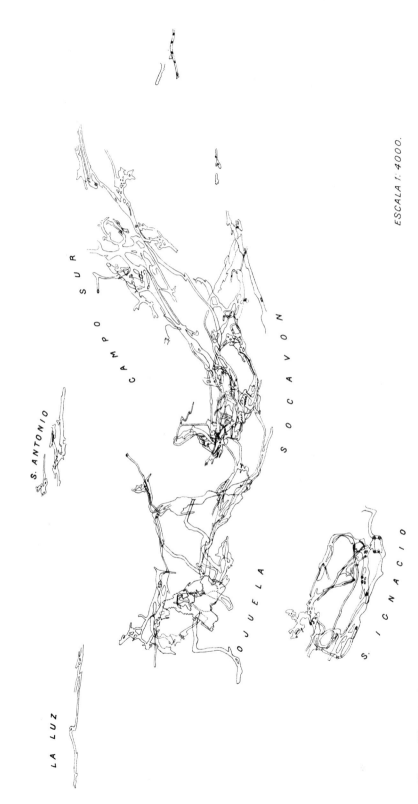

Figure 25. A plane view of the Ojuela, Socavon, San Ignacio, Campo Sur, San Antonio, and La Luz mines, Mapimi mining district, circa 1906. (Villarello, 1906, end page, copyright 1906 by Litog y Tipog. S. A.)

ESCALA 1: 4000.

where the solid sulfide ores are encountered. In the Ojuela Mine, the water table where the solid sulfide ores are encountered. In the Ojuela Mine, the water table occurs between the twelfth and thirteenth levels (Fig. 26).

There are four distinct zones of mineralization in the district, each with its own mineral assemblage: the copper-bearing contact zone; the zinc/lead zone; the silver/lead zone; and the barren carbonate zone. In general, the zones are easy to distinguish from each other, although some overlapping does occur. The copper-bearing contact zone principally contains calcium silicates, such as

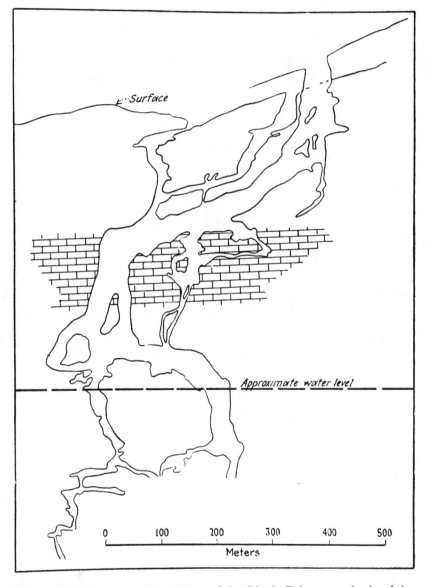

Figure 26. A cross-sectional view of the Ojuela-Paloma ore body of the Ojuela Mine. (Prescott, 1926, p. 289, copyright 1926 by Engineering and Mining Journal/Penoles)

amphibole, wollastonite, and pyroxene; also arsenopyrite, chalcopyrite, pyrite, and in lesser amounts galena, sphalerite, enargite, and molybdenite. Arsenopyrite is the principal sulfide in this zone and is important because its oxidation provides the needed arsenic to combine with the copper, lead, iron, and zinc to form the many rare arsenate minerals for which Mapimí is especially famous. the zinc/lead zone principally contains sphalerite and galena with lesser amounts of arsenopyrite and calcium silicates, and smaller quantities of pyrite, chalcopyrite, molybdenite, enargite, and marcasite, which forms psuedomorphs after enargite. The silver/lead zone mainly contains argentiferous galena and lesser amounts of sphalerite, pyrite, chalcopyrite, enargite, molybdenite, stibnite, and arsenopyrite; the arsenopyrite is now less than 1 percent. The barren carbonate is remarkable only for its many small veins of calcite, siderite, quartz, and stibnite. However, at times pyrargyrite occurred in quantity, which in the early years provided much of Mapimí's high grade silver ore. Of the four mineral zones, the oxidized silver/lead and zinc/lead zones are of primary interest to the mineral collector and have produced most of the fine mineral specimens from this district.

SOMBRERETE

Sombrerete, located in the northwestern part of the state of Zacatecas, must enter the history books not so much for the vast quantity of silver that was produced from its mines, but rather for the extreme richness of its silver ore bodies. To the collector, this mining district produced many of the famous silver specimens, which may be just simply labeled Zacatecas, and which in all likelihood were found in the Sombrerete mining district.

History

It is known that between 1530 and 1555, the Spanish maintained several detachments of troops in this region of Zacatecas. Silver was discovered in the mountains just to the north of Sombrerete in 1548. In 1555, Captain Juan de Tolsa discovered silver and began developing a mine at what in a few years was to become Sombrerete. By 1570, the convent of San Matero had been built and the mining camp elevated in status and was officially named, Villa de Yerena Real y Minas de Sombrerete.

The records of the first 100 years of operations at Sombrerete have been lost in antiquity. Because the silver was sent to Zacatecas for minting and taxation, these values were added to the mining district of Zacatecas and not credited to the district of Sombrerete. It was not until 1670 that official mining records separated the two. A royal mint was established in 1681 at Sombrerete to handle local silver, but because records were lost or destroyed it is difficult to place exact amounts of early production.

Early exploration took place on the two principal veins, La Veta Negra and El Pabellón. It was not until 1677, however, that Captain Francisco Costilla y Espinosa and his two associates began a major mining venture on the Veta Pabellón. In 1675, an astonishingly rich find of silver ore was made. It was hailed at the time as the richest silver strike in the entire world. A large mass of ore was uncovered in the Veta Pabellón composed entirely of light ruby silver or proustite, 32 meters (105 feet) long, 55 meters (180.5 feet) deep, and 1.2 meters (3.9 feet) wide, altogether a 2,112 cubic meters (73,914.8 cubic feet) mass of solid proustite. It took five years to mine out the ore, recover the silver, and mint it into coins. The official records placed a value of $30 million on this find which netted the king of Spain $6 million in royal taxes. The true amount returned to Costilla and his associates is not known, but the figure of $20 thousand income per day for five years, as stated in the mining records could be accurate. It also suggests that $6 million was unreported and some of this unreported profit realized from the bonanza was used to build the Church of San Juan Bautista in Sombrerete in 1679.

The official records that are available indicate that from 1675 to 1691, between $48 million and $60 million in silver had been produced within the district. During the 1680s and 1690s numerous mines were opened on the two veins, but in 1696, because of subterranean water flooding problems and disputes over ownerships, the mines began to close. Sombrerete almost became a ghost town. In the 1780s the Fagoaga family obtained ownership of the properties and resumed mining. What took place after resumption of operations is a truly remarkable tale of unbelievable riches, uncovered by persistence and luck. Don Jose Mariano Fagoaga, previously involved in mining ventures at nearby Fresnillo, visited the district on advice of his personal secretary, Señor Tarve. The latter persuaded his employer to risk $16 thousand in capital to work the Veta Negra, with Tarve becoming director of operations for one-fourth of the profits. Tarve hired Manuel Unzain, one of the best miners in the region, to manage the underground operations. Almost immediately a bonanza strike of silver ore resulted and in 1786 Tarve's quarter share of the profits netted him $360 thousand.

With enough money and knowledge of ore potentials in the district, Tarve decided to exploit the Veta El Pabellón on his own. In México City he settled up with the Fagoaga family and resigned his directorship of their mining operation on the Veta Negra. Unfortunately, he became ill and died at Zacatecas, but not before passing on his ideas concerning the El Pabellón potential to his friends, Don Jose Mariano Fagoaga and Don Juan Martin de Izmendi. Izmendi had the backing of the Fagoaga for mining the Veta Negra. The Fagoaga family had come upon hard times and development capital was exhausted. Don Jose Fagoaga gave the orders to stop mining. Izmendi disregarded the order and carried out Tarve's deathbed plans for running a cross-cut level from the Veta Negra to connect to the Veta El Pabellón just below where the great bonanza of 1675 had pinched out and disappeared.

When the San Rafael cross-cut was completed in 1792, it was 1 meter (3.28 feet) lower than planned, a most fortunate mistake as it turned out, because it just barely struck the top of the second major bonanza of proustite ore. This strike still remains the richest strike ever made at a single point of a vein ore body. The actual size of the proustite mass is not fully known, but it was said to be 30 meters (98.4 feet) long, 50 meters (164 feet) deep, and of unknown width. Because of the extreme richness of this discovery, records were poorly kept to avoid the heavy royal tax. The official records show the strike produced 5,000 metric tons of proustite ore that assayed in 1792 at not less than $2,300 per metric ton in silver. It required over eight months of steady work to excavate the ore, and from 1792 to 1811 to mill and mint the silver. The Fagoaga family's share of this bonanza came to almost $12 million. Don Jose Fagoaga, now extremely wealthy and influential, was made Marquis de Apartado by the king of Spain. In 1811, when the Fagoaga's quit mining at Sombrerete, they had mined over 244,944 kilograms (720,000 troy pounds) of silver at a profit greater than $40 million. Tarve's estate paid the municipal indebtedness of Sombrerete and established the first public school.

The outbreak of the War of Independence in 1810 coupled with increasingly lower grade ore combined with the increase of metallurgical problems the district's mines closed. The silver-rich ore found now was contaminated with arsenic, which came from the increasing quantities of arsenopyrite encountered in the lower levels of the district's mines. The arsenic prevented recovery of the silver from the ores. The best metallurgical scientists were at the University in México City, but despite their efforts they could not solve the silver recovery problem. The War of Independence (1810–1821) put an end to any further scientific investigations of the problem in the district.

During the War of Independence, the mines were worked by a cooperative of buscones (prospectors) gouging out small pockets of high-grade ore. In 1824, a British company entered Sombrerete and set up an efficient mining operation. High expenses, law suits, and domestic problems in México forced closure of the operations in 1832. The buscones again worked the mines for high-grade ore until 1880, when the Sombrerete Mining Company, backed by U.S. capital, resumed formal mining after erecting new buildings and installing hoists, pumps, and reduction works of the most modern design. With the completion of the railroad, passing 29 kilometers (18 miles) to the west, high transportation costs were greatly reduced. With the approach of the revolution in the early 1900s, mining again slowed and eventually stopped. In the 1930s, a Mexican

Figure 27. The Tocayos Mine is the largest mine now operating within the Sombrerete mining district.

subsidiary of the U.S. owned Sombrerete Mining Company took over mining, but on an intermittent basis.

Until the late 1890s, the ore within the district averaged a remarkable $330 per metric ton in silver. However, ore that assayed only $220 per metric ton or less was not mined and left in reserve. Mexican mining records report that until 1900, the mines of Sombrerete produced $300 million. This reconstructed figure is probably somewhat higher than the actual figures that were destroyed or lost during the two revolutions.

Mining at Sombrerete is presently conducted by Compañía Fresnillo's Unidad Tocayo (Fig. 27) and averages 220 grams (7.1 troy ounces) silver per metric ton. The company extracts 500 metric tons daily from its operations. The principal silver mineral is still proustite, but acanthite is gaining and almost equally important. The earlier metallurgical problems of arsenic in the deeper portions of the ore body have been overcome by modern milling and processing methods.

ZACATECAS

Zacatecas, the capitol of the state of the same name, is the center of one of the world's leading silver districts. It is of special interest and importance to the mineral collector. Zacatecas is located in the Mesa Central Mineral province 73 kilometers (45.4 miles) east of the Sierra Madre Occidentals mountain range in central México and 790 kilometers (491 miles) northwest of México City (Fig. 28).

History

On September 8, 1546, a detachment of soldiers, friars, and Indian scouts camped at the base of Cerro de la Bufa at a site that later would become the city of Zacatecas. The captain in charge, Juan de Tolosa, with the help of his scouts, convinced the local Indians that they meant no harm, and, in return, the Indians offered them a few pieces of silver. The Indians told them the place where the precious metal could be found. Tolosa convinced his friends, Captain Cristóbal de Onate, the acting Governor of Nueva Galicia, Diego de Ibarra, and Tenino de Bañuelas, to help finance his prospecting and mining ventures. They established the first mining camp on the northern frontier here at Zacatecas. The four partners began a silver aristocracy that changed life in New Spain. The silver that made them the richest men in the Americas was used to build churches and encourage missionary endeavors, which served as bases for frontier expansion.

Historians disagree on the specific dates when ore veins and mines were first discovered and worked. Church records in Spain allege that Tolosa discovered silver in 1546, probably at the place shown to him by the local Indians, and most likely the Veta Grande. According to other records, however, the Veta San Barnabe was the first vein to be discovered and the San Barnabe Mine the first mine to be worked beginning on June 11, 1548. Records in México City confirm the San Barnabe or the Veta de Malanoche as being first discovered sometime in 1548, and the Alvarado Mine as first formal operation, which commenced mining on March 21, 1548. The Alvarado Mine, however, is not located on the San Barnabe vein, but rather the Veta Grande. Although the dates and the

Figure 28. The small silver mining camp of Zacatecas has grown into a modern capital city, 1979.

names of the mines are a bit confused, the credit for all first discoveries remains with Tolosa and his associates.

High-grade silver ore was discovered in the San Barnabe Mine on the surface and first mined with a trench 732 meters (2,401.6 feet) long following the vein. As surface ore was exhausted the mine was developed in depth. During the early period of Zacatecas mining, a third major vein system was discovered, the Quebradilla or Veta de la Cantera that brought a big rush to the district. By 1550, 34 different companies were in operation, and by 1562, 35 Haciendas de beneficios or ore beneficiation patios were processing the ore. A steady supply of Negro slaves and Indians comprised the early labor force in the rapidly developing mining district.

Early Zacatecas ore bodies were often so rich that it was common to recover 18.7 kilograms (50 troy pounds) of silver from 45.4 kilograms (100 pounds) of ore. Between 46 and 73 meters (150.9-239.5 feet) below the surface, the ore changed in character from oxidized (colorados) to sulfides (negros) ore. Whereas the oxidized ore contained native silver, chlorides and bromides of silver, and iron oxide, the sulfide ore contained azurite in the upper zone, also silver sulfides, ruby silvers, pyrites, galena, and sphalerite. The mining district became famous for its specimens of dark ruby silver or pyrargyrite.

In 1588, the king of Spain, very pleased by the great quantities of silver sent to him from the district's mines, presented Zacatecas with an official coat of arms and named the city, The Very Noble and Loyal City of Our Lady of Zacatecas. By the early 1600s the city boasted of 300 large residences of wealthy Spaniards, but Negro and Indian miners lived outside the city in very poor conditions in the small town of Guadalupe. In 1606, a Royal Mint (Caja

Real or Casa de la Moneda) was established in the city as one of the first and finest outside of México City. Its 300 employees operated three die stamps which could mint six reales or pieces of eight ($48) per minute. The practice among the mine proprietors was to turn in their silver bars on the first of the week and by the last of the week pick up newly minted coins less the appropriate taxes for the king.

By 1629, Zacatecas became the third largest city in New Spain, after México City and Puebla de Los Angles. As it continued to grow, the city fathers added fountains and beautiful parks, built a large aqueduct to bring in better drinking water, and built many fine churches. In the early 1700s, the Franciscan Order built the College of Our Lady of Guadalupe in the village of Guadalupe expressly for the training of missionaries for the conversion of the Indians to the north. By 1707, it was serving as the base of all missions to the north, including missions in what would become the western United States.

From 1600 to 1750 Zacatecas mines alternated periods of good production with leaner times. Often, when anyone would least expect it, bonanza ore would be found and bring the district back to life. For example, during 1765 to 1782 a bonanza was uncovered by Jose de la Borda, a mine operator who was very familiar with bonanzas. La Borda had made and lost over $30 million in the silver and gold mines of Tlalpujahua and Taxco. When he arrived in Zacatecas, he only had the $100 thousand in capital that he received from the sale of the gold, silver, and diamond custodia. In 1765, he and Señor Ansas worked the mines on the claims of San Acasio, Alvarado, and San Francisco on the Veta Grande. A new shaft in the San Acasio Mine exploited a massive silver strike and over the next 17 years, $20 million in silver was recovered. La Borda's silver strike is still famous in the district, not only for its richness but also for the long period of production of rich ore. In 1775, la Borda opened the long idle Quebradilla Mine. After much expense in draining the older workings, he uncovered a small area of bonanza ore but decided not to push his good luck too far and retired with profits of $2 million. When he died, he left an estate worth over $3 million to his heirs.

The Fagoaga family, mentioned earlier in connection with their mining activities in Sombrerete, also appear in the historical records of Zacatecas. In the late 1700s and early 1800s members of the family worked the Veta Grande and in a few years made a princely fortune of almost $17 million. They purchased the Hacienda de la Savceda from the la Borda family and retired from mining.

As the mines grew in numbers, size, and depth along the three major vein systems, the older less efficient mining methods could no longer be tolerated; mine flooding was a major problem on two of the vein systems. The number of malacates (horse driven hoists) used to haul ore and water from the mines had to be greatly increased. The problem was so severe at the Quebradilla Mine that they increased its number of malacates to 16 and still could not keep the mine dry. By the early 1800s, 1,415 miners worked underground, but this crew required 1,135 workers and 800 horses on the surface to operate the malacates and other equipment. When the War of Independence (1810-1821) was over, the newly established Mexican government asked England to help solve its mining problems. In the late 1820s, the English firm of the Bolaños Company began working the Veta Grande. They brought with them the most up-to-date

European mining methods and solved the water drainage problem by installing massive beam engines (Cornish pumps). As a result, the need for large crews above and below the ground was greatly lessened and the Quebradilla Mine for example, reduced its work force from 2,400 to 550 employees. As the mines deepened, however, the ore increased in silica content that made the patio process less effective in recovering the precious metals. In 1889, smelters were installed by the American companies that were operating in the district and the patios abandoned. By the early 1900s the more economical cyanide process was in general use for precious metal recovery in the district. Keeping pace with production of lower grade ores were further cost-cutting measures, such as bringing in a railroad spur (1883) and cheaper hydroelectric power from outside the Zacatecas mining district (1906). Even with such steps many mines closed. They slowly filled with water; only the mines that produced good ore remained in operation. Shortly after the turn of the century, the American companies installed larger pumps and better steam hoists, which allowed for considerably deeper work. Now, in addition to the silver, copper was discovered, and in the early 1900s exceeded silver by a ratio of three to one in ore tonnage.

During the 1910s and 1920s, the district's mines were plagued with economic problems because of México's second major political revolution. They either closed or reduced production to very low levels. In the 1930s, when the American companies pulled out of the district, the mining operations were transferred to their Mexican subsidiaries or taken over by Mexican government-backed companies. At present, the old Spanish El Bote Mine, operated by government's Fomento Minera, still recovers silver and employs 100 miners (Fig. 28). Compañia Fresnillo still operates mines on the famous Veta Grande, while other small independent mines are operating on all three veins.

Zacatecas ore veins. The three major ore veins previously discussed are impressive, not only for their size, but also for their general richness (Fig. 29). The Nuestra Señora de Guadalupe de La Veta Grande, better known simply as Veta Grande is the longest; it is over 5,250 meters (17,224 feet) long, ranges in width from .3 meters to 28 meters (1-92 feet), and averages 9.1 meters (29.9 feet) wide for over 400 years of mining. It is the main vein of the District and one of the major veins in México. Its richest ore assayed as much as 24.0 kilograms (771.3 troy ounces) of silver per metric ton but the average was 2.3 kilograms (72.8 troy ounces) per metric ton in silver, with very little gold. Unlike the other two vein systems, underground water posed few problems in the Veta Grande.

Between the Veta Grande and the city of Zacatecas is the Descubridora or San Barnabe vein, which in later years was called the Malanoche Vein. It is more than 3,658 meters (12,001 feet) long, averages 4.6 meters (15.1 feet) in width, and carried an average assay of 27 kilograms (88.2 troy ounces) of silver per metric ton and a trace of gold. The ore occurs in soft slate and is the easiest to mine in the district, but there are water problems. In over 400 years of production, four major bonanzas have been encountered on this vein system, one which produced over $600 thousand in one week from one mine gallery alone. The third major vein system is the Cantera or Quebradilla Vein, located directly beneath the city of Zacatecas. It is the smallest of all, with a length of only 1,830 meters (6,004 feet) and ranges in width from 3.7 meters to 11 meters (12.1-36.1 feet). The Quebradilla Vein produced a large amount of low-grade silver ore and three major high-grade bonanzas. Major water problems made

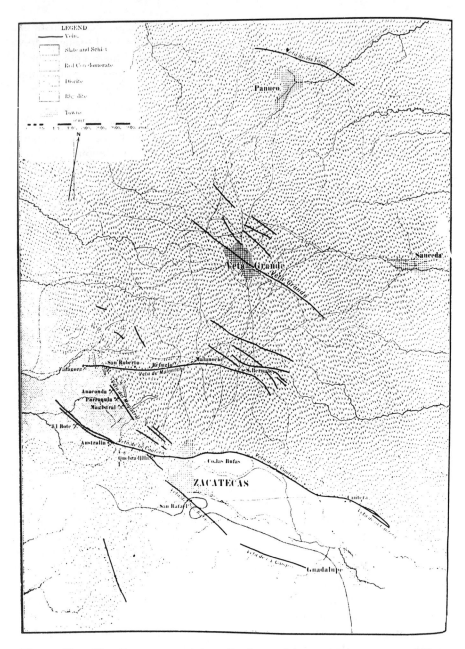

Figure 29. The Zacatecas mining district and its ore vein systems. (Rice, 1908, p. 404, copyright 1908 by Engineering and Mining Journal)

this vein the most expensive to exploit. It was not uncommon to spend over $400 thousands annually to drain the mines. In over 400 years of production, assays averaged 1.2 kilograms (38.6 troy ounces) of silver per metric ton.

According to the figures given to Von Humboldt in 1804, Spain shipped 576,283 kilograms (1,543,958 troy pounds) of silver annually to Europe and Asia from the ports of Veracruz and Acapulco. The mines of Zacatecas provided a major share of this quantity and Zacatecas was second only to Guanajuato as the leading silver producer in México and the world during the 1700 and 1800s. Using 1980 dollars the Zacatecas mining district in its 400 year career produced in excess of $2-3 billion in silver, gold, and copper.

Geology and Mineralogy

The rocks of the district are diorites capped in places with rhyolite. In the western portions of the district the older and deeper slates have been altered by the intrusions from below. The zone of oxidation in the district varies from 46 meters to more than 76 meters (150.4-249.3 feet) beneath the surface. The minerals from this zone include native silver, acanthite, chlorargyrite, bromargyrite, quartz, calcite, siderite, and barite. Beneath the oxidation zone is the sulfide zone that contains azurite, malachite, pyrargyrite, acanthite in the upper regions and galena, sphalerite, stibnite, chalcopyrite, quartz, and calcite in the deepest areas.

GUANAJUATO

Guanajuato, the capitol of the state of Guanajuato is located 274 kilometers (173 miles) northwest of México City. The city enjoys the title of King of the Mexican Mining Camps because of its mines producing billions of dollars in gold and silver in their 430 year history (Fig. 30).

History

Local Otomi Indians, under the control of the more powerful Tarascans, mined some surface silver and gold from the large outcrop of the rich Veta Peregriña in the hills above the Río Guanajuato. When the Spanish entered the region, all that remained of the outcrop was a trench 18 meters (59 feet) wide, 31 meters (101.7 feet) deep, and 457 meters (1,449.3 feet) long. The Spaniards became interested in the area when a silver strike was made near Cerro Cubilete in 1548. A group of mule skinners returning to the mines of Zacatecas from México City made the discovery. They noticed silver dripping from the red-hot rocks in their camp fire. They quickly located the spot where they found the rocks and with a little digging uncovered a vein of silver. This became the San Barnabe claim, whose richness set off a rush to where the mining camp of La Luz was established and the Veta La Luz was being developed. Further explorations uncovered rich silver ores in the hills above the Cañada de Marfil. A mining camp was established in 1554 in the valley below. In 1557, the growing mining camp was given the name of Santa Fe y Real de Minas de Quanaxhuato. A military garrison and four small forts were built to defend the developing mines and mining camps from the hostile Indians. On April 15, 1558, the Mellado Mine and the next day, April 16, the San Juan de Rayas Mine were begun. In 1619, the town was elevated in status to a Vílla. On December 8, 1741 King Felipe V granted the title, Very Noble Loyal City and renamed it Santa Fe de Gonnajoato.

As Guanajuato grew, buildings began to line the narrow twisting cobblestone

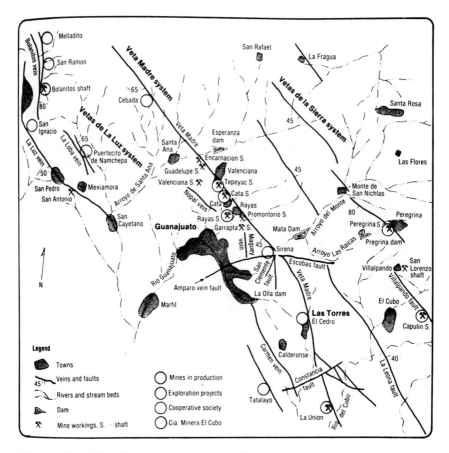

Figure 30. The Guanajuato mining district and its ore vein systems. (White, 1980, p. 157, copyright 1980 by Engineering and Mining Journal)

street that lead up the canyon to the main plaza and to the parish church. The Parroquia, as the church was named when it was dedicated in June of 1696, was renamed the Basilica of Nuestra Señora de Guanajuato. On its main alter is the most famous, and deeply venerated, statue of the Virgin of Guanajuato. The statue was made in the seventh century and presented to the city in 1557 by King Phillip II. It rests on a large base of pure silver, which was a gift from the miners of Guanajuato.

In 1809, the city built on a hill above the Río Guanajuato, a fort-like granary (Alhondiaga de Granaditas). During the Mexican Revolution of 1810, the granary served as a fort and prison. It became the focal point of the early days of the revolution, the first battle was fought here in September of 1810.

In 1812, a mint was established in Guanajuato to coin local silver. Because it was one of the most modern in all of México and the district mines continued to produce silver, it remained opened until the 1900s.

Discovery of the Veta Madre. The fabulously silver-rich Veta Madre or mother lode of the Guanajuato mining district was discovered in 1558. The vein, principally composed of amethyst and calcite, carries not only silver, but considerable amounts of gold too. The vein ranges from its narrowest width of 48 centimeters (18.9 inches) to over 49 meters (160.8 feet), often branching into three sub-veins with bonanza ore encountered at their junctions or where they narrowed. It is 12.9 kilometers (8 miles) long and has been worked to 640 meters (2,099.7 feet) below the surface with indications of greater depth. While the Veta Madre averaged only 2.42 kilograms (77.8 troy ounces) of silver per metric ton it was not uncommon for assays to run from 103 kilograms (3,311.5 troy ounces) to below 341 grams (11 troy ounces) of silver per metric ton. The Valenciana Mine, the major operating mine on the Veta Madre, graded its ore into four classes: (1) $6,600.00 silver per metric ton, (2) $220.00 silver per metric ton, (3) $132.00 silver per metric ton, and (4) $50.00 silver per metric ton. It was the practice in the district to leave ore in place when it assayed less than 330 grams (11 troy ounces) per metric ton, or to stockpile it for later extraction. Records shown to Alexander von Humboldt during his visit in 1803 indicated that over 21,723,000 kilograms (58.2 million troy pounds) of Guanajuato gold and silver had been produced in 245 years. Other vein systems were discovered in the district, but could never rival the Veta Madre's production records.

The first opening on the Veta Madre, the Mellado Mine, commenced on April 15, 1558, when a shaft was sunk where the surface silver was originally discovered. The mine owner, Don Matias de Bustos, also owned and developed the Cata Mine, a few kilometers northwest of the Mellado. The Mellado later passed into the Ruhl family who operated the mine until the mid 1800s, at which time the property became part of the Rayas Mine group. The latter, the San Juan de Rayas Mine, was the second major operation on the Veta Madre, thought to be named after its discoverer Juan Raya, who ran out of operating capital and sold the mine to Diego de Ahedo. He continued the mine development but sold the property to Don José de Sardeneta y Legaspi, an experienced and knowledgeable miner. The new owner struck rich ore and became a very wealthy person, receiving from the king of Spain the title of Marquis de Rayas. Vigorously pursuing his dream of striking a great bonanza, the Marquis acquired adjacent properties and expanded mining operations. He died, however, before he could realize his dream. His son, the second Marquis de Rayas, continued the operations and at depth discovered bonanza ore with a second and even larger bonanza being found soon after. Both bonanzas brought to the Sardeneta family over $11 million in profits. Gainful mining continued until 1780 when a flash flood in the mountains above filled the mine with water and killed over 100 of the miners. It took the next 19 years to drain the mine before ore could be produced once again.

In 1803, the famous octagonal shaft, Tiro General, was begun at the Rayas Mine. It measured 12.2 meters (40 feet) in diameter and at the time was the largest such shaft in the world. When the Mexican Revolution began in 1810, the shaft had reached 267 meters (876 feet) below the surface, far above the planned depth of 535 meters (1,755 feet). Problems in financing the mining operations caused Sardeneta to obtain funds from his cousin, Don Lucas Alaman, who at the time was director of a large English mining venture in Guanajuato. Because of the steep slope of the mountain where the Tiro General was being sunk, a 31 meter (101.7 foot) high wall with heavy buttresses was built. It was backfilled and leveled to provide room for the malacates (house powered hoists) needed near the edge of the shaft. Work resumed on the Tiro General, which when finished was served by eight malacates (Fig. 31 and 32) that allowed the exploitation of the Sangre de Cristo Labor silver bonanza. During its peak, the bonanza produced 516,196 kilograms (516 metric tons) per

Figure 31. The famous San Juan de Rayas Mine 1970.

day that assayed 1,028 grams (33.1 troy ounces) of silver per metric ton. Often large areas of this bonanza encountered ore that ran assays over 4.9 kilograms (157.5 troy ounces) of silver per metric ton.

The Valenciana Mine. The discovery of the Valenciana Mine began as a dream of Pedro Lucero Obregón y Alcocer. His life-long ambition was to own a silver mine. He searched along the Veta Madre and finally located a claim along a part of the vein thought by others to be totally barren of silver. In 1760, he started to sink the San Antonio shaft, but when his working capital was exhausted he attempted to get his friends to finance his mining. They called him the "Fool of Valenciana," saying there was no silver where he was mining. Finally, a store owner in Rayas, Don Pedro Luciano Otero, joined him in his mining venture. Shaft sinking continued and after 6 years of passing through barren ground they struck a great bonanza at 300 meters (984.3 feet). The "Fool of Valenciana" and his partner were now the richest men in México, and perhaps the entire world. The king of Spain, in 1780, was so impressed by the enormous silver strike that he awarded Obregón the title of Conde de Valenciana. Early years of mining in the Valenciana produced from the original bonanza and others that followed vast quantities of silver. From 1766 to 1803, the ore never dropped below 2.1 kilograms (67.5 troy ounces) of silver per metric ton and sometimes ran as high as 115.7 kilograms (3,719.8 troy ounces) of silver per metric ton. During this 38 year period the Valenciana Mine produced just over 4,255,000 kilograms (11.4 million troy pounds) of silver.

Other shafts, such as the hexagonal shaft of the Nuestra Señora de Guadalupe, were sunk to reach the various levels in the expanding Valenciana Mine. But these were still not enough to handle the vast amount or ore being produced. At the time the mine employed 1,800 miners and 1,300 workers on

Figure 32. The largest malacates in the world was at the San Juan de Rayas Mine, circa 1905. (Martin, 1906, p. 69, copyright 1906 by Cheltenham Press)

the surface. In 1801, the octagonal shaft of El Señor San José, 9.8 meters (32.1 feet) in diameter and 581 meters (1,906.2 feet) deep, was begun and completed just prior to the Mexican Revolution of 1810 at which time mining ceased and the mine filled with water and fell into disrepair.

In the early 1820s, an English firm began to work the water-filled mine. They used eight large malacates to hoist the water and slowly drain the mine. In 1825, active work began around the clock to drain the mine and recondition the mine workings. After 21 months, the water was lowered over 150 meters (492.1 feet) and ore was finally sent to the surface. The ore decreased in value with depth and only as the older upper levels were extended laterally was good ore recovered. Cornish steam pumps shipped from England were installed but even they could not handle the water flow as the El Señor San José (Tiro General) shaft was lowered to the 640 meter (2,099.7 foot) level. With all its great wealth, the Valenciana underground workings had been poorly organized. The shafts, drifts, and galleries were built too large, thus increasing operating costs greatly. It was not uncommon for the mine's crosscuts to be over seven meters (23 feet) high to aid ventilation and help lower the 32°C temperature (90°F) underground. The galleries or stopes were also enormous; this was thought necessary for good exploration. Even the inclined stairways were built to grand proportions and were often 7 to 10 meters (23-32.8 feet) wide. With the lowering ore values and high drainage costs, the English mining efforts were not successful. When profits fell, the Valenciana and other mines closed and the English left the district.

Even though the mine owners were anxious to make vast fortunes, most never forgot that God had watched over their destinies and consequently built large shrines, chapels, or churches. Conde de Valenciana was no exception and in 1765 began construction of a large church on a hill overlooking the Valenciana Mine. The church, San Cayetano, dedicated in 1788, was said to have been constructed with mortar mixed with finely crushed silver ore, Malaga wine, and water. The three altars inside the church were made of carved wood and gilded with gold and silver from the Valenciana Mine.

Stealing was such a constant problem at the Valenciana and other mines in the district, that mine owners hired people to search for stolen ore. Miners at Guanajuato as well as most mining districts in México stole not only ore, but tools, clothes, and candles. During the 1700s and 1800s, it was not uncommon to see the church alters ablaze with companies' candles during the local saint's day. The mine owners often asked the local parish priests to give an occasional sermon on stealing. It was always remarkable that on the day after such sermons, tools and other things that had disappeared mysteriously reappeared. The priests were well paid by the mine owners for their help.

In 1890, American interests investigated Guanajuato. They looked at the dumps, which contained ore that was considered too poor to process earlier. Such dumps assayed from 136.7 grams to 206.1 grams (4.4-6.6 troy ounces) silver per metric ton and a trace of gold. When the dumps were combined with the higher grade ore in the pillars and backfills in the mines, operations became profitable. Cheap electrical power arrived in 1889. The improvement in rail service caused operating costs to drop considerably. District mines were able to reopen and expand. New milling techniques and the use of cyanide for better metal recovery further lowered costs and enabled processing of abundant lower grade ores. Properties under the control of U.S. firms included many older and famous mines such as the Valenciana, Cata, Rayas, Mellado, and Purisima. They operated from as many as twenty shafts on the Veta Madre. The other two veins of the district, Veta de la Sierra and Veta La Luz, were also developed by U.S. capital. By the late 1920s, the silver and gold values in the ore greatly lessened and the veins began to be pinched off. Combined with falling metal prices, profitability decreased and mines began to close. It had always been thought that the Veta Madre was part of a larger and richer vein and excitement reached a new high in the district when an unknown section of the Veta Madre was discovered. Unfortunately, it was of such low grade ore and the ore was buried so deeply that it was not economical to mine. The election of President Cardenas in 1935 brought an end to foreign investment in México. The mines closed again and fell into disrepair. The Mexican Government established a cooperative to develop the mines of Guanajuato. Several of the older and better producing mines were reopened and operated on a limited basis, recovering what high grade ore that remained in the upper levels.

Guanajuato today.   In the late 1960s modern geophysical exploration techniques were used on the southern end of the Veta Madre and uncovered good ore. Also, the La Luz and Sierra vein systems were explored and ore was discovered there. In 1976, the mines reopened. Some openings were newly dug shafts, others were old original shafts. The ore was not as rich as in the past, but its volume resembles past reserves. The cooperative built a new headframe over El Señor San José (Tiro General) of the Valenciana Mine, improved the hauling system of the Rayas Mine and began to remove lower grade ore from the pillars and fill from the old galleries. These ores average about 110 grams (3.5

troy ounces) silver per metric ton, a far cry from the bonanza ores of years past. New mines were reopened in the mid 1980s while older ones closed.

The true production values for Guanajuato will never be known because early mine owners did not keep proper records in an effort to avoid the Royal tax imposed by the king of Spain. The early records that did exist for the most part were either lost or destroyed during periods of political unrest in México. Records in Europe, the United States, and México estimate the output up to 1985 from the district's mines at between $5 billion and $6 billion (1980 dollars and metal values) in silver and gold. The Valenciana Mine alone has produced over $1 billion of the district's total production. The values taken from the official tax records show the Valenciana Mine from 1771 to 1886 produced just over $700 million in silver and gold. The San Juan de Rayas from 1558 to 1886 had produced over $500 million dollars and its neighboring mine, the Mellado, produced during the same period $160 million dollars in silver and gold. These values are as reported in metal prices and dollar values of the period. It is estimated that even today, one-sixth of all the silver in use everywhere originally came from Guanajuato. The mines of Guanajuato and especially those on the Veta Madre have had a major impact on the world and its economy.

Geology and Mineralogy

The geology of the district is similar to the geology of other silver districts in the tableland of central México. Basement rocks are granite, upon which rest a layer of foliated slate capped with sandstone, limestone, and a porphyritic rock composed of syenite, serpentine, and hornblende. The ore veins of the district intrude these rocks, but are best developed in the foliated slate. Vein minerals are of two types: (1) oxidized (colorados) ores containing native silver, chlorargyrite, quartz, and calcite as the principle minerals; and (2) sulfide (negros) ores, containing principally pyrargyrite, argentiferous galena, acanthite, polybasite, stephanite, gold, chalcocite, pyrite, chalcopyrite, quartz, and calcite. Associated with both ore types are fluorite, barite, chalcedony, siderite, adularia. Rare minerals species found in the vein of Guanajuato and La Luz are guanajuatite, paraguanajuatite, aguilarite, and naumannite.

ZIMAPAN

Zimapán, the seat of government for the Municipio de Zimapán, is located in the western portion of the state of Hidalgo, 160 kilometers (99.4 miles) northwest of México City. The mines of the district have produced large amounts of ore, but on a whole, the district is little known to collectors. The mines themselves are located about 15 kilometers (9.3 miles) northwest of Zimapán and are situated in the extremely rugged and heavily vegetated Barranca de Toliman and several small side canyons.

History

Silver was discovered in 1632 by a local native, Lozenzo del Sabra, on the side of one of the canyons of the area. A narrow path zigzagged along the side of the sheer 914 meter (3,000 foot) cliff to the developing mine. The rich silver vein and the mine were named after the mountain on which they are located, the Lomo del Toro. The Veta Lomo del Toro (vein) and the Lomo del Toro Mine developed into the most important vein and mine in the district. The Veta Lomo del Toro and other silver-ore veins in the district, rich in lead, assayed 1.4-1.8 kilograms (45-57.9 troy ounces) of silver per metric ton. Because the ore could be directly smelted, the Conde de Regla, who operated the mine during its heyday, brought down thousands of tons each year to serve as fluxing ore in his smelters at the Hacienda de Regla near Pachuca de Soto, some 130 kilometers (81 miles) away. Many tunnels, drifts, and immense stopes or Claros were excavated in the ore bodies, which at the time were believed to be inexhaustible. Mining continued uninterrupted from 1632 to the beginning of the War of Independence in 1810.

A notable bonanza was struck in the Lomo del Toro Mine in 1793, yielding 5.2 million kilograms of lead/silver ore for Don Angel de Bustamante, Marquis de Batopilas, the owner at the time. He also operated the Santa Rita Mine in the district. Each year from 1791 to 1793 over $100 thousand in ore was mined from the Santa Rita. District mine records indicate that from 1785 to 1798 approximately $2 million worth of silver was produced with much of it being directly smelted locally. However, much ore was transported to other mining districts because of its value as a flux. During this period, much of the lead from the ores was recovered and used to make water pipes for México City. Fifty-two smelting furnaces were in operation in the district at the beginning of the War of Independence of 1810. From 1825 to 1828, Compañia Mineral Real del Monte y Pachuca mined the Lomo del Toro Mine and the Compañia Anglomexicana operated the La Cruz, Guadalupe, and San Fernando Mines. From 1830 to 1840, roughly $2 million was produced from the Lomo del Toro Mine. The la Cruz, San Fernando, and Guadalupe mines were also ore producers during this period. In 1840, just as the Lomo del Toro was going to close, a small bonanza was uncovered and quickly produced $40 thousand in profits. It appears that ore production slowed, and perhaps even stopped, between 1840 and 1879. Assay records, however, do show that during this period assays were running between $38.57 to 70.53 per metric ton in silver for whatever ore that was being mined. At the Centennial Exposition in Philadelphia in 1876, the ores and minerals from this district and the Lomo del Toro Mine won awards for their richness and beauty. From 1890 to 1901, two new adits, San Damian and San Guillermo were developed at the old Lomo del Toro Mine. The steep inclines or winzes of Los Bronces and San Agustin were driven from the old Spanish stope of Ave María to find and develop new bodies that produced a small amount of ore from 1901 to 1910. During this period, the La Luz Mine was also being developed.

The Mexican Revolution of 1910 caused the shut down of the Lomo del Toro Mine, which remained idle until 1928. Smaller operations, however, continued elsewhere within the district. In 1913, for example the Hidalgo Copper Mining & Smelting Company operating from the Concordia and several other mines smelted approximately 1.74 million kilograms (1,742 metric tons) of ore, assaying 21 percent lead and 806 grams (25.9 troy ounces) of silver per metric ton.

By the early 1920s, Compañía Fundidora y Minera de Zimapán, the Hidalgo

Copper Mining and Smelting Company, and Presisser family were operating within the district, but their operations were hampered by a shortage of working capital and extremely high freight costs. In 1928, the Lomo del Toro Mine reopened. Within the next year Compañía Minera Mexicana Negociación Minera La Aurora was operating other mines within the district. In 1945 exploration from the Lomo del Toro Mine toward the Los Balcones Mine discovered a large oxide ore body. The road into these two mines was improved and paved to help increase production. In 1948, Compañía Minera Fresnillo began to explore portions of the district and in 1949 began to ship ore. Compañía Minera La Llave was the only company smelting ore in the district, everyone else was shipping their ores to smelters outside the district. The total from the Lomo del Toro and Los Balcones Mines from 1929 to 1948 was about 3.2 million kilograms of lead/silver ore. At present, the Lomo del Toro and Los Balcones Mines are still in operation, as well as several of the older, smaller mines within the district. Most of the mines have been modernized and more efficient milling plants installed to economically process low-grade sulfide ores. Over its 355 year history, the Zimapán mining district, has produced an estimated $300 million (1980 dollars) in silver, gold, lead, and zinc.

Geology and Mineralogy

Area rock formations in the Zimapán mining district consist of shale, limestones, shaley limestone with sandstones and mudstones of Mesozoic and Jurassic ages, followed by fanglomerate, andesite, and basalt of Tertiary age. All are topped by another fanglomerate of Pleistocene age and covered with recent alluvium and terrace deposits. The entire rock sequence has been faulted, folded, and metamorphosed then heavily eroded. Ores occur in veins along the many faults, while replacement bodies in limestone are found forming typical mantos and chimneys for which México is famous. The chimneys, which are essentially vertical and have the general appearance of pipes, have been found up to 35 meters (114.8 feet) in diameter and 65 meters (213.3 feet) long; they are the principal source of ore in the district. The district's mantos have been found up to 150 meters (492.1 feet) in length, 75 meters (246.1 feet) in width, and 15 meters (49.2 feet) in thickness. The minerals at Zimapán are formed as either pyrometasomal (tactite) and mesothermal. The former is characterized by sphalerite, galena, pyrite, arsenopyrite, pyrrhotite, chalcopyrite, and jamesonite; the later by sphalerite, galena, pyrite, and calcite. Both types have been oxidized to a depth of more than 200 meters (656 feet). The most important ore mineral in the oxidized ore zone is plumbojarosite, which contains a high concentration of silver but is probably a mixture of both plumbojarosite and argentojarosite. It was mined in large quantities in the earlier years of the district. Other important oxide zone minerals are cerussite and anglesite. Sulfide ore contains pyrite, sphalerite, galena, chalcopyrite, arsenopyrite, and pyrrhotite. It is interesting to note that vanadinite was first identified from Zimapán by Andrés del Río in the 1790s. He received no credit, but was later credited with the identification of a new element, vanadium, from the district's ores (see vanadinite in catalog for the complete story). The Zimapán district boasts a list of ninety two different minerals.

TAXCO

Taxco, located halfway between Acapulco and México City in the state of Guerrero, is the oldest mining camp and district in México and in the Western Hemisphere.

History

Long before the arrival of the Spaniards in the early 1500s, the local Tlachuica Indians mined metals from small pits near their village of Tlaxco or Tlachco (known today as Taxco el Viejo). To obtain the gold, silver, and tin that was being mined and worked by the Tlahuicas, the Aztecs, under Itzcoatl, conquered them in 1445. Tlaxco became part of the Aztec empire and was forced to pay tribute to Moctezuma I in the form of gold bars.

Shortly after the Spanish Conquest in 1521, Hernán Cortés sent men into the region looking for tin that was desperately needed to make cannon bronze. He had seen tin disks being traded in the Aztec's large, open market at Tlaltelolco. Cortés had been told that the metal came from near Tlaxco. At first, the Spanish mined tin but soon discovered silver in the side of Cerro de Bermeja. In 1522, Cortés himself made a mining denouncement (claim) and began to excavate a mine tunnel called the Socavón del Rey or "Kings' Tunnel" beneath the area where the Tlahuicas had surface mined years before. In 1529, the mines were mentioned in historical documents as El Real de Tetelcingo. The small nearby mining camps were absorbed by the expanding district and by 1570 the entire district became known as the Real de Minas de Tasco. The Spaniards developed the silver lodes of La Argentita and La Marquesa.

Taxco continued to flourish and the now-famous silversmithing trade began. German historian and mining engineer Alexander von Humboldt reported that by the 1570s silver from the mines of Taxco had reached Europe. Local silvercrafting brought great fame to the city, a fame enjoyed throughout the colonial period and continues today. In 1581, the Spaniards renamed their developing mining camp, Tasco.

In 1716, French adventurer Joseph Laborda, known in New Spain as José de la Borda began working with his brother Francisco at his San Ignacio Mine. At first, their mining venture was unsuccessful and it was not until after his brother Francisco's death in 1743 that José discovered the fabulously rich San Ignacio vein. La Borda's annual share for the next twenty years was between 48,521 to 74,648 kilograms (129,996 and 199,994 troy pounds) of silver that yielded him $2-3 million yearly. La Borda stated that when God gives to Borda, Borda will give to God. Out of these words came the Church of San Sebastián y Santa Prisca in 1748. He personally supervised and employed a staff of European artisans to build the church to his triumph. It is a masterpiece of Baroque and Mexican Churrigueresque architecture, costing la Borda approximately $420 thousand when finished in 1758. After his great strike, however, the mines rapidly declined and la Borda was almost reduced to poverty. Through his son, the parish priest at Church of San Sebastián y Santa Prisca, he received permission from the archbishop of México to sell to the cathedral in México City the magnificent gold, silver, and diamond custodia, which la Borda had donated to the Tabernacle of the church. With the $100 thousand received from the sale, he and his wife left for the mining camp of Zacatecas. Once again la Borda would receive fame and fortune as a result of his mining activities.

On February 16, 1802, the district was devastated by a severe earthquake that disrupted and changed the flow of water from regional caves that was much needed to power the stamp mills processing silver. Without the water, the mills could not operate and new methods of powering the mills and mining equipment had to be brought into the district. In 1803, Alexander von Humboldt visited Taxco. He reported that since the end of the 1700s, the mines were producing just slightly more than 13,733 kilograms (36,900 troy pounds) of silver per annum.

During the Mexican Revolution of 1810, Taxco played an extremely important role. The Spanish forces that controlled Taxco were defeated by the Mexican rebel forces led by General Morelos; the General directed his forces from the safety and remoteness of Taxco. At the small convent of San Bernadino in Taxco, the Iguala Plan was written by the rebel leaders. The plan, a political philosophy for governing of México, united the rebel forces to defeat Spain and bring about independence to México in 1821. In 1872, by official decree, the Mexican government changed the earlier spelling of Tasco to Taxco.

With depth, ore values decreased and silver content lessened, while lead and zinc values increased. No longer were the veins yielding between 2.1 and 2.8 kilograms (67.5 and 90 troy ounces) per metric ton of silver. Because the new metals uncovered were of no interest, and the price of gold and silver was low, the mines were forced to close and the district remained inactive through the Mexican Revolution of 1910. Transportation, another problem which the district faced, was improved with the arrival in 1910 of the railroad. In 1928, when a modern highway reached Taxco, the mines were almost closed. However, in 1929, an American, William Spratling, settled in Taxco and reestablished the lost, local art of silversmithing. The city once again became famous for its silverware, jewelry, and other objects of art made from silver.

Prior to World War II, large scale mining was resumed in the district by American Smelting & Refining Company's ASARCO Mexicana, which later would become the Mexican owned Industrial Minera de México S. A. (IMMSA). Exploration found new fissure vein systems. Some had widths up to 10 meters (32.8 feet) and contained irregular ore bodies, which averaged only 137.8 grams (4.4 troy ounces) of silver per metric ton. The new emphasis was now on the recovery of lead and zinc. The silver that made the district famous was now of secondary importance. IMMSA expanded its operations in 1975 to mine and mill 2,400 metric tons per day from its El Pedregal, San Antonio, Jesús, and Guerrero mines. A new shaft, the El Solar, was sunk at the Guerrero Mine to a depth of 581 meters (1,906 feet) and developed by 60 meter (197 feet) separations on ten levels. Presently, this mine is the most important in the district. Complying with the Mexican government's orders, its buildings, shafts, and mills were colonial in style (Fig. 33). In 1981, IMMSA again expanded its mining and milling capacities to 3,575 metric tons per day.

In 1979 Taxco produced 80,000 kilograms (214,333 troy pounds) of silver; 16,120,000 kilograms (16,152 metric tons) of lead; and 29,680,000 kilograms

Figure 33. The old mining camp of Taxco and the Guerrero Mine in the distance, 1979.

(29,739 metric- tons) of zinc. Ore reserves have been increased to 8 billion kilograms (8 million metric tons) of proven ore that contains 137.8 grams (4.4 troy ounces) of silver per metric ton, 1.9 percent lead, and 4.2 percent zinc. A large amount of ore that has not been proven could more than double this figure. The mining future of Taxco is very promising. The mines are expanding and new high-grade ore has been uncovered that has caused a rebirth in mining within the district.

In the early 1980s, the Mexican government changed slightly the name of the city of Taxco. It is now known officially as Taxco de Alarcón with the municipio's and mining district's name remaining the same; Taxco.

Total silver production from Taxco's mines amounts to more than 3 million kilograms (21,433,333 troy pounds). Using 1980 metal values yields a total district production since the early 1500s of just over $1.8 billion in silver and when combined with all the other metals recovered, the district totals reach to almost $2 billion. This would place Taxco mining district as one of México's important mining areas.

Geology and Mineralogy

The base rocks of the district are granite, overlain mainly by limestone and micaceous slate, and the latter capped in places by a porphyry. Metalliferous veins intrude all rock strata, but are the richest in the limestone. The vein width ranges from 1.8 to 3.1 meters (5.9-10.2 feet) and may extend for lengths of 1,585 meters (5,200 feet). Fissure veins presently being worked, however, vary in length from 200 meters (656 feet) to more than 1 kilometer (3,280 feet). Some may be as wide as 1.5 to 10 meters (4.9-32.8 feet) and extend to a depth of 300 meters (984.3 feet). A common occurrence is that the silver content drops with depth, while the lead and zinc increases. Ore bodies have been found along with the veins as replacements in the limestone that take irregular courses, more or less parallel to the local limestone strata. Such ore bodies are about 20 meters (65.6 feet) wide but their metal content is much lower than in the fissure veins.

According to early reports, veins were composed of finely disseminated native silver, acanthite, and to a lesser amount of ruby silver. Von Humboldt reported seeing balls of silver up to 12 centimeters (4.7 inches) in diameter, which were composed of native silver, acanthite, ruby silvers, and quartz. He also reported that in the upper vein areas it was not uncommon to find crystallized selenite containing crystallized silver. Azurite crystals on pyrargyrite were commonly encountered in the upper mine levels. Fine specimens of smoky quartz with pyrite and marcasite have been found in the deeper portions of the veins and are usually associated with galena and sphalerite. Amethyst common in most silver deposits in México is found at Taxco as deep colored crystals of large size. Some of the crystal vugs have produced stout crystal points to just over 13 centimeters (5.1 inches) in length.

DISCOVERIES AND
DEVELOPMENTS IN MEXICO

20,000–10,000 B.C.	Evidence of human inhabitation in México
6,000 B.C.	Mining for materials to make tools and weapons; agricultural crops develop in the valley of Tehuacán, Puebla
2,000–1,500 B.C.	Olmecs develop concept and use of the zero. They begin carving jade and other gemstones
350 B.C.	Aztec's establish capital at Teotihuacán
300-900 A.D.	Mayans develop modern calendar
1,000 A.D.	Mixtec gold and silversmiths in the valley of Oaxaca develop the combined uses of metals and gemstones; gem carving develops
1325	Beginning of the Aztec civilization
1492	Columbus discovers America
1517	Francisco Hernández de Córdoba discovers the Yucatán peninsula
1518	Juan de Grijalva names the newly discovered land Nueva España (New Spain)
1519	Hernán Cortés lands at Veracruz and visits the Aztec capital of Tenochtitlán
1521	Cortés retakes and destroys Tenochtitlán, conquest complete, México City founded
1522	Tin mined at Taxco
1524	Silver discovered at Pachuca, Hidalgo
1525	First silver mines in Nueva España (New Spain); Taxco; Guerrero; Morcillo, Jalisco; and Espíritu Santo, Nayarit; and Sultepec, México
1528	Silver discovered at Real del Monte, Hidalgo
1530	Silver discovered at Mazapil, Zacatecas
1534	Silver mines of Taxco worked, Cortés constructs first mine tunnel
1535	First Viceroy appointed in México; first silver bonanza in

	México uncovered at Espíritu Santo; and first printing press in the Americas is established at México City
1536	First mint established at México City
1538	Silver veins of Leones and Santa Isabel discovered at Charcas, San Luis Potosí
1540	Conorado explores northward in the Tierra Incognita searching for Cibola
1544	Silver discovered at Santa Bárbara, Chihuahua
1546	Silver discovered at Zacatecas, Zacatecas
1548	Silver discovered at Sombrerete; Zacatecas; La Luz and; Guanajuato; mining begins at Vetagrande, Zacatecas and; at Zacatecas, Zacatecas; silver discovered at Bolaños, Jalisco; the silver vein of Los Tajos discovered at Panuco, Coahuila
1549	Silver discovered at Temascaltepec, México
1551	The Royal and Pontifical University of México is founded; mining begins at Real del Monte, Hidalgo
1552	Cerro de Mercado (iron) discovered in Durango; silver vein of Mololoa discovered, Hostotipaquillo, Jalisco
1553	Silver discovered at Guanajuato, Guanajuato; classes begin at the Royal and Pontifical University of México
1554	Medina invented and applied his "Patio Method" for processing silver ore at Pachuca (Purisima Grande); silver discovered at Fresnillo, Zacatecas
1555	Silver discovered at San Martín, Zacatecas
1558	Veta Madre discovered at Guanajuato; operations begin at San Juan de Rayas Mine, Guanajuato
1574	Mining begins at Charcas, San Luis Potosí
1583	Gold/silver discovered at Cerro de San Pedro, San Luis Potosí
1591	Silver discovered at Santa Eulalia, Chihuahua
1596	The city of Monterrey, Nuevo León is founded
1598	Silver/gold discovered at Mapimí, Durango
1600	Silver discovered at Hidalgo del Parral, Chihuahua
1604	Silver discovered at Villa de Ramos, San Luis Potosí
1606	Mint established at Zacatecas, Zacatecas
1615	Mercury discovered at Guadalcázar, San Luis Potosí
1628	Silver discovered at Guazapares, Chihuahua
1630	Silver discovered at Urique, Chihuahua
1632	Silver discovered at Zimapán, Hidalgo; silver discovered at Batopilas, Chihuahua
1650	Silver production declines
1655	Silver discovered at Rosario, Sinaloa
1666	Mining begins at the silver mines of Cusihuiriáchic, Chihuahua
1677	Mercury discovered at Chilapa, Guerrero
1680	Gold and silver discovered at Alamos, Sonora
1703	Mining begins at Santa Eulalia, Chihuahua
1714	Silver discovered and mining begins at Asientos, Aguascalientes
1725	Silver discovered at Planchas de Plata, Sonora
1740	Silver/gold discovered and work begins in Minas Prietas, Hermosillo, Sonora

1754	Gold/silver discovered at Tlatlaya, México
1755	Mining begins at Fresnillo, Zacatecas
1760	Mining begins at the Valenciana Mine, Guanajuato
1767	Jesuits expelled from México
1773	Silver discovered at Catorce, San Luis Potosí
1774	Silver discovered and mining begins at Mineral del Doctor, Querétaro
1775	Silver discovered and mining begins at San Antonio, Baja California Sur
1781	Alamos, Sonora becomes the silver capital of the world
1783	New mining ordinances provide for college mining in México; silver discovered at Guarisamy, Durango; Elhuyar discovers Tungsten
1785	Silver discovered and mining begins at Natividad, Oaxaca
1786	University of Guanajuato founded (School of Mines)
1790-1810	Friedrich Sonneschmid works at Royal Seminary of Mining, México City
1792	Royal Seminary of Mining (College of Mines, México City) held first classes, Fausto de Elhuyar is the director
1794	Don Andrés Manuel del Río arrives in México; silver discovered on the Sierra de Naica, Chihuahua
1795	First class in mineralogy taught by del Río at Royal Seminary; he wrote first book on mineralogy in México
1801	Del Río discovers the element vanadium at Zimapan, Hidalgo
1803	Silver discovered at Cuale, Jalisco
1803-1804	Alexander von Humboldt visits México
1810	Mexican Revolution begins with Father Hidalgo's Declaration of Mexican Independence
1820	Institute of Geology founded
1821	Mexican Independence from Spain
1828	Operations begin at Cuchillo Parado, Chihuahua
1830	Mining begins at Naica, Chihuahua
1835	Gold/silver discovered at Guadalupe y Calvo, Chihuahua
1846	War with the United States; U.S. invades México and captures México City
1853	School of Mines established at Fresnillo, Zacatecas, and the Gadsden Purchase
1855	The Opal is discovered in San Juan del Río, Querétaro
1860	Juárez becomes President
1861	France invades México
1864	Maximilian is emperor of México
1867	France leaves México
1868	Copper discovered at Santa Rosalía (Boleo) Baja California Sur
1870	Gold discovered and mining begins at Real de Castillo, Baja California; opal mining begins at Santa Maria Iris Mine, Cerro Caja de León, Querétaro
1873	Mercury discovered at Huitzuco, Guerrero
1876	Díaz becomes President and invites foreign investments
1877	Compressed air is first used in México's mines at Catorce, San Luis Potosí

1880	Coal discovered at Sabinas, Coahuila
1885	Mining begins at Santa Rosalia (Boleo) Baja California Sur
1886	Mexican Mining Society founded; Instutute of Geology becomes part of UNAM
1888	Copper discovered and mining begins at Cananea, Sonora; Gold discovered at Alamo, Baja California
1889	D. Antonio del Castillo writes "Catalogue Descriptf des Meteorites de Méxique"
1890	Bonanza uncovered at Cinco Señores Mine, Mineral de Pozos, Guanajuato; mining exhibit opens at the Municipal Palace at Zacatecas
1891	Geologic commission of México founded
1893	Antimony discovered and mining begins at Wadley, San Luis Potosí; MacArthur-Forrest cyaniding method first used in México; first mine electric plant established at Santa Ana, San Luis Potosí
1900	First coke fired blast furnace starts modern steel industry, Monterrey, Nuevo León; oil exloration in México
1904	Opening of the Museum of Minerals and Rocks at the Institute of Geology
1906	Geological Society of México founded; labor strikes at the Green Copper Company's Cananea Mine; building built for Institute of Geology
1907	Opening of Mineral Exposition at Chihuahua, Chihuahua
1909-1910	Maximum mining activities in México
1910	Beginning of Méxican Revolution
1916	Francisco (Pancho) Villa raids Columbus, New México; U.S. forces enter México after Pancho Villa
1917	Pershing's U.S. forces leave México; México adopts new constitution declaring minerals found in México belong to México
1919	U.S. forces enter México again
1920	End of Méxican Revolution
1923	Papal delegate expelled from México; Francisco (Pancho) Villa slain
1926	Religious exercises suspended by the Church in México
1928	President Obregón assassinated
1934	President Cardenas nationalizes foreign holdings
1937	Pope Pius 11 appoints first archbishop in eleven years
1938	President Cárdenas expropriates oil fields under foreign control
1939	Comisión Fomento Minero formed
1942	México declares war on the Axis
1954	The French leave Santa Rosalía (Boleo)
1964	Consejo de Recursos Minerales founded
1976	Institute of Geology moves into new building on the campus of the National University of México
1984	Sociedad Méxicana de Mineralogía, A. C. (Méxican Mineralogical Society) is founded

BIBLIOGRAPHY

Adams, W., 1825. *Actual State Of the Méxican Mines.* London.

Agricola, G., 1950. *Annotations.* H. C. Hoover and L. H. Hoover, eds. Trans. by De Re Matallica. New York: Dover Publications.

American Automobile Association, 1983. *Travel Guide to México.* Washington: American Automobile Association.

Anderson, C. A., 1941. Geology of the Gulf of California. *Geol. Soc. Am. Bull.* **52:**1888.

Anderson, C. E., 1926. Geology and Ore Deposits of the Asientos-Tepezala District, Aguascalientes, México. *Trans. Am. Inst. Min. Eng.* **76:**238-254.

Anonymous, 1843. *Life in México.* 2 vols. Boston: Little, Brown, & Co.

Anonymous, 1893. Mining at the Columbian Exposition. Number 13. *Eng. Min. Jour.* **56:**315.

Anonymous, 1946. Ancient Taxco Now a Base Metal Camp. *Eng. Min. Jour.* **147:**76.

Anonymous, 1946. A. S. & R.'s Modern Operation at Santa Eulalia. *Eng. Min. Jour.* **147:**77-78.

Anonymous, 1946. Parral—Big and Progressive. *Eng. Min. Jour.* **147:**78-80.

Anonymous, 1963. Santa Barbara Camp—Continues Active. Number 3. *Eng. Min. Jour.* **137:**132-136.

Anonymous, 1978. México. *Min. Ann. Rev.* 367-369.

Anonymous, 1979. México. *Min. Ann. Rev.* 363-365.

Anonymous, 1981. Eyes on México. *Min. Jour.* 369-370.

Aquilera, J. G., 1904. The Geographical and Geological Distribution of the Mineral Deposits of México. *Trans. Am. Inst. Min. Eng.* **XXXII:**497-520.

Arem, J. E., 1977. *Color Encyclopedia of Gemstones.* New York: Van Nostrand Reinhold Co.

Arenas, E. G., ed., 1979. *México Today—Imágenes de México 1979.* México: Compañía Periodistica Internacional, S. A.

Arizona-Sonora Desert Museum, 1980. *Catalog of Minerals.* Arizona: Arizona-Sonora Desert Museum.

ASARCO, updated report. *Mining Properties and Exploration in México.* New York: ASARCO.

ASARCO, 1979. *Mining Properties and Exploration in México.* New York: ASARCO.

Avino Mines & Resources LTD., 1983. *Annual Report* Canada: Avino Mines & Resources LTD.

Bagby, W. C., 1979. *Geology, Geochemistry and Geochronology of the Batopilas Quadrangle, Sierra Madre Occidental, Chihuahua, México.* Dissertation, University of California at Santa Cruz.

Bagg, R. M., 1905. The Sahuayacan District, México. *Eng. Min. Jour.* **79:**749-751.

Bagg, R. M., 1908. Geology of the Mining Districts of Chihuahua. *Min. Sci. Press* **97:**152-153, 189-197.

Bain, H. F., 1897. A Sketch of the Geology of México. *Jour. Geol.* **5:**384-390.

Baker, C. L., 1922. General Geology of the Catorce Mining District. *Am. Inst. Min. Eng.* **66:**42-48.

Bakewell, P. G., 1971. *Silver Mining and Society in Colonial México 1546-1700.* Number 15. Cambridge Latin Series. Cambridge: The University Press.

Bancroft, P., 1984. *Gem and Crystal Treasures.* Fallbrook: Western Enterprises/ Mineralogical Record.

Barbour, J., 1964. Black Opal at San Juan del Rio. *Gems & Min.* **326:**26-27.

Barbour, J., 1964. Mexican Topaz and Where It Comes From. *Gems & Min.* **316:**22-23, 61-63.

Baron, H. J., 1911. Chihuahua After the Revolution. *Eng. Min. Jour.* **92:**685-689.

Barron, E. M., 1958. The Gem Minerals of México. *Lapid. Jour.* **12:**4-60.

Benitez, A. T., 1922. The Camp of Guanaceri. *Eng. Min. Jour.* **114:**139-144.

Bennett, J. F., 1908. Diamond Drilling at Mapimi, *Eng. Min. Jour.* **86:**718-719.

Bernstein, M. D., 1964. *The Mexican Mining Industry 1890-1950.* Albany: State University of New York.

Blair, G., 1980. Memories of México. Part 1. Number 9. *Lapid. Jour.* **34:**1906-1914.

Blair, G., 1981. Memories of México. Part II. Number 10. *Lapid. Jour.* **34:**2106-2112, 2154-2157.

Blake, W. P., 1861. *Silver Mines and Silver Ores.* New Haven.

Blake, W. P., 1902. Notes on the Mines and Minerals of Guanajuato, México. *Trans. Am. Inst. Min. Eng.* **32:**216-223.

Blake, W. W., 1907. Books on Mining in México. Number 17. *Min. World* **27:**718-720.

Bleyleden, K., Anton, A., von Hassenpflug, W., and Klug, F. J., *Baedeker's México.* New Jersey: Prentice-Hall.

Bonillaes, Y. S., 1937. Geology of the Taxco Mining District, Guerrero, México. *Econ. Geol.* **32:**200.

Botsford, C. W., 1909. Geology of the Guanajuato District, México. *Eng. Min. Jour.* **87:**691.

Botsford, C. W., 1911. Southern Sonora and Chihuahua. *Eng. Min. Jour.* **92:**704-706.

Boulanger, R., 1968. Trans. by J. S. Hardman. *México.* Paris: Harchette.

Brenner, A., and Leighton, G. R., 1973. *The Wind That Swept México—the History of the Mexican Revolution of 1910-1952.* Austin: University of Texas Press.

Brodie, W., 1909. History of the Native Silver Mines of Batopilas. *Min. World* **30:**1105-1110.

Brodie, W., 1909. The Native Silver Mines of Batopilas-II. Number 26. *Min. World* **30:**1201-1208.

Brodie, W., 1911. The Milling of Batopilas Native Silver Ores. *Méx. Min. Jour.* 21-23.

Brodie, W. M., 1910. Native Silver in Southwestern Chihuahua. *Eng. Min. Jour.* 664-665.

Brodie, W. M., 1916. Metallurgy of Native Silver Ores of Southwestern Chihuahua. *Eng. Min. Jour.* **101:**297-301.

Brodie, W. M., 1917. Metallurgy of Native Silver Ores of Southwestern Chihuahua. Number 1. *Min. World* **31:**48-54.

Burton, G., 1976. Mapimi, México's Glory Hole. Number 6. *Lapid. Jour.* **30:**1542-1544.

Burton, L. W., 1981. The Opals of Queretaro. *Gems & Miner.* **521:**6.

Bush, K., 1980. *México XXIV.* Hanover: Nägeleu Obermiller.

Castillo, B. D., del, 1956. I. A. Leonard, ed. *The Discovery and Conquest of México.* New York: Farrar, Straus and Giroux.

Chamberlain, A., 1897. The Ancient Silver Mines of Zacatecas. *Min. Collect.* 88-93.

Chan, R. P., 1970. *A Guide to Mexican Archaelogy.* México: Minutiae Mexicana.

Christensen, A. O., 1911. La Noria Mine in Zacatecas. *Eng. Min. Jour.* 700-703.

Clark, K. F., and Fuente, L., de la *Distribution of Mineralization in Time and Space in the Sierra Madre Occidental Province, Chihuahua, México.* México: XXVth International Geological Congress.

Cortes, H., 1908. Trans. by F. A. MacNutt, ed. *Letters of Cortes.* Volume 1 and 2. New York: Putnam and Sons.

Cousen, T. W., 1956. Mexican Fire Opal. *Gems & Miner.* **232:**16-17, 68-71.

Cserna, Z., de, 1969. *General Geology of South-Central México, between México City and Acapulco.* Guidebook 9. México: The Geological Society of America.

Cumenge, E., and Bouglise, G. de la, 1885. *Etude sur le District Cuprifere de Boleo—Basse Californie.* Paris: Librairie Centrales des Chemins Fer.

Dahlgren, C. B., 1883. *Historic Mines of México.* New York: Private Printing.

Dahlgren, C. B., 1883. *Historic Mines of México—Handbook.* New Jersey: MacCrellism and Quigley Printers.

Dahlgren, C. B., 1887. *Minas Históricas de la República Mexicana.* México: Oficina Tipografica de la Secretaría de Fomento.

Dana, E. S., 1966. *A Textbook of Mineralogy.* Revised by W. E. Ford. 4th edition. New York: Wiley & Sons.

Dana, J. D., and Dana, E. S., 1944. Revised by C. Palache, H. Berman, and C. Frondel. *System of Mineralogy.* Volume 1. 7th edition. New York: Wiley & Sons.

Dana, J. D., and Dana, E. S., 1951 *System of Mineralogy.* Volume 2. New York: Wiley & Sons.

Dana, J. D., Dana, E. S., 1962. *System of Mineralogy.* Volume 3. New York: Wiley & Sons.

Davis, M. L., and Pack, G., 1963. *Mexican Jewelry.* Austin: University of Texas Press.

Desautels, P. E., Clarke, R. S., 1963. Re-Examination of Legrandite. *Am. Miner.* **48:**1258-1265.

Digby, A., 1972. *Maya Jades.* London: The Trustees of the British Museum.

Dolch, O., Jr., 1980. Ghost Mines of New Spain (Now Old México). Number 9. *Lapid. Jour.* **34:**2038-2040.

Dominquez, N., 1902. The District of Hidalgo del Parral, México in 1920. *Trans. Am. Inst. Min. Eng.* **XXXII:**459-477.

Drugman, J., and Hey, M. H., 1932. Legrandite, A New Zinc Arsenate. *Miner. Mag.* **32:**175-178.

Dulourcq, E. L., 1910. Minas Pedrazzini Operations near Arizpe, Sonora. *Eng. Min. Jour.* **90:**1105-1106.

Dumble, E. T., 1890. Notes on the Geology of Sonora, México. *Trans. Am. Inst. Min. Eng.* **49:**1-31.

Dunn, P. A., 1978. Gen Peridot and Enstatite with Spinel Inclusions from Chihuahua, México. Number 4. *Jour. Gemol.* **16:**236-238.

Dunning, C. H., and Peplow, E. H., Jr., 1959. *Rocks to Riches.* Phoenix: Southwest Publishing Company.

Edelen, A. W., and Lee, H. C., 1941. The Teziutlan Copper-Zinc Deposit, Teziutlan, Puebla, México. *Trans. Am. Instit. Min. Eng.* **144:**314-323.

Emmons, N. H., 2nd., 1902. The Values of Ores in México. *Trans. Am. Instit. Min. Eng.* **XXXII:**94-99.

Emmons, S. F., 1910. Cananea Mining District of Sonora, México. *Econ. Geol.* **5:**312-356.

Emmons, S. F., 1906. Pilares Mine, Nacozari, Sonora, México. *Econ. Geol.* **1:**629-643.

Fechet, E. O., 1893. The Miners of Sierra Mojada, México. Number 5. *Minerals* **III:**79-84.

Felger, R. S., 1980. *Vegetation and Flora of the Gran Desierto, Sonora, México.* Tucson: Arizona-Sonora Desert Museum.

Finkelman, R. B., 1974. A Guide to the Indentification of Minerals in Geodes from Chihuahua, México. *Lapid. Jour.* **27:**1742-1744.

Finkelman, R. B., Evans, H., and Matzko, J., 1974. Manganese Minerals in Geodes from Chihuahua, México. *Miner. Mag.* **39:**549-558.

Finkelman, R. B., Matzko, J. J., Woo, C. C., White, J. S., and Brown, W. R., 1972. A Scanning Electron Microscopy Study of Minerals in Geodes from Chihuahua, México. Number 5. *Miner. Rec.* **3:**205-212.

Finley, G. L., 1903. Geology of the San Pedro District, San Luis Potosí, México. *Columbia Univ. Sch. Mine Quart.* **25:**60-69.

Fishback, M., 1910. Mines of Zomelahuacan, Veracruz, México. *Eng. Min. Jour.* **90:**1017-1019.

Fleisher, M., 1983. *Glossary of Mineral Species.* Fourth edition. Tucson: The Mineralogical Record.

Fletcher, R. A., 1929. México's Lead-Silver Manto Deposits and Their origins. *Eng. Min. Jour.* 509-513.

Foshag, W. F., Carminite and Associated Minerals from Mapimi, México. *Am. Min.* **22:**474-484.

Foshag, W. F., 1927. The Minerals of México. *Explor. Field Work Smithson. Inst.* **2957**:21-26.

Foshag, W. F., 1927. The Selenite Caves of Naica, México. Number 6. *Am. Min.* **12**:252-256.

Foshag, W. F., 1928. Mineralogy and Geology of Cerro Mercado, Durango, México. *Proc. U.S. Natl. Mus.* **74**:1-27.

Foshag, W. F., 1934. The Ore Deposits of Los Lamentos, Chihuahua, México. Number 4. *Econ. Geol.* **29**:330-345.

Foshag, W. F., 1953. Mexican Opal. Number 9. *Gems & Gemol.* **7**:278-282.

Foshag, W. F., and Fries, C., Jr., 1942. *Tin Deposits of the Republic of México.* Bulletin 935-C. Washington: U.S. Geological Survey.

Foshag, W. F., and Gonzalez, R. J., 1956. *Birth and Development of Paricutin Volcano.* Bulletin 965-D. Washington: U.S. Geological Survey.

Fowler, G. M., Hernon, R. M., and Stone, E. A., 1944. The Taxco Mining District, Guerrero, México. *Econ. Geol.* **39**:93-94.

Fries, C., Jr., and Schmitter, E., 1945. *Scheelite Deposits in the Northern Part of the Sierra de Juarez Northern Territory Lower California, México.* Bulletin 946-C. Washington: U.S. Geological Survey.

Fries, C., Jr., and Schmitter, E., 1948. *Tin-bearing Placers Near Guadalcazar, State of San Luis Potosí, México.* Bulletin 960-D. Washington: U.S. Geological Survey.

Frye, K., ed., 1981. *Encyclopedia of Mineralogy.* Pennsylvania: Hutchinson Ross Publishing Co.

Gaines, R. V., 1970. The Moctezuma Tellurium Deposit. *Mineral. Rec.* Number 1. **2**:40-44.

Gallagher, D., 1952. *Geology of the Quicksilver Deposits of Canoas, Zacatecas, México.* Bulletin 975-B. Washington, D.C.: U.S. Geological Survey.

Gallagher, D., and Perez, R., 1946. *Geology of the Cuarenta Mercury District, State of Durango, México.* Bulletin 946-F. Washington, D.C.: U.S. Geological Survey.

Gallagher, D., and Perez, R., 1948. *Geology of the Huahuaxtla Mercury District, State of Guerrero, México.* Bulletin 960-E. Washington, D.C.: U.S. Geological Survey.

Garfias, V. R., 1937. Historical Outline of Mineral Production in México. *Trans. Instit. Min. Eng.* **CXXVI**:346-355.

Gary, M., McAfee, R., Jr., and Wolf, C., eds. 1974. *Glossary of Geology.* 3rd Printing. Washington: American Geological Institute.

Garza-Nietro, J. L., 1974. General Sketch of Old Mineral Districts of Gold and Silver in the Guanajuato State, México. Unpublished report. Guanajuato.

Gendrop, P., 1982. *A Guide to Architecture in Ancient México.* México: Minutiae Mexicana.

Gerhard, P., and Gulick, H. E., 1970. *Lower California Guidebook.* A. H. Clark Co.

Gierloff, H. G., 1964. *Die Halbinsel Baja California Ein Entwicklungsgebiet Mexikos.* Hamburg: Im Selbstverlag der Geolgraphischen Gellschaft.

Graham, R. T., 1907. The History of Mining Developments in México. Number 17. *Min. World* **27**:704-705.

Grau, J., 1978. *Acapulco.* New York: Crescent Books.

Griggs, J., 1907. *Mines of Chihuahua.* Chihuahua: Private Printing.

Griggs, J., 1909. Review of Mining and Metallurgy in México. *Min. World* **31:**48-63.

Guiza, R., 1956. *The Guanajuato Mining District.* Guidebook A-2. XXth International Geological Congress.

Hall, R. T., 1902. The Geographic and Geologic Features and Their Relations to Mineral Products of México. *Trans. Am. Instit. Min. Eng.* **XXXII:**163-178.

Halse, E., 1908. *A Dictionary of Mining, Metallurgical and Applied Terms.* London: Griffin & Co. LTD.

Haywood, M. W., and Triplett, W. H., 1931. *Occurrence of Lead-Zinc Ores in Dolomitic Limestones in Northern México.* Paper 442. AIME Tech.

Hazen, R. M., ed., 1979. *North American Geology — Early Writings.* Stroudsburg: Dowden, Hutchinson & Ross.

Heinrich, E. W., 1984. Some Unusual Examples of Mineralized Tactites in México. Paper read at MSA/FM Joint Symposium, 12 February, 1984, at Tucson, Arizona.

Hess, F. L., 1909. Graphite Mining Near La Colorado, Sonora. *Eng. Mag.* **28:**26-48.

Hewitt, W. P., 1968. Geology and Mineralization of the Main Mineral Zone of the Santa Eulalia District, Chihuahua, México. *Trans. Am. Instit. Eng.* 229-260.

Hewitt, W. P., 1943. Geology and Mineralization of the San Antonio Mine, Santa Eulalia District, México. *Geol. Soc. Am.* 173-204.

Hey, M. H., 1975. *An Index of Mineral Species and Varieties Arranged Chemically.* London: British Museum of Natural History.

Hey, M. H., 1963. *An Index of Mineral Species and Varieties Arranged Chemically.* Appendix to the Second Edition. London: British Museum of Natural History.

Hey, M. H., and Embrey, P. G., 1974. *An Index of Mineral Species and Varieties Arranged Chemically.* Appendix to the Second Edition. London: British Museum of Natural History.

Heylmun, E. B., 1983. Opal Localities in West Central México. Issue number 4. *Lapid. Jour.* **37:**598-602.

Hill, R. T., 1907. México: Its Geology and Natural Resources. Number 17. *Min. World* **27:**158-160.

Hill, R. T., 1907. The Santa Eulalia District (Chihuahua), México. *Eng. Min. Jour.* **83:**686-691.

Hoffman, V. J., 1968. *The Mineralogy of the Mapimi Mining District, Durango, México.* Thesis. Tucson: University of Arizona.

Horcasitas, A. S. and Snow, W. E., 1956. *Geological Resumé of the Santa Eulalia District, Chihuahua, México.* Guidebook A-2. XXth International Geological Congress.

Horsfall, H. A., 1889. Gold Discoveries in Batopilas. *Eng. Min. Jour.* **47:**5.

Howe, W., 1949. *The Mining Guild of New Spain and its Tribunal General, 1770-1821.* Cambridge: Harvard University Press.

Hubbell, A. H., 1936. Parral's Output — Mostly from Cerro de la Cruz. *Eng. Min. Jour.* **137:**127-130.

Hulin, C. D., 1929. Geology and Mineralization at Pachuca, México. *Bull. Geol. Soc. Am.* **40:**171-173.
Hunt, J. H., 1893. The Minerals of México. *Minerals* **III:**145-149.
Hutchinson, W. S., 1906. Mining in Western Chihuahua. *Eng. Min. Jour.* **81:**418.

Jackson, D. D., and Wood, P., 1975. *The Sierra Madre.* American Wilderness Series. Virginia: Time-Life Books.
Jennison, H. A. C., 1923. Mining History of México-I. *Eng. Min. Jour.* **8:**364-368.
Jenison, H. A. C., 1923. Mining History of México-II. *Eng. Min. Jour.* **115:**401-403.
Johnson, C. W., 1953. Notes on the Geology of Guadalupe Island, México. *Am. Jour. Sci.* **251:**231-236.
Johnson, P. W., 1962. Boleo Copper District-part 1. Numbers 7 and 8. *Mineralogist* **30:**4-9.
Johnson, P. W., 1962. Boleo Copper District-part 2. Numbers 9 and 10. *Mineralogist* **30:**8-12.
Johnson, P. W., 1963. Cerro de Mercado Mine. Number 1. *Mineralogist* **31:**4-6.
Johnson, P. W., El Fenomeno Mine. Number 5. *Mineralogist* **31:**4-10.
Johnson, P. W., 1963. The Minerals of Guanajuato. Number 2. *Mineralogist* **32:**9-11.
Johnson, P. W., 1964. Laguna Guerrero Negro. Number 2. *Mineralogist* **32:**14-17.
Johnson, P. W., 1965. *Field Guide to the Gems and Minerals of México.* Mentone: Gembooks.
Johnson, W. W., 1972. *Baja California—American Wilderness Series.* Virginia: Time-Life Books.
Jones, R. W., Mexican Wulfenite Discovery. Number 4. *Rock & Gem* **5:**36-42.
Jones, W., 1962. New Fire Agate Find at tule Hill. *Lapid. Jour.* 113.
Jones, W., 1962. The Guadalcazar-Realejo Collecting Area. *Lapid. Jour.* 282-283.
Just, E., 1936. Howe Sound Pioneers in The Sierra Madre. *Eng. Min. Jour.* 1.
Just, E., México is Strangling It's Mining Industry. Number 6. *Eng. Min. Jour.* **147:**75-76.
Just, E., Several Mines At Santa Barbara. Number 10. *Eng. Min. Jour.* **147:**76-77.

Keller, P. C., 1977. *Geology of the Sierra del Gallego Area, Chihuahua, México.* Thesis. Austin: University of Texas.
Keller, P. C., 1979. Quartz Geodes from near the Sierra Gallego area, Chihuahua, México. Number 4. *Mineral. Rec.* **10:**207-214.
Keller, P. C., 1984. Quartz Geodes of the Sierra de Gallego area, Chihuahua, México. Paper read at the MSA/FM Joint Symposium, 12 February 1984 at Tucson, Arizona.
Kelly, D. G., 1971. *Edge of a Continent.* American West Publishing Co.
Kelly, J., 1921. The Naica Controversy. Number 7. *Eng. Min. Jour.* **111:** 320-321.
Kemp, D. C., 1971. *Quicksilver to Bar Silver—Tales of México's Silver Bonanzas.* Pasadena: Socio-Technical Publications.
Kemp, J. F., 1906. Copper Deposits of San Jose, Tamaulipas, México. *Bull. AIME* **4:**178-203.
Kibbe, P., 1978. *A Guide to Mexican History.* México: Minutiea Mexicana.

Kimball, J. P., 1869. On Cretaceous Age of Silver Deposits in Chihuahua, México. *Proc. Am. Assoc. Adv. Scie.* **18:**120-179.

Koivula, J. I., Fryer, C., and Keller, P. C., 1983. Opal from Queretaro, México: Occurrence and inclusions. Number 2. *Gems & Gemol.* **19:**87-96.

Krieger, P., 1932. An Association of Gold and Uraninite from Chihuahua, México. Number 7. *Econ. Geol.* **XXVII:**651-660.

Krieger, P., 1935. Primary Native Silver Ores of Batopilas, México. *Am. Miner.* **20:**715-723.

Krieger, P., 1935. Primary Silver Mineralization at Sabinal, Chihuahua, México. Number 3. *Econ. Geol.* **XXX:**242-259.

Kunz, G. F., *New Observations on the Occurrences of Precious Stones of the Archaeological Interest in America.* Extrait de Memoires et Deliberations de XV Congres des Americanistes.

Kunz, G. F., 1883. Sapphire from México. *Am. Jour. Sci.* 26, 3rd ser., 151, p.75.

Kunz, G. F., 1887. On a Gigantic Jadite Votive Adze from Oaxaca. Private printing.

Kunz, G. F., 1888. Chalchihuitl: a note on the Jadeite discussion. *Sci.* **12:**192.

Kunz, G. F., 1890. *Gems and Precious Stones of North America.* New York: Scientific Publishing Co.

Kunz, G. F., 1902. Gems and Precious Stones of México. *Trans. Am. Instit. Min. Eng.* **XXXII:**55-93.

Kunz, G. F., 1906. *Gems and Precious Stones of México.* México: 10th International Geological Congress. pp. 1029-1080.

Kunz, G. F., 1907. *Precious Stones of México.* México: Secretaría de Fomento.

Kunz, G. F., 1928. *Precious Stones Used by the Prehistoric Residents of the American Continent."* Reprint: Proceeding of the 23rd International Congress of Americanists. pp. 60-66.

Lamb, M. R., 1908. Stories of the Batopilas Mines, Chihuahua. *Eng. Min. Jour.* **85:**689-691.

Lamb, M. R., 1910. Tales of Mountain Travel in México. *Eng. Min. Jour.* **90:**676-677.

Lee, M. L., 1912. A Geological Study of the Elisa Mine, Sonora, México. Number 4. *Econ. Geol.* **VII:**324-339.

Leiper, H., ed., 1961. The Agates of North America. *Lapid. Jour.* Number 1. **15:**95.

Lindgren, W., 1888. Notes on the Geology of Baja California, México. 2nd Series. *Proc. Calif. Acad. Sci.* **1.**

Linton, R., 1912. Geology of the Ocampo District. *Eng. Min. Jour.* **94:**653-658.

Los Angeles Board of Supervisors, 1954. *Southern California Mines Including Baja California—México.* Los Angeles: L. A. Baundaj Sup.

Los Angeles County Museum of Art, 1963. *Master Works of Mexican Art.* Los Angeles: Los Angeles County Museum of Art.

Lowther, G. K., and Marlow, G. C., 1956. *Geology of the Parral Area.* XXth International Geological Congress. Guidebook A-2. XXth International Geological Congress. 64-75.

Lyon, G. F., 1971. *Journal of a Residence and Tour in the Republic of México.* Volume 1 and 2. Port Washington: Kennikat Press.

Lyons, W. J., and Young, J. R., 1961. Colorful Agates of Northern México. *Lapid. Jour.* **264:**258-259, 264.

McAllister, J. F., and Hernandaz, D., 1945. *Quicksilver-Antimony Deposits of Hitzuco, Guerrero, México.* Bulletin 946-B. Washington: U.S. Geological Survey.

McConnell, D., 1933. Garnets from Sierra Tlayacac, Morelos, México. *Am. Miner.* **18:**25-29.

MacGillivray, W., 1839. *The Travels and Researches of Alexander von Humboldt.* New York: Harper Brothers.

McGowan, D., 1981. México's Historic Rancho La Arizona. Number 2. *Ariz. Highw.* **57:**30-35.

Megaw, P. K. M., 1984. Geology and Mineralogy of the San Antonio Mine, Chihuahua, México. Paper read at the MSA/FM Joint Symposium, Tucson, Arizona.

Malcolmson, J. W., 1902. The Sierra Mojada, Coahuila, México and Its Ore-Deposits. *Trans. Instit. Min. Eng.* **XXXII:**100-139.

Malcolmson, J. W., 1907. *The Mineral Resources of the Kansas City, México & Orient Railway.* El Paso: Private Printing.

Malte-Brun, V. A., 1864. *La Sonora.* Paris: Arthus Bertrand.

Mandarino, J. A., and Williams, S. J., 1961. Five New Minerals from Moctezuma, Sonora, México. *Science* **133:**2017.

Manning, L. J., 1969. San Francisco Mine, Sonora, México. Unpublished report. Vancouver: Ramada Resourses Ltd.

Manzano, J. P., 1902. The Mineral Zone of the Santa Maria del Oro, San Luis Potosí, México. *Trans. Instit. Min. Eng.* **XXXII:**478-483.

Marshall, E. M., 1941. Trailing Minerals in Old México. Number 2. *Mineralogist.* **IX:**47-48.

Martin, P. F., 1906. *México's Treasure-House-Guanajuato.* New York: Cheltenham Press.

Martinez-Carrillo, R., 1921. What Others Think. Number 4. *Eng. Min. Jour.* **111:**576-577.

Mayers, D. E., 1947. Mexican Black Opal. Number 11. *Gems & Gemol.* **5.**

Mendoza, A. de, 1907. Mineral Resources of Sonora. *Min. Sci. Press* **96:**33-40.

Merrill, F. J. H., 1906. Mining Camps of Sinaloa, México. *Eng. Min. Jour.* **82:**635-636.

Merrill, F. J. H., 1907. Mineral Resources of Sonora. *Min. Sci. Press* **96:**33-40.

Miller, Harry F., and Olson, R. L., 1966. Adventuring 'Off The Beaten Track' in México. Number 6. *Lapid. Jour.* **20:**700-709.

Miller, H. F., and Olson, R. L., 1966. Adventuring 'Off The Beaten Track' in México. Number 8. *Lapid. Jour.* **20:**924-936, 1008-1015.

Miller, H. F., and Olson, R. L., 1967. Adventuring 'Off The Beaten Track' in México. Number 10. *Lapid. Jour.* **20:**1208-1219.

Miller, H. F., and Olson, R. L., 1978. The Mineral Collecting Situations Today in México. Number 10. *Lapid. Jour.* **23:**1420-1424.

Mishler, R. T., 1921. Geology of the El Tigre District, Sonora. *Min. Sci. Press* **121:**583-591.

Mitchell, R. S., 1979. *Mineral Names— What Do They Mean.* New York: Van Nostrand Reinhold Co.

Modreski, P. J., 1984. Green Uranium-Activated Fluorescence of Adamite from the Ojuela Mine, Mapimi, México. Paper read at the MSA/FM Joint Symposium, 12 February at Tucson, Arizona.

Motten, Clement G., 1972. *Mexican Silver and the Enlightenment.* New York: Octagon Books.

Mrose, M. E., 1948. Notes by D. E. Mayers, and F. A. Wise. Adamite from the Ojuela Mine, Mapimi, México. *Am. Miner.* **33:**449-457.

Nelson, E. W., 1919. *Lower California and Its Natural Resources.* Washington: Manessier Publishing Co.

Nelson, E. W., 1921. Lower California and Its Natural Resources. Washington: Memoirs of the National Academy of Sciences, V. XVI.

Nelson, E. W., 1966. *Lower California and Its Natural Resources.* Washington: Manessier Publishing Co.

Newman, B., 1910. Mining and Smelting in Aquascalientes. *Eng. Min. Jour.* **90:**678-679.

Niven, W., 1910. Mineral Resources of the State of Guerrero. *Eng. Min. Jour.* **90:**672-674.

Oakman, M. R., Foord, E., and Maxwell, C. H., 1984. Durangite from Black Range, New México and Coneto, Durango, México. Paper read at the MSA/FM Joint Symposium, 12 February at Tucson, Arizona.

Ober, Frederick A., 1884. *Travels in México.* Boston: Estes and Lauriat.

Olson, R., and Arnold, J. C., 1961. The Story of an Adventure in Baja California. Number 1. *Lapid. Jour.* **15:**34-49.

Ordóñez, E., 1904. The Mining District of Pachuca, México. *Am. Instit. Min. Eng.* **XXXII:**224-241.

Panczner, C. S., and Panczner, W. D., 1983. Terra Incognita—The Search for Minerals in Spanish Arizona. Number 2. *Miner. Rec.* **14:**67-71.

Panczner, W. D., 1980. *The Baja Penninsula and the Gulf of California—Its Geologic History.* Tucson: Arizona-Sonora Desert Museum.

Panczner, W. D., 1984. Mapimi. Paper read at the Rochester Symposium, 14 April at Rochester, New York.

Panczner, W. D., 1983. México, a Cornucopia of Noted Mineral Locations. Paper read at the MSA/FM Joint Symposium, 14 February at Tucson, Arizona.

Panczner, W. D., 1984. México: Its Geographical and Mineralogical Provinces. Paper read at the MSA/FM Joint Symposium, 12 February at Tucson, Arizona.

Panczner, W. D., 1984. Notes from México—Benny J. Feen. *Miner. Rec.* **4:**239-240.

Panczner, W. D., 1985. Miquel Romero and the Romero Mineralogical Museum. Number 2. *Miner. Rec.* **16:**129-136.

Parker, M. B., 1979. *Mules, Mines and Me in México.* Tucson: University of Arizona Press.

Parliman, C. R., 1960. New Agate Fields in Old México. Number 1. *Lapid. Jour.* **14:**4-7.

Parsons, A. B., 1925. Penoles Company Expands Operations in México. *Eng. Min. Jour.* **119:**217-220.

Parsons, A. B., 1926. Revival of Zacatecas. *Eng. Min. Jour.* **122:**324-330.

Payne, H. M., 1921. Recent Developments in Mining in Jalisco and Nayarit, México. *Eng. Min. Jour.* **112:**1045-1048.

Pearce, W. D., 1905. Mapa del Estado de Chihuahua (Mining Districts). Number 12. *Min. World* **25:**320.

Pearce, W. D., 1907. Past Year's Progress in Mining Greatest in The State's History. *Min. World* Number 17. **27:**721-728.

Pearce, W. D., 1907. Review of Mining in Chihuahua. Number 17. *Min. World* **27:**721-731.

Peterson, F. A., 1959. *Ancient México An Introduction to the Pre-Hispanic Cultures.* London: George Allen & Unwin Ltd.

Petruk, W., and Owens D., 1974. Some Mineralogical Characteristics of the Silver Deposits in Guanajuato Mining District, México. *Econ. Geol.* **69:**1078-1085.

Pinch, W. W., 1983. The Mineralogy of Selenium. Paper read at the MSA/FM Joint Symposium, 14 February at Tucson, Arizona.

Place, A. E. and Elton, H. L., 1907. The Mineral Deposits of Oaxaca. Number 17. *Min. World* **27:**749-752.

Pough, F., 1960. *A Field Guide to Rocks and Minerals.* Boston: Houghton Mifflin Co.

Prescott, B., 1916. The Main Mineral Zone of the Santa Eulalia District, Chihuahua. Volume 101. *Trans. Am. Instit. Min. Eng.* **51:**57-99.

Prescott, B., 1926. The Underlying Principles of the Limestone Replacement Deposits of the Mexican Province I and II. *Eng. Min. Jour.* **122:**246-253, 289-296.

Prieto, C., 1970. M. D. Bernstein, ed. *Mining in the New World.* New York: McGraw-Hill Book Co.

Probert, A., 1951. Early Silver Capitals of México. Paper read at the 3rd Congress Panamerican Institute of Mining Engineers and Geologists, 4 November at México.

Probert, F. H., 1916. The Treasure Chest of Mecurial México. *Natl. Geogr.* Number 1. **XXX:**33-67.

Probst, A., 1969. Bartholome de Medina, the Patio Process and the Sixteenth Century Silver Crisis. *Jour. West* **II:**90-124.

Pye, Willard, 1972. *San Francisco Mine, Sonora, México.* Unpublished report. Tucson: Devex Corp.

Ray, H. C., 1944. Sultepec Silver Mines of México. *Rocks and Miner.* **19:**146.

Raymond, R. W., La Sierena, A Wonderful Deposit of Jamesonite. *Eng. Min. Jour.* **99:**9-10.

Rice, C. T., 1908. Mines of Penoles Company, Mapimi, México. Part I. *Eng. Min. Jour.* **86:**310-314.

Rice, C. T., 1908. Mines of Penoles Company, Mapimi, México. Part II. *Eng. Min. Jour.* **86:**373-374.

Rice, C. T., 1908. Mining and Transportation At Santa Eulalia. *Eng. Min. Jour.* **86:**33-36.

Rice, C. T., 1908. Ores and Mines of Santa Eulalia, México. *Eng. Min. Jour.* **85:**1283-1286.

Rice, C. T., 1908. The Ore Deposits of Santa Eulalia, México. *Eng. Min. Jour.* **85:**1229-1233.

Rice, C. T., 1908. Zacatecas, A Famous Silver Camp of México. *Eng. Min. Jour.* **86:**401-407.

Rice, P. C., 1980. *Amber—The Golden Gem of The Ages.* New York: Van Nostrand Reinhold Co.

Rickard, T. A., 1907. *Journeys of Observation.* San Francisco: Dewey Publishing Co.

Rickard, T. A., 1924. The Ahumada Lead Mine and The Ore Deposits of The Los Lamentos Range in México. *Eng. Min. Jour.* **118:**365-373.

Rickard, T. A., 1932. *Man and Metals: A History of Mining in Relation to the Development of Civilization.* Volume 1 and 2. New York: McGraw-Hill.

Rippy, J. F., *Latin America.* Ann Arbor: University of Michigan Press.

Roberts, W. L., Rapp, G. R., Jr., and Webber, J., 1974. *Encyclopedia of Minerals.* New York: Van Nostrand Reinhold Co.

Robertson, W. P., 1853. *A Visit to México. Volume 1 and 2.* London: Simpkin, Marshall & Co.

Roca, P. M., 1967. *Paths of The Padres Through Sonora.* Tucson: Arizona Pioneers' Historical Society.

Roe, A., and White, J., Jr., 1976. *A Catalog of the Type Specimens in the Collection, National Museum of Natural History.* Washington: Smithsonian Press.

Rogers, A. H., 1908. Character and Habit of the Mexican Miner. *Eng. Min. Jour.* **85:**700-702.

Romero, M., 1984. Memories of Eduardo Schmitter. Paper read at the MSA/FM Joint Symposium, 12 February, at Tucson, Arizona.

Romero Mineralogical Museum, 1978. *Catalog of Minerals.* 1978. Puebla: Romero Mineralogical Museum.

Rosenzweig, A., Taggart, J. E., Jr., 1984. Plattnerite Paramorphs After Alpha-Lead Dioxide from Mapimi, Durango. Paper read at the MSA/FM Joint Symposium on Minerals of México, 12 February, at Tucson, Arizona.

Royal Ontario Museum, 1979. *Catalog of Minerals.* Canada: Royal Ontario Museum.

Russell, B. E., 1908. Las Chispas Mines, Sonora, México. *Eng. Min. Jour.* **86:**1006-1007.

Russell, B. E., 1908. Nacozari Mining District, Sonora, México. *Eng. Min. Jour.* **86:**657-662.

Sackett, E., 1980. Hydrometallurgy May Rejuvenate Historic Boleo District. *Min. Eng.* 799-800.

Schmitt, H., 1931. Geologic Notes on the Santa Barbara Area in the Parral District of Chihuahua, México. *Eng. Min. Jour.* **126:**407-411.

Schmitt, H., 1931. Geology of the Parral Area of the Parral District, Chihuahua, México. General Volume. *Trans. Am. Instit. Min. Eng.* 268-290.

Seamon, W. H., 1910. Mining Operations in the State of Chihuahua. *Eng. Min. Jour.* **90:**654-657.

Segerstrom, K., 1961. *Geology of the Bernal-Jalpan Area, Estado Queretaro, México.* Bulletin 1104-B. Washington: U.S. Geological Survey.

Segerstrom, K., 1962. *Geology of South-Central Hidalgo and Northeastern México, México.* Bulletin 1104-C. Washington: U.S. Geological Survey.

Shannon, E. V., 1923. *Crystallographic Notes on Stephanite in a Silver Ore from México.* Number 2479. Washington: U.S. Printing Office.

Shedenhelm, W. R. C., 1980. *Rockhounding in Baja.* Glendale: La Siesta Press.

Shepherd, A. R., Jr., 1935. *A Summary of the Batopilas Native Silver Mines, Their Past Record of Production and Outlook for Future Yield.* Unpublished report.

Shepherd, G., 1938. *The Silver Magnet; Fifty Years In A Mexican Silver Mine.* New York: E. P. Dutton.

Simmons, F. S., and Mapes, V., Eduardo, *Geology and The Ore Deposits of the Zimapan Mining District, State of Hidalgo, México.* Professional paper 284. Washington: U.S. Geological Survey.

Simmons, W. B., Wayne, D., and Rog, A. M., 1984. Authigenic Adularia, Albite, and Pumpellyite from The Sierra Del Fraile, Near Monterrey, Nuevo Leon, México. Paper read at the MSA/FM Joint symposium, 12 February, at Tucson, Arizona.

Simonin, L., 1867. *La Vie Souterraine.* Paris: Librarie de L. Hachette et C.

Simonin, L., 1869. Trans. by H. W. Bristow, and D. Appleton. *Mines and Miners or Underground Life.* London: MacKenzie.

Sindeeva, N. D., 1964. *Mineralogy and Types of Deposits of Selennium and Tellurium.* New York: Wiley Interscience Publishers.

Sinkankas, J., Prehnite, Apophyllite, and Other Species from El Fenomeno Mine (Baja California, México). Number 6. *Mineralogist* **31:**12-14, 16.

Sinkankas, J., 1959. *Gemstones of North America.* Volume 1 and 2. New York: Van Nostrand Reinhold Co.

Sinkankas, J., 1962. Strawberry Quartz . . . What is it? Issue 9. *Lapid. Jour.* **16:**677-678.

Sinkankas, J., 1963. Chromian Sphene—A New find in Baja (California, México). *Lapid. Jour.* **17:**4-5.

Sinkankas, J., 1964. Gemstones and Minerals of Baja, California—An Annotated Directory. *Lapid. Jour.* **18:**48-63.

Sinkankas, J., 1964. *Mineralogy for Amateurs.* New York: Van Nostrand Reinhold Co.

Sinkankas, J., 1965. Spectacular Strike of Axinite in Baja California. *Lapid. Jour.* **19:**436-438, 440-447.

Sinkankas, J., 1966. Iris-Opal from México. *Jour. Gemol.* **10:**100-105.

Sinkankas, J., 1970. México's Mineral Classics. *Miner. Dig.* **1:**7-16.

Sinkankas, J., 1981. *Emerald and Other Beryls.* Pennsylvania: Chilton Book Co.

Sinkankas, J., 1982. *Gemstones of North America: Past, Present and Future. In International Gemological Symposium Proceedings.* D. M. Eash, ed., New York: Gemological Institute of America, 407-421.

Smith, W. C., 1950. Segerstrom, K., and Guiza, R., *Tin Deposits of Durango.* Bulletin 962-D. Washington: U.S. Geological Survey.

Smithsonian Institution, 1980. *Catalog of Minerals.* Washington, D.C.: Smithsonian Institution.

Soustelle, J., 1967. Trans by J. Hogarth. *México* New York: The World Publishing Co.

Stewart, W. O., 1939. Famous Localities of México. Number 6. *Mineralogist* **VII:**235-236, 243-245.

Stewart, W. O., 1939. Mineral Collecting in México. Number 5. *Mineralogist* **VII:**193-194, 210-212.

Stewart, W. O., 1938. Opals and Silver. Number 5. *Mineralogist* **VI:**3-4, 16-25.

Stewart, W. O., The Selenite Caves of Naica, México. Number 8. *Mineralogist* **VIII:**193-195.

Stone, J. B., 1956. Notes on Fresnillo Mining District, Zacatecas. Guidebook A-2. XXth International Geological Congress. pp. 111-113.

Stone, J. B., 1956. *Notes on the Sombrerete Mining District of Zacatecas.* Guidebook A-2. XXth International Geological Congress. p110.

Stout, J. A., 1981. *The Liberators.* Tucson: Westernlore Publications.

Swoboda, E., 1976. Boleo-A Classic Locality Reworked. Issue 10. *Lapid. Jour.* **30:**1814-1830.

Taggart, J. E., Foord, E., Rosenweig, A., 1984. Natural Occurrence of Alpha-Lead Dioxide at Bingham, New México, USA and Mapimi, Durango, México. Paper read at MSA/FM Joint Symposium on Minerals of México, 12 February, at Tucson, Arizona.

Tamayo, J. L., 1949. *Geografía General de México.* México.

Taylor, A. S., 1981. *A Historical Summary of Lower California.* Tucson: Westernlore Publications.

Tays, E. A. H., 1908. Mining in México, Past and Present. *Eng. Min. Jour.* **86:**665-667.

Thayer, W. H., The Physiography of México. *Jour. Geol.* **74:**72.

Thomas, C. S., Jr., 1911. Traveling in México. *Eng. Min. Jour.* 1201-1203.

Tod, S., 1907. History and Development of Batopilas Mine, México. *Min. World* 566-568.

Tod, S., The Mines of Batopilas. Number 7. *Min. World* **27:**667-794.

Todd, A. C., 1977. *The Search for Silver—Cornish Miners in México 1824-1947.* Cornwall: The Lodenek Press.

Toothaker, C. R., 1941. Crystals with Rhomboid Tubes from Guanajauto, México. *Am. Miner.* **26:**733-735.

Touwaide, M. E., 1930. Origin of the Boleo Copper Deposit, Lower California, México. *Econ. Geol.* **25:**113-114.

Treadwell, J. C., 1905. The Sahuayacan District, México. *Eng. Min. Jour.* **80:**1213-1216.

Valentine, W. G., 1936. Geology of the Cananea Mountains, Sonora, México. *Bull. Geol. Soc. Am.* **47.**

Valerio-Ortega, M., 1902. The Patio Process for Amalgamation of Silver Ore. *Trans. Am. Instit. Min. Eng.* **XXXII:**276-284.

Van Horn, F. R., 1912. Occurrence of Silver, Copper, and Lead Ores at the Veta Rica Mine, Sierra Mojada, Coahuila, México. AIME.

Vaupell, C. W., 1937. The Huitzuco Mercury Miners, Guerrero, México. *Econ. Geol.* **32:**196.

Vaupell, C. W., 1941. Mercury Deposits of Huitzuco, Guerrero, México. *Trans. Am. Instit. Min. Eng.* **114:**300-313.

Villarello, J. D., 1909. The Mapimi District. *Min. World* **31:**62-63.

von Humboldt, A., 1811. *Political Essay on the Kingdom of New Spain.* London.

von Humboldt, A., 1824. Notes by J. Taylor. *Selections from the Works of the Baron de Humboldt, Relating to Climate, Inhabitants, Productions, and Mines of México.* London: Longman, Hurst, Rees, Orme, Brown, and Green.

Wallace, L. 1867. The Mines of Santa Eulalia, Chihuahua. *Harpers* **210:**681-702.

Wampler, J., 1978. *México's 'Grand Canyon.'* Berkley.

Wandke, A., and Martinez, J., 1928. The Guanajuato Mining District, México. *Econ. Geol.* **23:**1-44.

Ward, H. G., *México.* 2nd Edition. Volume 1 and 2. London: Henry Colburn.

Webster, R., 1975. *Gems, Their Sources, Descriptions, and Identification.* London: Butterworths.

Weed, W. H., 1902. Notes on Certain Mines in The States of Chihuahua, Sinaloa, and Sonora, México. *Trans. Am. Inst. Min. Eng.* **XXXII**:396-443.

Weigand, P. C., 1978. The Mines and Mining Techniques of the Chalchihuites Culture. *Am. Antiq.* **33**:45-61.

Wenrich, K. J., Wenrich, P., Modresky, J., and Zielinski, R. A., 1983. A New Mineral of Hydrothermal Origin from the Peña Blanca Uranium District, México. Paper read at the MSA/FM Joint Symposium, 13 February, at Tucson, Arizona.

Wentworth, I. H., The San Nicolas Mining District, San Nicolas, Tamaulipas, México. *Trans. of the Am. Instit. of Min. Eng.* **43**:304-313.

Whitaker, A. P., 1951. The Elhuyar Mining Missions and the Enlightenment. *Hisp. Am. Hist. Rev.* **XXI**:558-585.

White, D. E., 1947. *Antimony Deposits of Soyatal District State of Queretaro, México.* Bulletin 960-B. Washington: U.S. Geological Survey.

White, D. E., and Gonzalez-Reyna, J., 1946. *San Jose Antimony Mines Near Wadley, State of San Luis Potosí, México.* Bulletin 946-E. Washington: U.S. Geological Survey.

White, D. E., and R. Guiza, Jr., 1949. *Antimony Deposits of El Antimonio District Sonora, México.* Bulletin 962-B. Washington: U.S. Geological Survey.

White, D. E., and R. Guiza, Jr., 1947. *Antimony Deposits of the Tejocotes Region, State of Oaxaca, México.* Bulletin 953-A. Washington: U.S. Geological Survey.

White, J. S., Jr., 1972. Extreme Symmetrical Distortion of Pyrite from Naica, México. Number 6. *Mineral. Rec.* **4**:267-271.

White, L., 1980. Mining in México, Feeding Needed Raw Materials To a Rapidly Expanding Economy. *Eng. Min. Jour.* **181**:62-194.

Wiese, J., and Cardenas, S., 1945. *Tungsten Deposits of the Southern Part of Sonora, México.* Bulletin 946-D. Washington: U.S. Geological Survey.

Wiesenthal, M., 1978. *Yucatán and the Maya Civilization.* New York: Crescent Books.

Wilcox, R. E., 1954. *Petrology of Paricutín Volcano, México.* Bulletin 965-C. Washington: U.S. Geological Survey.

Wilkerson, G., 1983. *Geology of the Batopilas Mining District, Chihuahua.* Dissertation. El Paso: University of Texas.

Wilson, W. E., 1979. The Collector's Library, Part III: Minerals of México. *Mineral. Rec.* Number 10. **3**:169-171.

Wilson, W. E., 1980. Famous Mineral Localities: Los Lamentos, Chihuahua, México. *Mineral. Rec.* Number 11. **11**:277-286.

Wilson, W., and Panczner, C. S., Famous Mineral Localities: The Batopilas Silver Mines, Chihuahua, México. *Mineral. Rec.* Number 1. **17**:61-80.

Wilson, I. F., 1956. *The Naica Mining District, Chihuahua, México.* Guidebook 2-A. XXth International Geological Congress. pp. 46-61.

Wilson, I. F., and Rocha, V. A., *Geology and Mineral Deposits of the Boleo Copper District, Baja California, México.* Professional paper 273. Washington: U.S. Geological Survey.

Wilson, I. F., and Veytia, M., 1949. *Geology and Manganese Deposits of the*

Lucifer District Northwest of Santa Rosalia, Baja California, México. Bulletin 960-F. Washington: U.S. Geological Survey.

Winchell, H. V., 1921. Geology of Pachuca and El Oro, México. *Trans. Am. Inst. Min. Eng.* **66**:27-40.

Wisser, E., 1954. Geology and Ore Deposits of Baja California, México. *Econ. Geol.* **49**:44-76.

Wisser, E., 1966. The Epithermal Precious-Metal Province of Northwest México. Volume 13. *Nev. Bur. Mines Rep.* **13**:63-92.

Wooton, P., 1907. Rich in Minerals but Handicapped by Lack of Transportation. Number 17. *Min. World* **27**:746-749.

Young, O. E., Jr., 1970. *Western Mining.* Norman: University of Oklahoma Press.

Zodac, P., 1946. Potosí Mine of México. *Rocks Min.* **21**:573.

CITATIONS AVAILABLE FROM SOURCES IN MEXICO

Aquilar y Santillán, R., 1889. *Bibliografía Geológica y Minera de la República Mexicana.* Boletín 10. México: Instituto de Geológico de México. Universidad Nacional Autónoma de México.

Aquilar y Santillan, R., 1904. *Bibliografía Geológica y Minera de la República Mexicana, completada Hasta.* Boletín 17. México: Instituto Geológico de México. Universidad Nacional Autónoma de México.

Aquilera, J. G., 1898. *Catálogos Sistematico y Geografico de las Especies Mineralogicas de la República Mexicana.* Boletín 11. México: Instituto de Geologico de México. Universidad Nacional Autónoma de México.

Anonymous, updated. *Las Minas de Guanajuato.* Guanajuato: Private Printing.

Anonymous, 1970. *Minería Prehispanica en la Sierra de Queretaro.* México: Secretaría del Patrimonio Nacional.

Anonymous, 1977. Catálogo de Informes Geológico Mineros Existenes en al Archivo Tecnico del Consejo de Recursos Minerales. *Geomimet* **89**:38-76.

Anonymous, 1977. Catálogo de Informes Geológico Mineros Existenes en al Archivo Tecnico del Consejo de Recursos Minerales. Part 2. *Geomimet* **91**:49-90.

Barrera, Madame, de, 1871. Joyas y Alhajos. O Sea Su Historia en Relacíon con la Politica, la Geografia, la Mineralógia, la Quimica. *Cron. Hisp. Am.* **25**:7-21.

Barrera, T., 1927. *Informe Geologico del Criadero y Mina de Santa Rosa, Pertenecientes a la Santa Rosa Mining Co., Dist. de Mazapil, Estado de Zacatecas.* Bolétin 46. México: Instituto Geológico de México.

Barrera, T., 1931. *Zonas Mineras de Los Estados de Jalisco y Nayarit.* Boletín 51. México: Instituto de Geológico de México.

Barrera, T., and Segura, D., 1927. *Itinerarios Geologicos en Estado de Michoacan.* Boletín 46. México: Instituto Geológico de México.

Beal, C. H., 1922. Reconocimiento Geologia de las Districtos de Hermosillo y Guaymas del Estado de Sonora, México. *Bol. Pet.* **13**:263-291.

Beteta, J. L., 1978. Breve Historia Minería de México. *Geomimet* **91**:37-43.

Bose, P. E., 1905. *Geologia de Chiapas y Tabasco*. Boletín 20. México: Instituto Geológico de México. Universidad Nacional Autónoma de México.

Bracho, A. M., 1984. Situacion General de la Minería. *Caminex Min.* **5:**5-19.

Brodie, W. M., 1889. El Socavon, 'Porfirio Diaz' et Batopilas, Estado de Chihuahua. Number 19. *Min. Méx.* **34.**

Burckhardt, C., 1906. *Geologie de la Sierra de Mazapil.* XXIV (exursions de nord). México: Xth International Geológical Congress.

Burckhardt, C., and Scalia, S., 1906. *Geologie des Environs de Zacatecas.* XIV (excursions du nord). México: Xth International Geológical Congress.

Cabrera-Ipina, O., 1975. *El Real de Catorce*. San Luis Potosí: Sociedad Potosina de Estudios Históricos.

Cabrera-Ipina, O., 1962. San Luis Potosí y su Territorio. San Luis Potosí: Private printing.

Cardenas, S., and Perez, F. M., 1947. *Los Yacimientos Argentiferos de Temascaltepec, Estado de México.* Boletín 12. México: Comité Directivo para la Investigacíon de los Recursos Minerales de México.

Cárdenas-Vargas, J., and Castillo-Garcia, L., del, 1964. *Yacimientos de Hierro de la Parla y La Negra, Municipio de Camargo, Chihuahua.* Boletín 69. México: Consejo de Recursos Naturales no Renovables.

Castillo, A., del, 1864. Catologo de las Especies Minerales y de sus Variedades que se Encuentran en México. Number 8. *Bol. Soc. Mex. Geogr. Estad.* **10:**464-471.

Castillo, A., del, *Minerales de la República*. México: Boletín de Siglo XIV.

Castrejón-Diez, R. N., and Castrejón-Diez, J., *Una Pequeña Historia de Taxco*. México: Santa Prisca.

Clarke, K. F., Damon, P. E., Schutter, S. R., and Shaffiqullah, M., 1979. Magmatismo en el Norte de México en Relacíon a los Yacimientos Metaliferos. *Assoc. Ing. Minas, Met. Geol. Mex. Mem. Tec.* **XVIII:**8-57.

Dana, E. S., 1976. *Tratado de Mineralogía*. 4th edition. México: Compañía Editorial Continental, S. A.

Desconocido, 1984. *Atlas de Carreteras Basicas de la República Mexicana*. México: Desconocido.

Echavarri-Prez, A., Saitz-Sau, O. A., and Salas-Piza, G. A., 1977. Mapa Metalogenético de Sonora. *Geomimet* **88:**33-64.

Elhuyar, Fausto de, 1835. *Memoria Sobre el Influjo de la Minería en la Agricultura, Industria, Poblacíon y Civilizacíon de la Nueva España en sus Diferentes Épocas, Con Varios Disertaciones Relativos a Puntos de Economía Pública Conexos Con el Propio Ramo.* Madrid: Inprenta de Amartia.

Fabregat-Guinchard, F. J., 1962. *Bibliografía Mineralogica de México*. Anales, Número 20. México: Instituto de Geologia. Universidad Nacional Autónoma de México.

Fabregat-Guinchard, F. J., 1963. *La Boleita*. Boletín 66. México: Instituto de Geologia. Universidad Nacional Autónoma de México.

Flores, T., *Estudio Geologico-Minero de Los Distritos de El Oro y Tlalpujahua*. Boletín 37. México: Instituto de Geológico de México. Universidad Nacional Autónoma de México.

Flores, T., 1946. *Geologia Minera de la Region NE del Estado de Michoacan.* Boletín 52. México: Instituto de Geológico, Universidad Nacional Autónoma de México.

Flores, T., *Granates Turmalinas, Micas y Feldspatos des Distrito Norte de la Peninsula de la Baja Californie. Tomo 4. México: Instituto de Geológico de México. Universidad Nacional Autónoma de México.*

Foshag, W. F., and Fries, C., Jr., 1946. *Los Yacimientos de Estano de la República Mexicana.* Boletín 8. México: Comité Directivo para La Investigacíon de Los Recursos Minerales de México.

Foshag, W. F., Gonzalez, R. J., and Perez, S. R., 1946. *Los Depositos de Fluorita del Distrito Minero de Taxco, Estado de Guerrero.* Boletín de Minas y Petroleo. Secretaría de Económica Nacional Direccion General de Minas y Petroleo.

Foto Picture Books, *Via Taxco en Color.* México: Foto Picture Books.

Fries, C., Jr., 1948. *Los Yacimientos de Calcita Optica de la República Mexicana.* Boletín 16. México: Comité Directivo para la Investigacíon de los Recursos Minerales de México.

Fries, C., Jr., and Schmitter, E., 1945. *Yacimientos de Scheelita en la parte norte de la Sierra de Juarez, Distrito Norte de la Baja California.* Boletín 2. México: Comité Directivo para la Investigacíon de los Recursos Minerales de México.

Fuente-La Valle de la, and Fernando E., 1969. *De la Mina El Potosí Distrita Minera de Santa Eulalia, Chihuahua.* Thesis, Universidad Nacional autónoma de México.

Gaines, R. V., *Mineralization de Telurio en la Mina Moctezuma, Cerca Moctezuma, Sonora.* Boletín 75. México: Instituto de Geológico de México. Universidad Nacional Autónoma de México.

Galvez, V., 1919. *Apuntes Sobre El Mineral de Puerto de Nieto E. of Guanajuato.* Boletín 6. México: Instituto Geológico de México.

Garcia, T., 1970. *Los Mineros Mexicanos.* 3rd Edition. México: Editorial Porrua, S. A. Universidad Nacional Autónoma de México.

Garfias, V. R., and Chapin, T. C., 1949. *Geologia de México.* México: Editorial Jus.

Gonzalez-Reyna, J., 1946. *La Industria Minera en el Estado de Chihuahua.* Boletín 7. México: Comité Directivo para la Investigación de los Recursos Minerales de México.

Gonzalez-Reyna, J., 1946. *La Industria Minera en El Estado de Zacatecas.* Boletín 4. México: Comité Directivo para la Investigación de los Recursos Minerales de México.

Gonzalez-Reyna, J., 1946. *Los Criaderos de Uranio y Oro en Placer de Guadalupe y Puerto del Aire Estado de Chihuahua.* Boletín 5. México: Comité Directivo para la Investigación de los Recursos Minerales de México.

Gonzalez-Reyna, J., 1946. *Los Depositos de Magnesita de la Porcion Central de Isla Margarita, Baja California.* Boletín 10. México: Instituto Nacional para la Investigación de Recursos Minerales.

Gonzales-Reyna, J., 1947. *Los Yacimientos Argentiferos de Batopila Estado de Chihuahua.* Boletín 11. México: Instituto Nacional para Investigación de Recursos Minerales, México.

Gonzalez-Reyna, J., 1949. *Geologia, Paragenesis y Recursos de los Yacimientos de Plomo y Zinc de México.* Boletín 26. México: Comité Directivo para la Investigación de los Recursos Minerales de México.

Gonzalez-Reyna, J., 1956. *Memoria Geological-Minea del Estado de Chihuahua (Minerales Metalicos).* XXth International Geological Congress.

Gonzalez-Reyna, J., 1956. *Riqueza Minera y Yacimientos Minerales de México.* 3rd Edition. XXth International Geological Congress.

Gonzalez-Reyna, J., 1958. *Erupción del Volcán de Fuego de Colima.* Berlin: Akademie Verlag.

Instituto de Geologico de México, 1897. *El Mineral de Pachuca.* Boletín 7, 8, and 9. México: Instituto de Geológico de México. Universidad Nacional Autónoma de México.

Jimenez, L. G., Wittch, E., and Griggs, J., 1978. Cronologia sobre el Desarrollo de la Minera en México. *Geomimet* **96:**54-60.

Karsten, D. L. G., 1804. *Tablas-Mineralogicas.* México: Seminario de Minería.

Kunz, G. F., 1889. Sur une Hache Votive Gigantesque en Jadeite de l "Oaxaca, et Sut Unpectoral en Jadeite de Guatemala." *Compte Rendu. Congr. Int. Anthro.* 517-523.

Maffei, F. D., 1970. *Biblioteca Espanola.* La Minería Hispana e Iberomericana. Volume I and 2. Leon: Bibligrafia VI congress Internacional de Minería, Departamento de Publicación, Catedra De San Isidoro.

Maffei, F. D., and Figueroa, R. D., 1872. *Biblioteca Espanola.* Volumes 1 and 2. Madrid: J. M. Lapuente.

Maldonado-Espinosa, D., 1979. *Geologia de la Mina San Antonio, Unidad Minera Santa Eulalia, Chihuahua, México.* Thesis. Universidad Autónoma de San Luis Potosí.

Maldonado-Espinosa, D., and Megaw, P. K. M., Geology of the Santa Eulalia Mining District, Chihuahua, México. Unpublished Report. Industrial Minera de México S. A.

Mendoza, A. de., 1536. *Ordenanzas de . . . Para el Tratamiento de los Indios de las Minas de Plata, Preganadas en México.* 30 June. México: Bibliot de la Acad de la Historia.

Miranda, E. G., de, and Gyves, Z. F., de, 1984. *Atlas-Nuevo Atlas Porrua de la Repüblica Mexicana.* 6th Edition. México: Editorial Porrua, S. A.

Muñoz, C. F., 1978. Estudios Petrografico y Mineralogico de las Minas San Antonio y Buena Tierra, Unidad Santa Eulalia. Unpublished Report. Industrial Minera de México S. A.

Ordóñez, E., 1904. El Mineral de Angangueo, Michoacan. *Instit. Geol. México* **3:**1-23.

Ordóñez, E., and Rangel, M., 1899. *El Real del Monte.* Boletín 12. México: Instituto de Geologico de México. Universidad Nacional Autónoma de México.

Orozco, R., 1921. *La Industria Minera de México-Distrito de Guanajuato.* México: Secretaría de Educacion Publica.

Ortiz-Asiain, R., 1956. *Notes on Cerro de Mercado.* Guidebook A-2. XXth International Geological Congress.

Pareja, J. S., 1883. *Resena Historia de Batopilas.* Alamos: J. M. Murillo.

Pesquera-Velazquez, R., 1978. Cronologia Sobre el Desarollo de la Mineria en México. *Geomimet* **96:**54-60.

Pesquera-Velazquez, R., 1978. Principales Minas Antiquas Inactivas o Parcialmente Trabajando Téchnico del Consejo de Recursos Minerales en México. Part 2. *Geomimet* **96:**43-53.

Prado, J. J., 1954. *La Fluorita.* Boletín 1-E. México: Instituto Nacional Para la Investigación de Recursos Minerales.

Ramdohr, P., 1948. *Las Especies Mineralogicas Guanajuatita y Paraguanajuatita.* Boletín 20. México: Instituto Nacional para la Investigación de Recursos Minerales.

Ramírez, S., 1884. *Riqueza Minera de México.* México: Oficina Tipográphica de la Secretaría de Fomento.

Ramírez, S., 1891. *Biografia del Sr. D. Andres del Rio: Primer Catedratico de Mineralogía del Colegio de Mineral.* México: Imprenta del Sagrado Corazon de Jesus.

Río, A. M., del, 1795. *Los Elementos de Orictognosia, o Del Conocimiento de los Fósiles.* México: Mariano Joseph de Zuniga.

Río, A. M., del, 1797. Discurso que Presencia del Real Tribunal de Minería Pronuncio . . . Catedratico de Minerlogia. Supliemento. Número 30. *Gas. Mex.* **III.**

Río, A. M., del, 1799. Discurso Sobre las Volcanes. *An. Hist. Nat.* **II:**335-348.

Río, A. M., del, 1803. Descripcion de una Piedra Perlada. *An. Hist. Nat.* **VI:**363-367.

Río, A. M., del, 1804. Observaciones de . . . Sobre un Tratado de Minas. *An. Hist. Nat.* **VII:**17-29.

Río, A. M., del, 1805. *Los Elementos de Orictognosia, o Del Conocimiento de los Fósiles-Part 2.* México: Mariano Joseph de Zuniga.

Río, A. M., del, 1810. *Discurso Sobre la Ferreria de Coalcoman.* Supplement. México: Diario de México.

Río, A. M., del, 1814. Carta Dirigida al Sr. Baron de Humboldt. *Mercur. Esp.* 169-176.

Río, A. M., del, 1827. *Nuevo Sistema Mineral Del Sr. Bercelio.* México: Instituto Mexicano.

Río, A. M., del, 1832. *Los Elementos de Orictognosía, o Del Conocimiento de los Fósiles Segun el Sistema de Bercelio.* Philadelphia: J. F. Hurtel.

Río, A. M., del, 1846. *Los Elementos de Orictognosía, o Del Conocimiento de los Fósiles Segun el Sistema Del Baron Bercelio.* 2nd Edition. México: R. Rafael.

Río, A. M., del, 1848. *Supplemento de Adiciones Correcciones de mi Mineralogía-1832.* México: Instituto Nacional de Francia y de Otra Acadamias Sociedadas Científicas.

Rogers, C. L., Cserna, Z., de, Amezcua, E. T., and Ulloa, S., 1957. *Geologia General y Depositos de Fosfatos Del Distrito De Concepcion Del Oro, Estado De Zacatecas.* Boletín 38. México: Instituto Nacional para La Investigación de Recursos Minerales.

Rueles, T., Miquel, A., 1973. Ojuela Pueblo Fantasma, Fue uno do Las Lugares Mas Ricos de México. 28 July. *El Siglo de Torreón.*

Salas, P. G., 1981. El Potencial de Los Recursos Minerales Del Estado de Guanajuato. Number 110. *Geomimet* 55-67.

Salazar-Salinas, L., 1923. *Catálogo Geografico de las Especies Minerales de México.* Boletín 41. México: Instituto Geológico de México, Universidad Nacional Autónoma de México.

Salazar-Salinas, L., 1923. *Catálogo Sietematico de Especies Minerales de México y sus Aplicaciones Industriales.* Bóletín 40. México: Instituto de Geológico de México. Universidad Nacional Autónoma de México.

Salazar-Salinas, L., 1923. *El Cerro de Mercado, Durango.* Boletín 44. México: Instituo de Geológico de México. Universidad Nacional Autónoma de México.

Santiago De La Laguna, Conde de, 1732. *Descripcion Breve de la Muy Noble y Real Ciudad de Zacatecas.* México.

Shepherd, A. R., 1900. Las Minas de Batopilas, Chihuahua, México. Number 19 and 20. *Min. Méx.* **36**.

Shepherd, G., 1978. *De Magnate de Plata.* Trans. Camu, Concepcion Montilla. México: Centro Librero la Prensa.

Sonneschmidt, F., 1825. *Tratado de Amalgamacion de Nueva Espana 1805,* Sacado a Luz D. J. M. F. México: Libreria de Bossange (Padre) Antoran y Cia.

Torres, A., Muñoz, J. M., and Muñoz, C. F., 1978. *Mineragrafia de las Calizas de Mina Vieja de Santa Eulalia, Chihuahua, México.* Unpublished Report. Industrial Minera de México S. A.

Vega, S. E., de la, 1980. Algunos Datos para la Cronologia de la Minería y Geologia en México. *Geomimet* **104:**63-72.

Vega, S. E., de la, 1975. Antecendentes Históricos de las Esculas de Minas y Geologia en Tecnico del Consejo de Recursos Minerales de México. Part 2. *Geomimet* **73:**38-45.

Vega, S. E., de la, 1981. Fresnillo en la Minería. *Geomimet* **113:**69-78.

Velasquez de Léon, J., 1884. Elógio Funebre del Sr. D. Andres del Río. El Minero Mexicano.

Villarello, J. D., 1906. *Le Mineral d'Aranzazu.* XXth International Geological Congress, XXV.

Villarello, J. D., 1906. *La Mineral de Mapimi.* XXth International Geological Congress, XVIII.

Weidner, F., 1858. *El Cerro de Mercado de Durango.* México: Impreso de Andrade y Escalante.

Zubiría y Campa, L., 1919. Bibliografía Minería, Geológica y Mineralogica del Estado de Durango. *Mem. Soc. Cient.* **38:**177-198.

CATALOG

The catalog is arranged alphabetically according to mineral and
further alphabetized within each
entry by state and municipio.
Mineral information is presented
as follows:

MINERAL

Chemical formula. Brief description of environment of formation
and/or rock type found in.

MEXICAN STATE

Municipio

Location: Distribution. (Mineral
variety). Mine or mines. Crystal
size. Associations. [Author's
comments.]

In a few cases, distribution
information was obtained from
documents that are over several
hundred years old. Current
conditions may differ. Crystal
sizes are based on the longest
recorded sizes for that location.

A

ACANTHITE

Ag_2S. Formed both in hydro-thermal and secondary enrichment mineral deposits. Dimorphous with argentite.

AGUASCALIENTES

Municipio de Asientos

Asientos: Moderately distributed within the Veta de los Pilares. Noted from the Cinco Señores, El Orito, San Francisco, Santo Cristo, and Descubridora mines.

Municipio de Tepezalá

Tepezalá: Moderately distributed.

BAJA CALIFORNIA

Municipio de Ensenada

Isla de Cedros: Limited distribution. Macara de Hierro Mine.

San Fernando: Moderately distributed. Chalcocita Mine.

BAJA CALIFORNIA SUR

Municipio de San Antonio

San Antonio: Moderately distributed. Noted from the Comstock and Reforma mines.

CHIHUAHUA

Municipio de Aquiles Serdán

Francisco Portillo: Moderately distributed. Noted from the Velardena and Inglaterra mines. Etched crystals to 3 cm. Associated with native silver, quartz, and calcite.

San Antonio el Grande: Moderately distributed. San Antonio Mine. Lustrous crystals to 2 cm. Associated with native silver and quartz.

Municipio de Ascención

Sabinal: Moderately distributed.

Municipio de Batopilas

Alisos: Moderately distributed. Noted from the Santa Sophia and Socorro mines.

Batopilas: Moderately distributed. Noted from the San Nestor, New Nevada, and Santo Domingo mines. Crystals to 2 cm. Associated with native silver, proustite, and calcite.

Jesús María: Moderately distributed. Yedras Mine. Associated with proustite and pyrargyrite.

Satevó: Limited distribution. San Miguel Mine.

Municipio de Coyame

Las Plomosas: Moderately distributed. El Lago Mine.

Municipio de Cusihuiriáchic

Cusihuiriáchic: Widespread. Most noted from the Princesa, Buenos Aires, and El Burro mines. Associated with quartz. [The mines of this district produced such a large volume of silver that the Spanish established a mint in this remote area of Chihuahua.]

Municipio de Guazapares

Guazapares: Moderately distributed. Agua Nueva Mine.

Municipio de Hidalgo del Parral

Hidalgo del Parral: Widespread.

Municipio de Jiménez

Huejuquilla: Moderate distribution. Noted from the Adargas and Saltillera mines.

Municipio de Julimes

Julimes: Moderate distribution. Noted from the Chanate, Guadalupe, and San Antonio mines.

Municipio de Morelos

Zapural: Limited distribution. Associated with native silver and stephanite.

Zapuri: Moderately distributed, noted from the San Rafael Mine.

Municipio de Moris

Sahuayacón: Moderately distributed. Santa Teresa Mine.

Municipio de Nuevo Casas Grandes

San Pedro Corralitos: Moderate distribution. Candelaria Mine.

Municipio de Ocampo

Pinos Altos: Moderately distributed. Pinos Altos Mine.

Municipio de Santa Bárbara

Santa Bárbara: Widespread. [Silver was discovered here in 1544 and Santa Bárbara became the capital of New Biscay from 1580 to 1638. A territory stretching northward from what was to become California to Arizona, New Mexico and Texas.]

Municipio de Saucillo

Naica: Moderately distributed. Etched crystals to 2 cm. Associated with native silver, calcite, quartz, and galena.

Municipio de Urique

Piedras Verde: Moderately distributed. Piedras Verde Mine.

Urique: Moderately distributed. Rosario Mine. Associated with chalcopyrite, pyrite, galena and quartz.

COAHUILA

Municipio de Múzquiz

La Encantada: Moderate distribution. La Encantada Mine.

Municipio de Sierra Mojada

Sierra Mojada: Moderately distributed. Veta Rica Mine. Associated with barite, native silver, proustite, and erythrite.

DURANGO

Municipio de Canatlán

Mejamen: Moderately distributed. Eureka Mine.

Municipio de Canelas

Birinuca: Moderately distributed.

Municipio de Cuencamé

Velardeña: Moderately distributed. Noted from the Velardena
and Socavón Hay mines.

Municipio de Guanaceví

Guanaceví: Widely distributed.
Noted from the Guadalupe and
Farmy mines. Associated with
native gold.

Municipio de Indé

Indé: Moderately distributed. Los
Colorados Mine.

Municipio del Oro

Santa Maria del Oro: Widely
distributed. Noted from the
La Candela, La Reina, La Predilecta, and La Princesa mines.

Municipio de Pánuco de
Coronado

Pánuco de Coronado: Moderate
distribution. Alvino Mine. [This
old location dates back into
the 1500s and was originally
known as San José de Avino.]

Municipio de Poanas

Cerro de Sacrificio: Moderately
distributed. Most noted from the
El Sacramento, San Carlos, and
El Rosario mines.

Municipio de Pueblo Nuevo

El Salto: Limited distribution.
las Animas Mine. Crystals to 1
cm. Associated with polybasite,
pyrargyrite, and calcite.

Municipio de San Dimas

Tayoltita: Widely distributed.
Associated with stephanite.

Municipio de San Pedro
del Gallo

Peñoles: Moderately distributed.
San Rafael Mine.

Municipio de Topia

La Portilla: Moderately distributed. Noted from the La Portilla
Mine.

Pilones: Moderately distributed.
La Primavera and Calera mines.

Topia: Moderately distributed.
Noted from the La Adrugada Mine.

GUANAJUATO

Municipio de Guanajuato

Guanajuato: Widespread. Most
noted from the Cata, El Cubo,
La Valenciana, and the San Juan
de Rayas mines. Sharp brilliant
crystals to 8 cm. Associated
with stephanite, polybasite,
pyrargyrite, galena, chalcopyrite, calcite, and quartz.
[During the 1970s and 1980s,
several large crystal specimens
have been found at the San Juan
de Royas Mine. The cubic crystals
all had brilliant and sharp
crystal faces. The crystals were
up to 8 cm in size and are the
largest known from México.] See
introduction, Guanajuato.

La Luz: Widespread. Noted from
the La Luz and Santa Rita
(Bolanitos) mines. Associated
with calcite and galena. [The La
Luz Mine has produced specimens
of acanthite replacing stephanite.]

Municipio de San Diego
de la Unión

Providencia: Limited distribution. Providencia Mine.

Municipio de San Luis
de La Paz

Mineral de Pozos: Moderately
distributed. Noted from the Santa
Brígida and La Joya mines.

Municipio de Xichú

Xichú: Moderately distributed.

GUERRERO

Municipio de Arcelia

Campo Morado: Moderate distribution. Noted from the El Naranjo and Reforma mines.

Municipio de Buenavista de Cuéllar

Buenavista de Cuéllar: Widespread. Most noted from the Pinahua, Aguacate, and Natividad mines.

Municipio de Chilpancingo

Chilpancingo de los Bravos: Moderately distributed. Noted from the Delfina and Porvenir mines.

Municipio de San Miguel Totolapán

La Coronilla: Moderate distribution. Noted from the Grande, Peregrina, and Pilar mines.

Municipio de Taxco

Taxco de Alarcón: Widespread. Noted from the Chontapan Mine. Crystals to 2 cm. Associated with native silver, pyrargyrite, proustite, azurite, and gypsum (selenite). [One of the oddities found were spheres to 6 cm composed of native silver, acanthite, pyrargyrite, and proustite. Also found were, selenite crystals containing native silver and acanthite crystals of acanthite with azurite crystals.]

Municipio de Tetipac

Cuadrilla Poder de Dios: Moderately distributed.

Municipio de Tlalchapa

Puerto del Oro: Limited distribution. Garduno Mine.

HIDALGO

Municipio del Arenal

Tepenene: Moderately distributed. Noted from the Ernestina, Podor de Dos, and Santo Niño mines.

Municipio de Mineral del Chico

Mineral del Chico: Widespread.

Municipio de Mineral del Monte

Mineral del Monte: Widespread. Noted from the Doloras Mine. Crystals to 3 cm. Associated with calcite and quartz.[Silver was discovered here in 1528 and is the location better known as Real del Monte.]

Municipio de Pachuca

Pachuca de Soto: Widespread. Most noted from the San Rafael, Entroetida, Jacal, Encino and Moctezuma mines. Crystals to 3 cm. Associated with native silver, calcite, quartz, and gypsum selenite. [Acanthite on native silver within selenite crystals have been found at the Encino Mine. Silver was discovered here in 1524. This location is better known as Pachuca.]

Municipio de Zimapán

Zimapán: Widespread. Associated with native silver.

JALISCO

Municipio de Ahualulco de Mercado

Caballo Muerto Cocultecos: Moderately distributed.

Municipio de Bolaños

Bolanos: Moderate distribution. Sante Fe Mine.

Municipio de Exatlán

Exatlán: Limited distribution. Cancinene Mine.

Acanthite, approx. 10 cm × 14 cm, from Las Chipas Mine, near Arizpe, Sonora. Formerly in the Pinch collection, but now in the Sams collection, Houston Museum of Natural History, Houston, Texas. *(Photograph by Wm. Pinch.)*

Adamite (cuprian) on goethite from the Ojuela Mine near Mapimí, Durango, in the Panczner collection, Tucson, Arizona. The sphere of crystals is approx. 6 cm in diameter.

Adamite (cuprian) on goethite, approx. 6 cm × 9 cm, from Ojuela Mine near Mapimí, Durango, in the William Larson collection, Fallbrook, California.

Azurite replacing atacamite associated with quartz and atacamite from Aranzazu Mine, Aranzazu, Zacatecas, in the Benny Fenn collection, Colonia Juárez, Chihuahua. The crystal is approx. 5 cm long.

Fluorite on calcite, with crystals up to approx. 12 cm long, from Naica, Chihuahua, in the Peter Megaw collection, Tucson, Arizona.

Gold on hematite, approx. 8 cm × 12 cm, from Monterde Mine, San José, Chihuahua, in the Wm. Pinch collection, Rochester, New York. *(Photograph by Wm. Pinch, courtesy of the Pinch Mineralogical Museum.)*

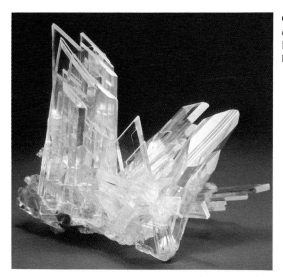

Gypsum (selenite), approx. 18 cm × 12 cm overall with crystals to 12 cm long, from the Santo Domingo Ramp, Francisco Portillo, Chihuahua, in the William Larson collection, Fallbrook, California.

Gypsum (selenite), with crystals up to 12 cm long from Naica, Chihuahua, in the Panczner collection, Tucson, Arizona.

Hemimorphite, with crystal sprays approx. 4 cm long, on goethite from Ojuela Mine near Mapimí, Durango, in the Panczner collection, Tucson, Arizona.

Legrandite "Aztec Sun" associated with goethite and scorodite in a complex specimen from Ojuela Mine near Mapimí, Durango, in the collections of the Museo Mineralógico de Romero, Tehuacán, Puebla. The legrandite crystals are approx. 24 cm long.

Legrandite, with crystals up to approx. 5 cm long, on goethite from Ojuela Mine near Mapimí, Durango, in the Wm. Pinch collection, Rochester, New York. *(Photograph by Wm. Pinch, courtesy of the Pinch Mineralogical Museum.)*

A complex twinned crystal of ludlamite with pyrite, approx. 11 cm long, from San Antonio Mine, San Antonio el Grande, Chihuahua, in the Sams collection, Houston Museum of Natural History, Houston, Texas.

Malachite replacing azurite associated with quartz from El Cobre Mine, El Cobre, Zacatecas, in the collections of Museo Mineralógico de Romero, Tehuacán, Puebla. Largest crystal on this specimen is approx. 12 cm long.

Municipio de Hostoti-
paquillo

Hostotipaquillo: Moderately
distributed. Mololoa Mine.
Associated with covellite and
jalpaite.

Municipio de Mascota

Natividad: Limited distribution.
Descubridora Mine.

Municipio de San Sebastián

El Bosque: Moderately distrib-
dora and Socorro mines.

Municipio de Talpa de
Allende

Guale: Moderate distribution.
Noted from the Ojo de Agua and
San Vicente mines.

Municipio de Tequila

Tequila: Widespread. Española
Mine.

MEXICO

Municipio del Oro

El Oro de Hidelgo: Widespread,
Vetas San Rafael. Noted from
the El Oro, Esperanza, México,
Nolan, and Dos Estrellas mines.
Crystals to 3 cm. Associated
with quartz, calcite, and pyrite.

Municipio de Ixtapan del
Oro

Ixtapan del Oro: Moderately
distributed.

Municipio de Sultepec

Sultepec de Pedro Ascencio de
Alquistras: Widespread. Crystals
to 3 cm. Associated with stephanite and quartz.

Municipio de Temascaltepec

Temascaltepec de González:
Widespread. Noted from the Veta
Chica and Protectora mines.

Municipio de Zacualpan

Zacualpan: Widespread. Noted
from the Carboncillo, Coronas,
and Durazno mines. Associated
with pyrite.

MICHOACAN

Municipio de Acuitzio

Curucupaseo: Moderately distributed.

Municipio de Carácuaro

Carácuaro de Morelos: Moderate
distribution. Descubridora Mine.

Municipio de la Huacana

La Huacana: Moderately distributed. Noted from the San Cristóbal and Inguaran mines. Lustrous
crystals to 3 cm.

Municipio de Tacámbaro

Huaungucha: Limited distribution.
La Luz Mine.

Municipio de Tlalpujahua

Tlalpujahua: Widespread, Vetas
Las Coronas and La Borda.
Most noted from the La Borda
and Los Zapateros mines. Associated with polybasite, pyrargyrite, native silver, and
quartz.

MORELOS

Municipio de Tlaquiltenango

Huautla: Moderately distributed.
Noted from the Santa Ana and
Peregrina mines. Associated with
calcite.

NAYARIT

Municipio de Amatlan de
Canas

Barranca del Oro: Moderately
distributed.

Municipio de Compostela

Miravalles: Moderate distribution. Noted from the Miravalles and Guadalupe mines.

Municipio de la Yesca

La Yesca: Moderately distributed. Noted from the Los Tajos, La Morada, and La Cabrera mines.

Municipio de Santa Maria del Oro

Santa Maria del Oro: Moderately distributed.

NUEVO LEON

Municipio de Mina

Villa de Mina: Moderately distributed. Cerro de Enmedio.

OAXACA

Municipio de Ocotlán de Morelos

Ocotlán de Morelos: Widespread. Noted from the El Consuelo Mine. Associated with coarse wires of native silver.

Municipio de San Jerónimo Silacalloapilla

San Jerónimo Silacalloapilla: Moderate distribution. Noted from the Guadalupe, Providencia, and San Rafael mines.

Municipio de San Jerónimo Taviche

San Jerónimo Taviche: Moderate distribution. Benjamín Mine.

Municipio de San Miguel Peras

Canadá de Peras: Moderately distributed. Rebeca Mine.

QUERETARO

Municipio de Cadereyta

Cadereyta de Monte: Moderate distribution. Noted from the Las Azulitas and Santa Ines mines.

El Doctor: Widespread. Noted from the San Juan Nepomuceno and Trinidad mines.

Maconi: Moderately distributed in the Cerro de San Nicolás. Noted from the San Nicolás and La Dificultad mines.

Municipio de Amoles

Pinal de Amoles: Widespread.

Municipio de Toliman

Toliman: Moderate distribution. Noted from the Animas and Providencia mines.

SAN LUIS POTOSI

Municipio de Catorce

Catorce: Widespread. Concepción Mine.

Municipio de Cerro de San Pedro

Cerro de San Pedro: Widespread. Crystals to 2 cm. Associated with native silver and quartz.

Municipio de Charcas

Charcas: Moderately distributed on Cerro de Santa Ines.

Municipio de Guadalcázar

Guadalcázar: Moderately distributed.

Municipio de Villa de Ramos

Ramos: Moderately distributed.

SINALOA

Municipio de Concordia

Pañuco: Widespread. Noted from the Cata Rica, El Porvenir, El Toro, Gran Captain, Restauracion, and San Antonio mines. Associated with native silver and calcite.

Municipio de Cosala

Cosala: Moderately distributed. Noted from the Nuestra Señora and La Candelaria mines.

Municipio de Rosario

El Rosario: Moderate distribution. El Tajo Mine. Associated with gold, quartz, and sphalerite.

SONORA

Municipio de Alamos

Alamos: Widespread. Etched crystals to 2 cm. Associated with native silver.

Municipio de Altar

Altar: Moderate distribution. Noted from the La Esteril and La Tigre mines.

Municipio de Arizpe

Arizpe: Moderate distribution. Las Chispas Mine. Brilliant crystals to 5 cm. (Some of the crystals from here appear to have undergone a slight melting after formation). Associated with polybasite, stephanite, pearceite, native silver, bromian chlorargyrite, pyrite, and quartz. [In the 1880s, the Santa María Mining Company began a small mining operation south of Arizpe, but because of poor management and the high grading by the miners of the rich gold/silver ore the operations closed. The mine's clerk, John Pedrazzini, paid the back taxes and took over the ownership in lieu of back pay. In 1907, he formed the Pedrazzini Gold and Silver Mining Company and started mining on the rich Veta Las Chispas. He named the mine the Las Chispas after the beauty and richness of the chispas or crystals of the silver minerals. The ore was found in stringers or lenses in a series of 14 parallel fissure veins in the local rhyolitic rocks. The ore was extremely rich; first class ore assayed better than 15.6 kilos (500 troy ounces) in silver and 125 grams (4 troy ounces) in gold per metric ton. In 1907, just over 73 kilos (160 pounds) of crystals of acanthite alone were mined and sent to the smelter. Stealing again became a major problem for the company and steps were taken to prevent it, but to no avail. Ore buyers would meet the miners in great secrecy to buy the high-grade ore.]

Sinquipe: Limited distribution. Graciosa Mine.

Municipio de Atil

Atil: Moderate distribution. Los Tajos Mine.

Municipio de Bacerac

Bacerac: Moderately distributed. Noted from the Dolores, La Vieja, and San Ignacio mines.

Municipio de Bavispe

Bavispe: Moderate distribution. San José Mine.

Municipio de Caborca

Heróica Caborca: Widespread. Noted from the Mina Grande and Oro Blanco mines.

Municipio de Hermosillo

Hermosillo: Moderately distributed.

Municipio de Huépac

Huépac: Limited distribution. La Lajuela Mine.

Municipio de Tubutama

Tubutama: Moderately distributed. Grand Republic Mine.

ZACATECAS

Municipio de Concepción del Oro

Aranzazú: Moderately distributed. Aranzazú Mine.

Mazapil: Widespread.

Municipio de General Francisco Murguía

Nieves: Moderate distribution. Santa Catarina Mine.

Municipio de Noria de Angeles

Noria de Angeles: Widespread. Noted from the Cata Rica, Cata Urista, San Franciso, and Cinco Estrellas mines. [This location is being developed into the world's largest open pit silver mine, the Real de Los Angeles Mine. The mine, when opened in the 1980s, is expected to produce 7 million troy ounces of silver per year. The low-grade ore reserve is calculated to contain 135.3 million troy ounces of silver!]

Municipio de Pinos

Pinos: Widely distributed. Most noted from the The María, Candelaria, Purísima, and San Ramón mines. Associated with wires and crystals of native silver and azurite.

Municipio de Sombrerete

Sombrerete: Moderately distributed. Noted from the Veta Negra, El Pabellón and Tocayos mines. Dull crystals to 2 cm. Associated with quartz, proustite, and pyrargyrite. See Introduction, Sombrerete.

Municipio de Vetagrande

Vetagrande: Widespread. Crystals to 3 cm. Associated with quartz, proustite, pyrargyrite, native silver, and azurite. See Introduction, Zacatecas.

Municipio de Zacatecas

Zacatecas: Widespread. Noted from the El Bote Mine. Crystals to 5 cm. Associated with wires of native silver, quartz, and calcite. See Introduction, Zacatecas.

ACTINOLITE

$Ca_2(Mg,Fe)_5Si_8O_{22}(OH)_2$. A member of the amphibole mineral group that forms in low-grade, regionally metamorphosed rocks, particularly limestones, gneisses, and schists.

AGUASCALIENTES

Municipio de Tepezalá

Tepezalá: Widespread. Santa Bárbara Mine. Dark green crystal sprays to 4 cm.

BAJA CALIFORNIA

Municipio de Ensenada

Rosa de Castilla: Widespread.

Municipio de Tecate

Tecate: Moderately distributed. Associated with quartz.

BAJA CALIFORNIA SUR

Municipio de Mulegé

Isla de Santa Margarita: Moderately distributed. Dark green crystals to 4 cm.

CHIHUAHUA

Municipio de Aquiles Serdán

Francisco Portillo: Moderately distributed.

San Antonio el Grande: Limited distribution. San Antonio Mine's skarn zone. Associated with epidote, hedenbergite, grossular, andradite, sphalerite, and galena.

Municipio de Saucillo

Naica: Moderately distributed. Maravilla Mine's skarn zone. Associated with dipside, calcite, grossular, and quartz.

DURANGO

Municipio de Mapimí

Mapimí: Moderately distributed. Ojuela Mine's skarn zones.

HIDALGO

Municipio de Zimapán

Zimapán: Moderate distribution.

SAN LUIS POTOSI

Municipio de Guadalcázar

Guadalcázar: Widespread.

VERACRUZ

Municipio de Las Minas

Las Minas: Widespread. Noted from the La Purísima, Ampliación, and San Valentín mines.

ZACATECAS

Municipio de Concepción del Oro

Concepción del Oro: Moderately distributed. Associated with quartz.

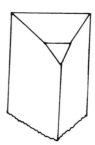

A typical 1 cm long adamite crystal from the Ojuela Mine, Mapimi, Drango. (Modified by Shawna Panczner after Sinkankas, 1964)

ADAMITE

$Zn_2(AsO_4)(OH)$. A rare secondary mineral found in the oxidized ore zones of arsenic rich base metal deposits.

DURANGO

Municipio de Mapimí

Mapimí: Moderate distribution. Ojuela Mine; 5th to the 15th level. Crystals to 12 cm. Associated with hemimorphite, calcite, smithsonite, goethite, and lotharmayerite. [Most noted areas of the Ojuela Mine for this mineral include: 5th level, Santo Domingo stope; 11th level of the Las Palomas ore body; 12th to 14th levels, San Judas stope, and; 14th level, San Juan Poiente stope. The color varies from colorless to white, through shades of yellow, green to blue, violet, and red. Occasionally it is strongly fluorescent because of trace amounts of uranium. Mayers and Wise wrote of their discovery that it was then the largest pocket of adamite known: "In June 1946, the writers had the good fortune to discover a remarkable pocket of adamite in a small manway of the Las Polomas ore body just above the 11th level. Enroute to a stope containing fine specimens of wulfenite and green mimetite, our lamps fell upon a pocket in the limestone. We saw a miniature

grotto, some four feet in diameter and as many deep, its interior carved with smoothly undulating waves of sparkling yellow crystals; it was a glorious sight, as though we were gazing upon a mineral specimen of unimaginable splendor."]
See Introduction, Mapimí.

HIDALGO

Municipio de Zimapán

Zimapán: Limited distribution. Miguel Hidalgo Mine. Pale yellow to yellow-green crystals to 1 cm.

Aguilarite from Guanajuato, Guanajuato, in the Pinch Collection, Pinch Mineralogical Museum, Rochester, New York.

AFWILLITE

$Ca_3Si_2O_4(OH)_6$. An uncommon mineral found in metatmorphic rocks and forms from the alteration of spurrite.

MICHOACAN

Municipio de Zitácuaro

Susupuato: Moderately distributed within the Cerro Mazahua skarn. Associated with spurrite.

AGUILARITE

Ag_4SeS. An extremely rare mineral found in very few base metal deposits.

GUANAJUATO

Municipio de Guanajuato

Guanajuato: Limited distribution. San Carlos Mine. Crystals to 3 cm. Associated with calcite, quartz, proustite, pearceite, and chalcopyrite. [This is the type location for this mineral. Named after P. Aguilar, the San Carlos Mine superintendent.]

La Luz: Limited distribution. Santa Rita (Bolanitos) Mine. Crystals to 1 cm.

Nayal: Limited distribution. Nino Perdido Mine.

Rayas: Limited distribution. Flores de Maria Mine.

GUERRERO

Municipio de Taxco

Taxco de Alarcón: Limited distribution. Chontalpan Mine. Associated with calcite, chalcopyrite, and sphalerite.

AIKINITE

$PbCuBiS_3$. A rare mineral found in some copper/lead deposits.

GUERRERO

Municipio de Apaxtla de Castrejon

Undisclosed location: Limited distribution.

AKAGANEITE

FeO(OH,Cl). A rare mineral found
in the oxidized zone of a few
metal deposits.

SONORA

Municipio de Trincheras

La Mur: Limited distribution. Las
Animas Mine. Associated with
argenosiderite, beudantite, and
olivenite. [This location has
been referred to as Benjamin
Hill, which is located south and
east of the mine.]

ALABANDITE

MnS. An uncommon primary mineral
found in vein deposits of
manganese and epithermal vein
deposits.

DURANGO

Municipio de Topia

Topia: Limited distribution.
Santa Cruz Mine.

HIDALGO

Municipio de Pachuca

Pachuca de Soto: Limited distribution. El Rasario Mine.

MICHOACAN

Municipio de Angangueo

Mineral de Angangueo: Moderately
distributed. Santa Bárbara Mine.

OAXACA

Municipio de San Miguel
Quezaltepec

San Miguel Quezaltepec: Moderate
distribution.

PUEBLA

Municipio de Tepeyahualco

Cerro Tlachiaque: Widespread near
Alchichica. Preciosa Sangue de
Cristo Mine.

Sierra de Tepeyahualco: Limited
distribution. La Prieta Mine.

Municipio unknown

La Ilucha: Moderately distributed.

ALAMOSITE

$PbSiO_3$. A rare secondary lead
mineral found in a few lead
deposits.

HIDALGO

Municipio de Zimapán

Zimapán: Limited distribution.
San Pascual Mine's "Salon Grande
de los Carbonatos" stope.
Associated with cerussite,
anglesite, and wulfenite.

SONORA

Municipio de Alamos

Alamos: Limited distribution.
Crystals to 2 mm. Associated with
wulfenite and leadhillite. [This
is the type location for which
this mineral was named.]

ALBITE

$NaAlSi_3O_8$. Occurs in igneous
rocks such as granite, diorite,
pegmatites and granular limestones, and marbles.

BAJA CALIFORNIA

Municipio de Ensenada

Ojos Negros: Widespread. La Verde
Mine. Crystals to 4 cm.

Municipio de Tecate

Tecate: Widespread, 40 km east. Associated with orthoclase, topaz, mica, and quartz.

OAXACA

Municipio de San Francisco Telixtlahuaca

San Franciasco Telixtlahuaca: Widespread. Santa Ana Mine. Associated with biotite, quartz, beryl, and orthoclase.

ALLANITE

$(Ce,Ca,Y)_2(Al,Fe)_3(SiO_4)_3(OH)$. A member of the epidote group and found occurring as an accessory mineral in granites, pegmatites, schists, and as a detrital mineral.

OAXACA

Municipio de San Francisco Telixtlahuaca

San Francisco Telixtlahuaca: Moderately distributed. Noted from the el Muerte, Huitzoita and Santa Ana Mines. Associated with zircon, epidote, orothoclase, beryl, biotite, and muscovite.

ALLOPHANE

$Al_2Si_5O \cdot nH_2O$. A common constituent of clay.

BAJA CALIFORNIA SUR

Municipio de Mulegé

Santa Rosalía: Widespread. Massive.

QUERETARO

Municipio de Tolimán

Soyatal: Widespread.

ALMANDINE

$Fe_3Al_2(SiO_4)_3$. A member of the garnet group found in regional metamorphosed argillaceous sediments, contact metamorphic environments, and occasionally in igneous environments.

BAJA CALIFORNIA

Municipio de Ensenada

Trinidad: Widespread. Crystals to 1 cm.

Municipio de Mexicali

Mexicali: Widespread, south of Mexicali. Crystals to 2 cm.

Santa Rosa: Moderately distributed. Crystals to 1 cm.

Santo Tomás: Moderatelty distributed. Crystals to 2 cm.

BAJA CALIFORNIA SUR

Municipio de Comondú

Loretto: Moderately distributed. Crystals to 1 cm.

CHIHUAHUA

Municipio de Chihuahua

Rancho Corralitos: Moderately distributed. Crystals to 1 cm.

OAXACA

Municipio de San Sebastián Abasolo

San Sebastián Abasolo: Limited distribution, Cresta de la Campana.

ALPHA-LEAD DIOXIDE (UNNAMED)

αPbO_2. A rare secondary mineral found in oxidized lead deposits. It is a paramorph of plattnerite.

DURANGO

Municipio de Mapimí

Mapimí: Limited distribution. Ojuela Mine. Deep orange known tabular crystals to 1 mm. Associated with plattnerite, rosasite, hemimorphite, and goethite.

ALTAITE

PbTe. A rare mineral found in a lead/tellurium deposits.

MEXICO

Municipio de Tlatlaya

Tlatlaya: Limited distribution. La Fama Mine.

ALUMINITE

$Al_2(SO_4)(OH)_4 \cdot 7H_2O$. Formed from the alteration of marcasite or pyrite on aluminum silicates.

MEXICO

Municipio de Temascalcingo

Solís: Moderately distributed.

ALUNITE

$KAl_3(SO_4)_2(OH)_6$. A common mineral found in the wall rocks of sulfide ore bodies, which formed from hydrothermal activities.

HIDALGO

Municipio de Tula de Allende

El Salto: Moderately distributed.

MEXICO

Municipio de Temascalcingo

Solis: Moderately distributed.

SONORA

Municipio de Naco

La Morita: Limited distribution.

ALUNOGEN

$Al_2(SO_4)_3 \cdot 17H_2O$. A common secondary mineral that forms from the decomposition of pyrite or in fumerolic environments.

CHIHUAHUA

Municipio de Aquiles Serdán

Francisco Portillo: Moderately distributed.

San Antonio el Grande: Moderately distributed. San Antonio Mine.

HIDALGO

Municipio de Tecozautla

Tecozautla: Modertely distributed.

ZACATECAS

Municipio de Fresnillo

Fresnillo de González Echeverría: Moderately distributed in the Cerro de Proano.

AMALGAM

Ag,Hg. A rare mineral found in base metal deposits.

CHIHUAHUA

Municipio unknown

Location Unknown.

JALISCO

Municipio de Chiquistlán

Chiquistlán: Limited distribution. San Antonio Mine. Massive.

MICHOACAN

Municipio de Coalcomán de
Matamoros

Coalcoman de Matamoros: Limited
distribution, Sierra de la
Aguililla. Massive.

AMESITE

$Mg_2Al(SiAl)O_5(OH)_4$. A member of
the kaolinite-serpentine group.
Forms as a product of hydrothermal alteration of ferromagnesian
silicates found in metamorphic
rocks.

HIDALGO

Municipio de Pachuca

Pachuca de Soto: Moderately
distributed.

AMPHIBOLE

A group name, not a specific
mineral.

ANALCIME

$NaAlSi_2O_6 \cdot H_2O$. A common mineral
formed in mafic and intermediate
igneous rocks and by the results
of hydrothermal conditions.

BAJA CALIFORNIA

Municipio de Tijuana

Rancho Ballistero: Moderately
distributed, 25 km south of
Tijuana, west of Rosarito.
Associated with stilbite.

HIDALGO

Municipio de Zimapán

Zimapán: Limited distribution.
Lomo del Toro.

JALISCO

Municipio de Tonalá

San Gaspar: Moderately distributed.

MORELOS

Municipio de Ayala

Xalostoc: Moderate distribution.
San Feliepe Mine.

ANATASE

TiO_2. An uncommon secondary
mineral found in veins and
cavities in schists and gneisses
forming from hydrothermal
solutions leaching the titanium
from local country rocks.

SONORA

Municipio de Alamos

Alamos: Limited distribution.
Associated with brookite crystals.

ANDALUSITE

Al_2SiO_5. An uncommon mineral
formed in regionally metamorphosed rocks especially slates,
schists, and gneisses; rarely in
granites.

BAJA CALIFORNIA

Municipio de Ensenada

Rancho Santa Clara: Moderately
distributed, 45 km east of
Ensenada.

MEXICO

Municipio de Ixtapan del Oro

Ixtapan del Oro: Widespread.

SONORA

Municipio de Guaymas

San Marcial: Moderately distributed, arroyo Escondido in the Valle de San Marcial. Crystals.

ANDESINE. See **PLAGIOCLASE**

ANDRADITE

$Ca_3Fe_2(SiO_4)_3$. A common member of the garnet group. Found in contact metamorphic rock and skarn deposits.

BAJA CALIFORNIA

Municipio de Encenada

Alamos: Widespread. Crystals to 1 cm.

CHIAPAS

Municipio de Pichucalco

Pichucalco: Widespread. Santa Fe Mine. Crystals to 3 cm. Associated with wollastonite and calcite.

CHIHUAHUA

Municipio de Aquiles Serdán

San Antonio el Grande: Moderate distribution. San Antonio Mine. Crystals to 2 cm. Associated with grossular, hedenbergite, galena, and sphalerite.

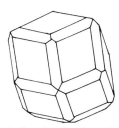

A 4 cm crystal of andradite from Trapichilles, Colima. (Modified by Shawna Panczner after Sinkankas, 1964)

Municipio de Julimes

Lázaro Cárdenas: Limited distribution. Ojos Espanoles Mine. Brilliant black crystals to 5 cm. Associated with quartz.

Municipio de Saucillo

Naica: Widespread. Olive green crystals to 4 cm. Associated quartz, calcite, chlorite, barite, fluorite, selenite, and anhydrite.

COAHUILA

Municipio de Monclova

Monclova: Moderate distribution. El Cinabrio Mine.

COLIMA

Municipio de Colima

Trapichillos: Moderately distributed, 10 km east. Bright, olive green crystals to 20 cm.

GUERRERO

Municipio de Zumpango del Río

Mescala: Moderate distribution.

HIDALGO

Municipio de Pachuca

Cerro de San Cristóbal: Limited distribution.

Municipio de Zimapán

Zimapán: Moderate distribution. Crystals. Associated with diopside, epidote, sphene, and axinite.

MICHOACAN

Municipio de Zitácuaro

Susupuato: Moderately distributed, Cerro Mazahua. Associated with diopside, merwinite, calcite and diopside.

MEXICO

Municipio de Sultepec

Sultepec de Pedro Ascencio de Alquisiras: Moderate distribution. El Porvenir Mine.

MORELOS

Municipio de Ayala

Ayala: Moderate distribution. San Luis Mine.

Jalostoc: Moderate distribution. San Felipe Mine. Crystals. [This location was spelled Xalostoc in old literature.]

NUEVO LEON

Municipio de Lampazo de Naranjo

Sierra del Carrizal: Widespread.

SAN LUIS POTOSI

Municipio de Guadalacázar

Guadalacázar: Moderately distributed. Associated with fluorite and quartz.

SONORA

Municipio de Alamos

Tepueste Ranch: Moderately distributed, 16 km northwest. Dark green crystals to 5 cm. Associated with calcite and malachite. [This area is often mislabled as Chihuahua.]

VERACRUZ

Municipio de Professor Rafael Rameriz

Piedras Parada: Moderately distributed. Greenish brown to bright green crystals to 3 cm. Associated with calcite, quartz, grossular, and schorlomite. [The crystals from here have been found both as singles and in large sheets of crystals up to 20 cm long. The andradite has been also referred to as demantoid garnets. The name of this municipio was recently changed from Las Vigas.]

ZACATECAS

Municipio de Concepción del Oro

Concepción del Oro: Widespread. Crystals to 2 cm.

ANGLESITE

$PbSO_4$. A common mineral in oxidized lead deposits.

BAJA CALIFORNIA SUR

Municipio de Mulegé

Santa Rosalía: Limited distribution. Noted from the Amelia and Curuglu mines. Crystals to 6 cm. Associated with boleite, gypsum selenite, quartz, and atacamite.

CHIHUAHUA

Municipio de Ahumada

Los Lamentos: Limited distribution. Ahumada/Erupción Mine. Crystals to 3 cm. Associated with goethite, sulfur, galena, and gypsum.

Municipio de Aquiles Serdán

Franscisco Portillo: Limited distribution. Buena Tierra Mine. Associated with galena, and pyrite.

San Antonio el Grande: Limited distribution. San Antonio Mine's oxide zone. Associated with cerussite, plumbojarosite, galena, and pyrite.

Municipio de Camargo

Sierra de Encinillas: Moderately distributed.

Municipio de Cusihuiriachic

Cusihuiriáchic: Moderately distributed. Princesa Mine. Associated with calcite.

Municipio de Manuel Benavides

Manuel Benavides: Limited distribution. Los Dos Marias Mine. Associated with minium.

Municipio de Santa Bárbara

Santa Bárbara: Moderately distributed. Santa Bárbara Mine.

Municipio de Saucillo

Naica: Moderately distributed. Associated with galena, chal-copyrite and calcite.

COAHUILA

Municipio de Múzquiz

Melchor Múzquiz: Widespread. Noted from the Cedral, Rosario, and San Francisco mines. [This location is often referred to simply as Múzquiz.]

DURANGO

Municipio de Mapimí

Mapimí: Moderate distribution. Ojuela Mine. Associated with goethite and galena.

Municipio del Oro

Santa Maria del Oro: Moderately distributed. Santa Ana Mine.

Municipio de San Pedro del Gallo

Penoles: Moderately distributed.

GUERRERO

Municipio del Zumpango del Rio

Xochipila: Moderately distrib-uted.

HIDALGO

Municipio de Pachuca

Pachuca de Soto: Widespread.

Municipio de Zimapan

Zimapan: Moderately distributed. Noted from the Lomo de Toro, Preciosa Sangre, San Pascual, and the La Cruz mines. Associated with galena, plumbojarosite, and gypsum.

MEXICO

Municipio de Zacualpan

Zacualpan: Moderate distribution. Noted from the Santa Ines and Guadalupe mines.

NUEVO LEON

Municipio de Monterrey

Monterrey: Moderately distrib-uted. Colorado Mine. Associated with galena.

SAN LUIS POTOSI

Municipio de Cerro San Pedro

Cerro San Pedro: Moderately distributed. Juarz Mine.

Municipio de Guadalcazar

Guadalcazar: Moderately distrib-uted.

Municipio de Villa de la Paz

La Paz: Moderate distribution. Noted from the Providencia and Plomosa mines.

SONORA

Municipio de Aconchi

San Felipe Canyon: Limited distribution. Los Lavaderos Mine. Cream crystals to 3 cm. Associ-ated with goethite, pyromorphite, and manganese oxide. [A large vug lined with crystals was mined out in the mid 1970s, which produced many fine specimens.]

Municipio de Arizpe

Arizpe: Moderately distributed.

Municipio de Cucurpe

Sierra Prieto: Moderate distribution. San Fancisco Mine. Massive. Associated with galena.

Municipio de Onavas

Onavas: Moderate distribution. San Javier Mine.

ZACATECAS

Municipio de Mazapil

Mazapil: Moderately distributed. Noted from the Cajon, Santa Ines, San Angel, La Vibora, and La Bohemia mines.

ANHYDRITE

$CaSO_4$. Forms in sedimentary environments and as a gangue mineral in sulfide deposits.

BAJA CALIFORNIA SUR

Municipio de Mulegé

Santa Rosalía: Moderately distributed.

CHIHUAHUA

Municipio de Aquiles Serdán

Francisco Portillo: Limited distribution. El Potosí Mine's 10th level. Associated with rhodochrosite, arsenopyrite, and pyrrhotite.

Municipio de Saucillo

Naica: Widespread. Light blue crystals to 15 cm. Associated with galena, chalcopyrite, quartz, calcite, andradite and gypsum. [In 1981, a large water-course was discovered on the Naica fault that contained crystals up to 15 cm with mostly flat terminations. A few double terminated crystals were found inside water-clear gypsum (selenite) crystals.]

GUERRERO

Municipio de Huitzuco

Huitzuco de los Figueroa: Moderately distributed. La Cruz Mine. Associated with gypsum, sulfur, and livingstonite.

ANKERITE

$Ca(Fe,Mg,Mn)(CO_3)_2$. A member of the dolomite group and found in metamorphosed environments and in metallic veins.

CHIHUAHUA

Municipio de Aquiles Serdan

Francisco Portillo: Moderately distributed. Buena Tierra Mine. Crystals to 3 cm. Associated with marcasite, quartz, and sphalerite.

DURANGO

Municipio de Mapimí

Mapimí: Limited distribution. Ojuela Mine. Associated with goethite.

GUANAJUATO

Municipio de Guanajuato

Guanajuato: Moderately distributed. San Juan Rayas Mine. Crystals.

ZACATECAS

Municipio de Concepción del Oro

Concepción del Oro: Moderately distributed. Associated with calcite and pyrite.

ANNABERGITE

$Ni_3(AsO_4)_2 \cdot 8H_2O$. A rare mineral of secondary origin found in nickel/arsenic deposits.

SINALOA

Municipio de Badiraguato

Alisito: Limited distribution. Gloria Mine. Associated with gersdorffite.

ANORTHITE

$CaAl_2Si_2O_8$. An uncommon mineral occurring in basic igneous rocks, granular limestones of contact deposits, and volcanic ejected rocks.

CHIHUAHUA

Municipio de Nuevo Casas Grandes

Undisclosed location.

ANTHOPHYLLITE

$(MgFe)_7Si_8O_{22}(OH)_2$. A member of the amphibole group, found in schists, gneisses, and contact metamorphic rocks.

HIDALGO

Municipio de Pachuca

Pachuca de Soto: Moderately distributed.

ANTIGORITE

$(Mg,Fe)_3Si_2O_5(OH)_4$. A member of the kaolinite-serpentine mineral group forming in metamorphic environments.

HIDALGO

Municipio de Pachuca

Pachuca de Soto: Moderately distributed.

ANTIMONY

Sb. A rare mineral element. Occurs in metal-bearing veins with silver and arsenic ores.

CHIHUAHUA

Municipio de Uruáchic

Arechuybo: Limited distribution. Massive. Associated with stibiconite, valentinite, kermesite, and cervantite.

GUANAJUATO

Municipio de Guanajuato

Guanajuato: Limited distribution.

SONORA

Municipio de Tepache

Nuevo Tepache: Limited distribution.

ANTLERITE

$Cu_3(SO_4)(OH)_4$. A rare secondary mineral which forms in the oxidized zones of copper deposits.

COAHUILA

Municipio de Sierra Mojada

Sierra Mojada: Limited distribution.

APATITE. See **FLUORAPATITE**

APHTHITALITE

$(K,Na)_3Na(SO_4)_2$. An uncommon
mineral formed by the action
of sulfuric acid vapors on
volcanic rocks and in saline
deposits were it crystallized
from solution.

MICHOACAN

Municipio de Uruapán

Volcan Paricutín: Limited
distribution.

APJOHNITE

$MnAl_2(SO_4)_4 \cdot 22H_2O$. An uncommon
member of the halotrichite
group. Occurs usually as an
efflorescence in protected
places.

MEXICO

Undisclosed location

APOPHYLLITE. See **HYDROXAPO-
PHYLLITE**

ARAGONITE

$CaCO_3$. An uncommon mineral
formed in spring deposits,
sulfate-bearing saline solutions,
and in limestone caves at
temperatures above 27oC.

BAJA CALIFORNIA SUR

Municipio de Mulegé

Santa Rosalía: Widespread.

CHIHUAHUA

Municipio de Aquiles serdán

Francisco Portillo: Moderately
distributed. White to clear
needles to 3 cm. Associated with
calcite. [In the mid 1980s, over
6,000 kilos (6 metric tons) were
removed from a cave in the Santa
Rita Mine. Crystals were up to
16 mm long.]

San Antonio el Grande: Moderately
distributed. San Antonio Mine.
White to clear needles to 5 cm.
Associated with calcite.

DURANGO

Municipio de Mapimí

Buruguilla: Moderately distributed. White to clear crystals to
4 cm. Associated with calcite.

Mapimí: Moderate distribution.
Ojuela Mine. Crystal needles to 3
cm. Associated with calcite.

GUANAJUATO

Municipio de Guanajuato

Guanajuato: Widespread. Associated with calcite and quartz
(amethyst).

GUERRERO

Municipio de Huitzuco

Huitzuco de los Figueroa: Limited
distribution. La Cruz Mine.

Municipio de Pilcaya

Grutas de Cacahuamilpa: Widespread. Associated with calcite.

HIDALGO

Municipio de Pachuca

Pachuca de Soto: Moderately
distributed. El Rosario Mine.

PUEBLA

Municipio de Zapotitlán

San Antonio Tezcala: Moderately distributed. Snow white botroyadal kidney shaped masses to 30 cm.

TLAXCALA

Municipio de Calpulalpan

Nanacamilpa: Moderately distributed.

ZACATECAS

Municipio de Fresnillo

Fresnillo de Gonzalez Echeverría: Moderately distributed. Jesús María Mine.

ARGENTOJAROSITE

$AgFe_3(SO_4)_2(OH)_6$. A rare secondary silver mineral found in the oxidized zone of silver deposits.

CHIHUAHUA

Municipio de Aquiles Serdán

Francisco Portillo: Limited distribution.

San Antonio el Grande: Limited distribution. San Antonio Mine.

COAHUILA

Municipio de Melchor Múzquiz

La Encantada: Moderate distribution. La Encantada Mine.

HIDALGO

Municipio de Zimapán

Zimapán: Moderate distribution.

ARGYRODITE

Ag_8GeS_6. A very rare silver mineral found in very few base metal deposits.

GUANAJUATO

Municipio de Guanajuato

Guanajuato: Limited distribution. Associated with canfieldite.

ARSENIC

As. A rare mineral (element) found in hydrothermal veins associated with cobalt, nickel, and silver ores.

CHIHUAHUA

Municipio de Batopilas

Batopilas: Limited distribution. Massive. Associated with native silver, and calcite.

GUANAJUATO

Municipio de Guanajuato

Guanajuato: Limited distribution. Massive.

ARSENIOSIDERITE

$Ca_3Fe_4(AsO_4)_4(OH)_6 \cdot 3H_2O$. A rare mineral found in base metal deposits.

DURANGO

Municipio de Mapimí

Mapimi: Limited distribution. Ojuela Mine. Associated with scorodite and carminite.

SONORA

Municipio de Trincheras

La Mur: Limited distribution. Las
Animas Mine. Crystals to 2 mm.
Associated with carminite,
beudantite, and akaganenite.
[This location is often referred
to as Benjamín Hill, which is
south and east of the mine.]

ZACATECAS

Municipio de Mazapil

Mazapil: Limited distribution.
Jesús María Mine. Associated with
calcite and quartz. [When this
mineral was first discovered here
it was named mazapilite, but
later was found to be arseniosiderite.]

ARSENOBISMITE

$Bi_2(AsO_4)(OH)_3$. A rare mineral
found in bismuth/arsenic deposits.

DURANGO

Municipio de Mapimí

Mapimí: Limited distribution.
Ojuela Mine. Associated with
carminite.

ARSENOPYRITE

FeAsS. A common mineral formed
under a variety of conditions:
high-temperature gold quartz
veins, contact metamorphic
sulphide deposits, pegmatites,
and low temperature ore veins.

BAJA CALIFORNIA SUR

Municipio de San Antonio

El Triunfo: Widespread.

San Antonio: Widespread.

CHIHUAHUA

Municipio de Aquiles Serdán

Francisco Portillo: Moderate
distribution. El Potosí Mine's
silicate ore body. Crystals to 4
cm. Associated with rhodochrosite.

San Antonio el Grande: Moderate
distribution. San Antonio Mine.
Brillant crystals to 6 cm often
with epitaxial growth. Associated
with galena, quartz, calcite,
pyrite, and siderite.

Municipio de Hidalgo del Parral

Hidalgo del Parral: Widespread.
El Tajo Mine. Brilliant crystals
to 6 cm. Associated with sphalerite, quartz, chalcopyrite,
pyrite, and jamesonite. [In the
early 1970s several pockets of
fine crystallized specimens were
recovered.]

Municipio de Nuevo Casas Grandes

San Pedro Corralitos: Moderate
distribution. San Pedro Mine.

Municipio de Santa Bárbara

Santa Bárbara: Widespread.

Municipio de Saucillo

Naica: Moderate distribution.
Naica Mine. Crystals to 2 cm.
Associated with quartz, calcite,
pyrite, and sphalerite.

DURANGO

Municipio de Canelas

Canelas: Moderately distributed.
Noted from the Trinidad, La
Calerad, and Dulces Nombres
mines.

Municipio de Mapimi

Mapimi: Limited distribution.
Ojuela Mine. Associated with
pyrite, quartz, and chalcopyrite.

Municipio de Poanas

Cerro de Sacifricio: Moderate distribution. Noted from the San Carlos, San Cristóbel, and El Carmen mines.

GUANAJUATO

Municipio de Guanajuato

Guanajuato: Moderately distributed. Crystals to 1 cm. Associated with acanthite, quartz, and calcite.

GUERRERO

Municipio de Arcelia

Campo Morado: Limited distribution. El Naranjo and Reforma mines.

Municipio de Taxco

Taxco de Alarcón: Moderately distributed. Crystals to 1 cm.

HIDALGO

Municipio de Zacualtipán

Zacualtipán: Limited distribution. Chacuaco Mine.

Municipio de Zimapán

Zimapán: Widespread. Noted from the El Carmen and Lomo del Toro San Damian Stope mines. Needle-like crystals to 1 cm. Associated with galena, and sphalerite.

MEXICO

Municipio de Temascaltepec

Culcapán: Moderate distribution.

Municipio de Tlatlaya

Rancho Los Ocotes: Moderate distribution. Noted from the La Esperanza, La Fama, La Pinta, Bella Mañana, and San Pedro mines.

Municipio de Zacualpan

Zacualpan: Moderate distribution. Guadalupe and San Juan Bautista mines.

MICHOACAN

Municipio de Angangueo

Mineral de Angangueo: Moderately distributed.

Municipio de Cuidad Hidalgo

Chapatuato: Limited distribution. El Grande Mine.

Municipio de La Huacana

Inguaran: Widely distributed.

MORELOS

Municipio de Ayala

Jalostoc: Moderately distributed on Cerro de Galvín.

NAYARIT

Municipio de Compostela

Compostela: Limited distribution. El Cajón Mine.

PUEBLA

Municipio de Libres

Temextla: Moderately distributed.

QUERETARO

Municipio de Cadereyta

Congregación de San Juan Tetla: Limited distribution. La Dificultad Mine.

El Doctor: Limited distribution. Las Aguas Mine.

SAN LUIS POTOSI

Municipio de Catorce

Catorce: Moderately distributed. San Pedro Mine.

Municipio de Guadalcázar

Guadalcázar: Limited distribution. La Trinidad Mine.

SINALOA

Municipio de Cosalá

Cosalá: Moderately distributed. El Refugio Mine.

SONORA

Municipio de Onavas

La Barranca: Moderate distribution. Belém Mine.

Municipio de Rayón

Rayón: Limited distribution. El Socorro Mine.

Municipio de San Javier

San Javier: Moderately distributed. Las Animas Mine.

ZACATECAS

Municipio de Concepción del Oro

Bonanza: Moderately distributed. Bonanza Mine. Crystals to 2 cm. Associated with bournonite, pyrite, and jamesonite.

Concepción del Oro: Moderate distribution. El Cobre Mine. Crystals to 2 cm. Associated with pyrite.

Municipio de Mazapil

Noche Buena: Noche Buena Mine. Crystals to 2 cm. Assoicated with pyrite, galena, and bournonite.

Municipio de Noria de Angeles

Noria de Angeles: Limited distribution. Cubierta Mine.

Municipio de Sombrerete

Sombrerete: Moderately distributed. Crystals to 1 cm.

Municipio de Zacatecas

Zacatecas: Moderately distributed. Associated with pyrite, quartz, and sphalerite.

ASBESTOS. See **SERPENTINE**

ASBOLAN

$(Co,Ni)_{1-y}(MnO_2)_{2-x}(OH)_x(OH)_{2-2y+x}nH_2O$. A rare mineral found in the oxidized ore zone of cobalt/nickel deposits.

PUEBLA

Municipio de Santiago Miahuatlán

Santa Ana: Limited distribution.

QUERETARO

Municipio de San Juan del Río

Ahuacatlán: Limited distribution. Mina de Grande de Escalena.

SONORA

Municipio de Hermosillo

El Zubiate: Limited distribution. Guadalupana Mine.

ATACAMITE

$Cu_2Cl(OH)_3$. A rare secondary mineral formed from the oxidation of other secondary copper minerals under arid and saline conditions.

BAJA CALIFORNIA SUR

Municipio de Mulegé

Mulegé: Limited distribution. Associated with chrysocolla.

Santa Rosalía: Moderately distributed. Noted from the El Toro, Amelia and Curuglú mines. Crystal to 1 cm. Associated with sphaeorocobaltite, gypsum, boleite and anglesite. [This area is the noted Boleo Copper District.]

TAMAULIPAS

Municipio de Villagran

San José: Moderate distribution. Noted from the Mina Verde, Santo Domingo, San Carlos, Imán, and La Priedra mines.

ZACATECAS

Municipio de Concepción del Oro

Aranzazú: Limited distribution. Aranzazú Mine. Crystals to 1 cm. Associated with malachite, azurite, and goethite. [Pseudomorphs after azurite to 5 cm have been found here.]

AUGITE

$(Ca,Na)(Mg,FeAl,Ti)(Si,Al)_2O_6$. A common rock forming mineral of the pyroxene group found in a wide variety of mafic igneous rocks, gabbro, diabase, and basalt.

DURANGO

Municipio de Durango

Victoria de Durango: Limited distribution. Cerro de Mercado Mine. [This location is better known as Durango.]

MICHOACAN

Municipio de Tlapujahua

Tlapujahua: Widespread. Associated with quartz.

OAXACA

Municipio de La Pe

La Panchita: Limited distribution, west of Ayoquiztco and north of La Panchita. La Panchita mine. Crystals to 5 cm. Associated with scapolite.

Municipio de San Martín Lachilá

San Martin Lachila: Moderately distributed.

AURICHALCITE

$(Zn,Cu)_5(CO_3)_2(OH)_6$. An uncommon secondary mineral formed in the oxidized zones of lead and copper deposits.

CHIHUAHUA

Municipio de Aquiles Serdán

Francisco Portillo: Limited distribution. El Potosí Mine. Associated with goethite.

San Antonio el Grande: Limited distribution. San Antonio Mine. Associated with goethite.

Municipio de Chihuahua

Labor de Terrazas: Moderately distributed. Associated with goethite. [This location is east of the main north-south highway about 17 km north of Chihuahua near the old ruins of the Hacienda de Torreón.]

Municipio de Janos

Rancho San Pedro: Moderate distribution. Noted from the El León and La Inmensidad mines.

Municipio de Saucillo

Naica: Moderately distributed. Associated with calcite and goethite.

DURANGO

Municipio de Mapimí

Mapimí: Moderate distribution. Ojuela Mine. Crystal tuffs to 3 cm. Associated with goethite, calcite, plattnerite, fluorite, hemimorphite, barite, and the newly discovered alpha lead oxide still unnamed. [Cavities have been found up to over 1 meter in diameter and 6-7 meters long lined with this mineral.]

HIDALGO

Municipio de Zimapán

Zimapán: Moderate distribution. Noted from the Lomo del Toro, Santa Luisa stope. Crystal tuffs and blades to 1 cm. Associated with malachite and goethite.

JALISCO

Municipio de Tapalpa

Sierra de Tapalpa: Moderately distrubuted.

NUEVO LEON

Municipio de Lampazos de Naranjo

Lampazos: Limited distribution. Flor de Peña Mine. Associated with goethite and calcite.

Municipio de Monterrey

Monterrey: Moderately distributed.

ZACATECAS

Municipio de Concepción del Oro

Bonanza: Moderately distributed.

AUSTINITE

$CaZn(AsO_4)(OH)$. A rare secondary mineral formed in the oxidzed zones of base metal deposits.

DURANGO

Municipio de Mapimí

Mapimí: Limited distribution. Ojuela Mine. Crystals to 5 mm. Associated with calcite and mimetite.

AUTUNITE

$Ca(UO_2)_2(PO_4)_2 \cdot 10-12H_2O$. An uncommon secondary uranium mineral formed from the alteration of other uranium minerals and in oxidized zones of hydrothermal deposits.

DURANGO

Municipio de Bermejillo

Sierra de Bermejillo: Moderately distributed.

SONORA

Municipio de Alamos

Alamos: Moderately distributed. Crystals to 5 mm. Associated with meta-autunite.

AXINITE. See **FERROAXINITE**

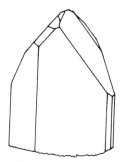

A typical 1 cm long azurite crystal from Aranzazu, Zacatecas. (Modified by Shawna Panczner from Anothony, Bideaux, and Williams, 1977)

AZURITE

$Cu_3(CO_3)_2(OH)_2$. A rather common secondary mineral found in the oxidized zones of copper deposits.

AGUASCALIENTES

Municipio de Asientos

Asientos: Moderately distributed. Noted at the Santa Francisca, Mercedes, El Cristo, Descubridora, and Cinco Señores mines. Crystals to 1 cm. Associated with malachite and goethite.

El Cobre: San Bartolo Mine.

Municipio de Tepezalá

Tepezalá: Moderately distributed. Don Federico Mine. Associated with malachite and goethite.

BAJA CALIFORNIA

Municipio de Ensenada

San Fernando: Limited distribution. San Fernando Mine. Associated with malachite.

Sauzalito: Limited distribution, 16 km northeast of El Aguajito. Associated with malachite and chrysocolla.

Municipio de Tecate

Rosa de Castilla: Limited distribution. El Fenomeno Mine. Associated with malachite.

BAJA CALIFORNIA SUR

Municipio de La Paz

Cacahilas: Limited distribution. Esperanza Mine. Associated with malachite.

Municipio de Mulegé

Santa Rosalía: Limited distribution. Santa Rita Mine. Crystals to 3 cm. Associated with malachite and goethite. [Most of the azurite was mined in the late 1890s and early 1900s.]

Municipio de San Antonio

El Triunfo: Moderately distributed. Associated with malachite.

San Antonio: Moderately distributed. Associated with malachite and goethite.

CHIAPAS

Municipio de Pichucalco

Pichucalco: Limited distribution. Santa Fe Mine. Associated with malachite and goethite.

CHIHUAHUA

Municipio de Aquiles Serdán

Francisco Portillo: Moderately distributed. Associated with malachite and goethite. [The Santa Eulalia mining district produced this mineral in its early years of operation.]

San Antonio el Grande: Limited distribution. San Antonio Mine. Associated with malachite and goethite.

Municipio de Chihuahua

Labor de Terrazas: Moderate distribution. Río Tinto Mine. Associated with malachite, copper, cuprite, and goethite.

Municipio de Cuidad Camargo

Sierra de las Encinillas: Limited distribution. Chanato Mine. Associated with malachite and goethite.

Municipio de Coyame

Cuchillo Pardo: Moderately distributed on Cerro de la Espumosa. Associated with malachite, cuprite, tenorite, and goethite.

Las Vigas: Moderately distributed on Sierra de la Boquilla. Associated with malachite and cuprite.

Municipio de Cusihuiriáchic

Cusihuiriáchic: Moderately distributed. Noted from the Princesa, San Miguel, San Bartolo, Grande and La Palma mines. Associated with crystals of acanthite.

Municipio de Hidalgo del Parral

Hidalgo del Parral: Moderate distribution on Cerro de la Cruz. El Tajo Mine. Associated with malachite and goethite.

Municipio de Manuel Benavides

Sierra Rica: Limited distribution. Crystals to 2 cm. Associated with quartz. [This location has been listed incorrectly as the San Carlos Mine. The San Carlos Mine has never reported azurite from its workings. This location is across the valley from San Carlos and is a series of small, shallow prospects.]

Municipio de Nuevo Casas Grandes

San Pedro Corralitos: Moderately distributed.

Municipio de Ojinaga

Magistral: Moderate distribution. Noted from the El Pelón, La Vieja, and Marcia del Velle mines. Associated with malachite.

Municipio de Saucillo

Naica: Moderate distribution. Crystals. Associated malachite and goethite.

Municipio de Urique

Urique: Moderate distribution. Noted from the San Antonio, Patrona, San Miguel, San Rafael, Santo Niño, Cuchila, Melchor Ocampo, and Santa Clara mines.

COAHUILA

Municipio de Candela

Pánuco: Moderate distribution. Escondida Mine. Associated with malachite and goethite. [This mine has also been listed as the Panuco Copper Mine.]

Municipio de Castaños

Castaños: Limited distribution. Mercedes Mine. Associated with malachite.

Municipio de Melchor Múzquiz

La Encantada: Limited distribution. La Encantada Mine.

Municipio de Sierra Mojada

Sierra Mojada: Moderate distribution. Noted from the Veta Rica and Dionea mines. Associated with goethite and malachite.

DURANGO

Municipio de El Oro

Santa Maria del Oro: Moderate distribution. Noted from the Santa Niño, La Cruz Negra, Reina Victoria, El Gran Suceso, Promontorio, and Celestina mines. Associated with malachite and goethite.

Municipio de Panuco de
Coronado

Panuco de Coronado: Moderate
distribution. Avino Mine.

Municipio de Guanaceví

Guanaceví: Moderate distribution.
Noted at the Rosario and Capuzal-
la mines.

Municipio de Mapimí

Buruguilla: Moderate distribution, 35 km northwest of Mapimi,
2 km west-northwest of Buruguilla. Descubridora Mine.
Associated with malachite and
goethite.

Mapimí: Moderate distribution.
Ojuela Mine. Crystals to 3 cm.
Associated with malachite and
goethite.

Municipio de San Dimas

Tayoltita: Moderately distributed. Associated with malachite
and goethite.

GUANAJUATO

Municipio de Guanajuato

Guanajuato: Limited distribution.
San Juan Rayas Mine. Associated
with malachite on acanthite.
[In the early years of mining
this mineral was reportedly
found, but not at the present
time.]

La Luz: Limited distribution.
Noted from the La Luz and San
Bernabé mines. Crystals to 1cm.
Associated with crystals of
pyrargyrite.

Municipio de San Luis
de la Paz

Mineral de Pozo: Moderate
distribution. Noted from the
Santa Brígida and Trinidad mines.
Associated with malachite and
goethite.

Municipio de Allende

Mineral de Puerto de Nieto:
Moderate distribution. La
Argentina Mine. Associated with
malachite and goethite.

GUERRERO

Municipio de Taxco

Taxco de Alcarón: Limited
distribution. Crystals to 1 cm.
Associated with acanthite
pyrargyrite, silver and polybasite. [This location is better
known as just Taxco.]

Municipio unknown

Las Fraguas: Las Minitas Mine.
Crystals.

Municipio de Zumpango del
Rio

Mezcala: Moderate distribution.
San Carlos Mine. Spherical
crystal masses to 8 cm, single
crystals to 1 cm. Associated with
malachite.

HIDALGO

Municipio de Mineral del
Monte

Mineral del Monte: Moderate
distribution.

Municipio de Pachuca

Pachuca de Soto: Moderate
distribution.

Municipio de Zimapán

Zimapán: Limited distribution.
Associated with malachite and
goethite.

JALISCO

Municipio de Bolaños

Bolaños: Moderate distribution.
Noted from the Santa Fe, Picacho,

Barranco, Descubridora, Iguana,
and Verde mines. Associated with
malachite and goethite.

Municipio de Chiquilistlán

Chiquilistlán: La Sobrina Mine.

MEXICO

Municipio de Tlatlaya

Rancho de los Ocotes: Limited
distribution. La Fama Mine.

MICHOACAN

Municipio de La Huacana

Inguaran: Widespread.

Oropeco: Moderate distribution.
Noted from the San Cristobal and
China mines. Associated with
malachite.

PUEBLA

Municipio de Tehuitzingo

Atopoltitlán: Limited distribution on Cerro del Guajolote.

SAN LUIS POTOSI

Municipio de Catorce

Catorce: Limited distribution.
Concepción Mine. Associated with
malachite and goethite.

Municipio de Cerro de San
Pedro

Cerro de San Perdro: Moderately
distributed. Noted from the
Victoria and Gogorran mines.

Municipio de Guadalcázar

Guadalcázar: Limited distribution.

Municipio de Villa de Ramos

Ramos: Moderate distribution.
Noted from the Cocinera and La
Purísima mines. Associations:

Cocinera Mine, malachite and
goethite; La Purísima Mine,
malachite, pyrargyrite, and
goethite.

SINALOA

Municipio de Badiraguato

Soyatita: Moderate distribution.
San Antonio Mine. Crystal druzes.
Associated with goethite and
malachite.

SONORA

Municipio de Alamos

Alamos: Limited distribution.
Druzy crystals. Associated with
malachite, smithsonite and
goethite.

San Antonio: Moderately distributed, 21 km east of Alamos. Noted
from the Malaquita, Aduanas, and
Chilla mines. Associated with
malachite and goethite.

Municipio de Arivechi

Arivechi: Moderate distribution.
Noted from the La Toribia and
Cobriza mines.

Municipio de Cananea

Cananea: Moderately distributed.
Crystals to 2 cm and drusy
masses. Associated with malachite
and goethite.

Municipio de Hermosillo

Hermosillo: Limited distribution.
Associated with malachite and
goethite.

Municipio de Imuris

Imuris: Moderate distribution.
Bonanza, Cerro Blanco and Santa
Apolonia mines. Associated with
malachite and goethite.

Municipio de Moctezuma

Moctezuma: Limited distribution.
Associated with malachite.

Municipio de Nacozari de Garcia

Nacozari de García: Moderate distribution. Pilares de Nacozari Mine. Crystals to 1 cm and drusy masses. Associated with malachite.

Municipio de Soyopa

San Antonio de la Huerta: Moderate distribution. Noted from the Santas Encinas, Cuchilla, and Ladrillera mines. Associated with malachite.

Municipio de Suaqui Grande

Suaqui Grande: Moderate distribution. Catalina, San Ignacio, La Blanca, Esperanza, and La Juana mines. Associated with malachite and goethite.

Municipio de Yécora

Yécora: Limited distribution. Crystal roses to 8 cm, single crystals to 3 cm. Associated with malachite.

TAMAULIPAS

Municipio de Villagran

San José: Limited distribution.

VERACRUZ

Municipio de Profesor Rafael Ramírez

Tenexpanoya: Limited distribution. Associated with malachite. [The name of this municipio was recently changed from Las Vigas.]

ZACATECAS

Municipio de Concepcion del Oro

Aranzazú: Limited distribution. Santa Elegio Mine. Brilliant crystalsd to 2 cm. Associated with malachite and goethite.

Bonanza: Limited distribution. Bonanza Mine. Crystals to 2 cm. Associated with malachite and goethite.

Concepción del Oro: Limited distribution. El Corbre Mine. Associated with malachite.

Municipio de Mazapil

Mazapil: Moderate distribution. Noted from the San Pedro, La Cruz, Pedernal, Animas, Cajón, Jesús María, San Carlos and Nieva mines. Crystals to 3 cm. Associated with malachite and goethite. [The San Carlos Mine in the late 1950s and early 1960s produced thousands of specimens, most of which were Zacatecas.]

Municipio de Noria de Angeles

Noria de Angeles: Limited distribution. Noted from the La Leona, Madrono, and Verde mines.

Municipio de Pinos

Pinos: Widespread. Crystals to 2 cm. Associated with wires and crystals of native silver. [This mineral was found in the earlier years of mining.]

Municipio de Zacatecas

Zacatecas: Moderate distribution. El Bote Mine. Associated with malachite and goethite.

B

BAKERITE

$Ca_4B_4(BO_4)(SiO_4)_3(OH)_3 \cdot H_2O$. A rare mineral found in very few deposits.

SAN LUIS POTOSI

Municipio de Charcas

Charcas: Moderate distribution. Noted from the La Bufa, San Bartolo and San Sebastián mines. Associated with calcite, stilbite, datolite, danburite, and quartz.

BAMBOLLAITE

$Cu(Te,Se)_2$. An extremely rare mineral found in copper/tellurium deposits.

SONORA

Municipio de Mactezuma

Moctezuma: Limited distribution. Noted at the La Bambolla and Moctezuma mines. Associated with selenium, klockmann, and chalcomenite. [The mineral was named after the mine where it was discovered, which is the type location for the mineral.]

BARITE

$BaSO_4$. A common mineral that forms in moderately low-temperature hydrothermal veins, in replacements, and in cavity fillings in sedimentary rocks by meteoric or hypogene solutions.

BAJA CALIFORNIA SUR

Municipio de Mulegé

Santa Rosalía: Widespread.

CHIAPAS

Municipio de Pichucalco: Pichucalco

Pichucalco: Limited distribution. Santa Fe Mine.

CHIHUAHUA

Municipio de Aquiles Serdán

San Antonio el Grande: Limited distribution. San Antonio Mine. Associated with fluorite, quartz and calcite. [This is a late vein deposition at this location.]

Municipio de Camargo

Sierra de Las Encinillas: Moderate distribution. Noted from the

Santa Rita and El Carmen mines.
Crystals to 5 cm.

Municipio de Coyame

Cuchillo Pardo: Widespread.

Municipio de Hidalgo del Parral

Hacienda de Valsequillo: Moderate
distribution.

Municipio de Meoqui

Sierra del Carrizo: La Plomosa.
Widespread.

Municipio de Nuevo Casas Grandes

San Padro Corralitos: Moderately
distributed. Crystals to 1 cm.

Municipio de Saucillo

Naica: Moderately distributed,
Naica fault. Noted from the Naica
Mine. White, greenish brown, and
blue crystals to 6 cm. Associated
with calcite, wollastonite,
anhydrite and andradite.

COAHUILA

Municipio de Sierra Mojada

Sierra de Mojada: Moderate
distribution. Veta Rica Mine.
Associated with native silver,
acanthite, proustite and erythrite.

COLIMA

Municipio de Minatitlán

Salto del Mamey: Moderately
distributed, 700 m northwest.

DURANGO

Municipio de Cuencamé

Cuencamé de Ceniceros: Moderate-
ly distributed, southwest.
[This location is better known as
Cuencamé.]

Municipio de Durango

Victoria de Durango: Moderate
distribution on Cerro de Mercado.
Cerro de Mercado Mine. [This
location is better known as
Durango.]

Municipio de Indé

Indé: Limited distribution. La
Sirena Mine.

Municipio de Mapimí

Mapimí: Widespread. Noted from
the Ojuela and El Burro mines.
White, golden brown, and blue
crystals to 5 cm. Associated with
mimetite, wulfenite, aurichalcite, goethite, calcite, plattnerite, murdockite, fluorite, and
hydrozincite.

GUERRERO

Municipio de Chilpancingo de Los Bravo

Cerro de San Vicente: Limited
distribution.

Municipio de Huitzuco

Huitzuco de las Figueroa:
Moderately distributed.

Municipio de Taxco

Taxco de Alarcón: Moderately
distributed. Blue to white
crystals to 6 cm. Associated with
quartz, pyrite, and marcasite.

HIDALGO

Municipio de Mineral del Monte

Mineral del Monte: Moderately
distributed. Dificultad Mine.

Municipio de Pachuca

Pachuca de Soto: Moderately
distributed. Barron Mine.

Municipio de Zimapán

Zimapán: Moderate distribution.
Pamplona Mine.

JALISCO

Municipio de Tapalpa

Realito: Moderately distributed.

MEXICO

Municipio de Zacualpan

Zacualpan: Limited distribution.
La Cuchara Mine.

MICHOACAN

Municipio de Angangueo

Minerod de Angangueo: Moderately
distributed. El Carmen Mine.
Clear, white and blue crystals to
5 cm. Associated with pyrite,
marcasite, sphalerite, quartz and
calcite.

NAYARIT

Municipio de Ruiz

Zopilote: Restauradora Mine.

NUEVO LEON

Municipio de Cerralvo

Cerralvo: Moderate distribution.
Refugio and Purísima mines.

PUEBLA

Municipio de Tehuacán

Tehuacán: Moderately distributed.

Municipio de Teziutlán

Teziutlán: Moderately distributed.

SAN LUIS POTOSI

Municipio de Cerro de San
Pedro

Cerro de San Pedro: Moderately
distributed.

Municipio de Charcas

Charcas: Widely distributed.
Colorless, white to pink crystals
to 2 cm. Associated with calcite,
pyrite, and sphalerite.

SINALOA

Municipio de Cosalá

Cosalá: Moderately distributed.

SONORA

Municipio de Alamos

Alamos: Moderately distributed.
Tesache Mine.

Municipio de Villa Pesquería

Villa Pesquería: Widespread. Tan
to white masses and crystals.
[This location is developing into
the largest barite mining area of
México.]

ZACATECAS

Municipio de Fresnillo

Fresnillo de González Echeverría:
Widespread on the Cerro Proano.

Municipio de General
Francisco Murguía

Nieves: Moderately distributed.
Santa Rita Mine. White to black
crystals to 10 cm. Associated
with jamesonite, pyrite, and
stibnite.

Municipio de Mazapil

Noche Buena: Moderate distribution. Noche Buena Mine.

Municipio de Noria de
Angeles

Noria de Angeles: Moderate
distribution. Noted from the
Cubierta and Anexas mines.

Municipio de Sombrerete

Sombrerete: Moderately distributed. Tocayos mine. White
crystals to 2 cm.

BARTHITE. See **AUSTINITE**

BAYLDONITE

$PbCu_3(AsO_4)_2(OH)_2 \cdot H_2O$. A rare mineral found in the oxide zones of lead/copper deposits.

DURANGO

Municipio de Mapimí

Mapimí: Limited distribution. Ojuela Mine. Crystals to 4 mm as druzes. Associated with wulfenite and goethite.

BEIDELLITE

$(Na,Ca_{0.5})_{0.33}Al_2(Si,Al)_4O_{10}(OH)_2 \cdot nH_2O$. A member of the smectite group. Common in hydrothermally altered mineralized areas and a questionable mineral.

CHIHUAHUA

Municipio unknown

Namiquipa: Moderately distributed. Princessa Mine.

BERTHIERITE

$FeSb_2S_4$. A rare mineral found in the sulfide ore zones of metal deposits.

BAJA CALIFORNIA SUR

Municipio de San Antonio

Triunfo: Moderately distributed. Most noted from the Comstock, Fortuna, Espinosena, Gobernadora, Humboldt, Nacimiento, Reforma, San Pedro, San Nicolás, Soledad, San Antonio, and San José mines. Associated with jamesonite, stibnite, pyrite, and galena.

DURANGO

Municipio de Indé

Indé: Limited distribution. Matracal Mine.

ZACATECAS

Municipio de Noria de Angeles

Noria de Angeles: Limited distribution. Noted from the Cubierta and Anexas mines.

BERTRANDITE

$Be_4Si_2O_7(OH)_2$. An uncommon mineral found in granite pegmatites.

COAHUILA

Municipio de Acuña

Sierra de Aguachile: Widespread (Pico Etereo district). Associated with fluorite. [This is the largest known deposit of bertrandite in the world.]

BERYL

$Be_3Al_2Si_6O_{18}$. An important ore mineral of beryllium, found in granite, pegmatite, mica schists, and dark limestones.

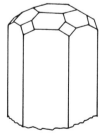

A typical 1 cm crystal of Beryl var. Aquarmarine from Ranchó Viejo, Baja California. (Modified by Shawna Panczner from Sinkankas, 1964)

BAJA CALIFORNIA

Municipio de Tecate

El Mesquite: Moderately distributed, 28 km south of the El Condor turnoff on Highway 2. Pegmatite located between Jasay and El Topo. Smooth and etched greenish yellow crystals, often tapered, to 8 cm. Associated with quartz, muscovite, albite, and fluorapatite.

La Jollita: Limited distribution (goshenite). Corroded golden yellow crystals to 2 cm. Associated with blue topaz, smoky quartz, fluorapatite, and elbaite.

Rancho Viejo: Moderately distributed. (aquamarine). Noted from the Socorro, Delicias, Aquamarina claims. Associated with tourmaline, sphene, quartz, and feldspar.

CHIHUAHUA

Municipio de Aldama

Placer de Guadalupe: Moderately distributed. (emerald).

HIDALGO

Municipio de Mineral del Monte

Mineral del Monte: Moderately distributed.

OAXACA

Municipio de San Francisco Telixtlahuaca

Santa Ana: Moderately distributed. Associated with biotite, muscovite, orthoclase, albite, spodumene, ilmenite, and dravite.

SAN LUIS POTOSI

Municipio de Guadalcázar

Realejo: Limited distribution, (Variety aquamarine). In the Arroyo de Los Arcosin on the Cerro de San Cristóbal. Crystals to 7 cm. Associated with various feldspars.

SONORA

Municipio de Moctezuma

Sierra de Oposura: Moderately distributed. Pale blue crystals to 10 cm. Associated with scheelite, biotite, dravite, albite, and orthoclase.

BERZELIANITE

Cu_2Se. A rare mineral found in few copper deposits.

DURANGO

Municipio de Mapimí

Mapimí: Limited distribution. Monterrey Mine. Small grains. Associated with umangite, chalcopyrite, bismuth, sphalerite, pyrrhotite and clausthalite.

BETAFITE

$(Ca,Na,U)_2(Ti,Nb,Ta)_2O_6(OH)$. A rare mineral and member of the pyrochlore group found in pegmatites.

OAXACA

Municipio de San Francisco Telixtlahuaca

Huitzo: Limited distribution. Muerto Mine. Crystals to 12 cm.

BEUDANTITE

$PbFe_3(AsO_4)(SO_4)(OH)_6$. A rare mineral of secondary origins formed by the alterations of lead minerals.

DURANGO

Municipio de Mapimí

Mapimí: Limited distribution. Ojuela Mine.

SONORA

Municipio de Herocia Carboraca

El Antimeneo: Limited distribution. El Antimeneo Mine. Associated with carminite and corkite.

Municipio de Trincheras

La Mur: Limited distribution. Las Animas Mine. Associated with arsenosiderite, akaganeite, and carminite. [This location is often referred to simply as Benjamín Hill, which is south and east of the mine.]

BIANCHITE

$(Zn,Fe)(SO_4) \cdot 6H_2O$. An uncommon secondary mineral and member of the hexahydrite group, which is found in zinc deposits.

DURANGO

Municipio de Mapimí

Mapimí: Limited distribution. Ojuela Mine.

BINDHEIMITE

$Pb_2Sb_2O_6(O,OH)$. A common mineral of the stibiconite mineral group found in oxidized antimonial lead deposits.

BAJA CALIFORNIA SUR

Municipio de San Antonio

Triunfo: Moderate distribution. La Gobernadora Mine. Associated with jamesonite, stibnite, pyrite, and galena.

Municipio de Los Cabos

San Jose del Cabo: Moderately distributed on the Rancho de San Felipe.

DURANGO

Municipio de Mapimí

Mapimí: Moderately distributed. Ojuela Mine's San Pointe stope, 14th level. Massive. Associated with mimetite, adamite, hedyphane, goethite, and jamesonite.

SAN LUIS POTOSI

Municipio de Guadalcázar

Arroyo Las Papas: Moderately distributed.

SONORA

Municipio de Cananea

La Morita: Limited distribution.

BIOTITE

$K(Mg,Fe)_3(Al,Fe)Si_3O_{10}(OH,F)_2$. The most common of the mica group and an important and abundant rock forming mineral. Forms under wide varieties of conditions and found in most types of rocks.

BAJA CALIFORNIA

Municipio de Ensenada

Pino Solo: Widely distributed, 25 km south. Associated with quartz, tourmaline, and various feldspars.

COAHUILA

Municipio de Candela

Pánuco: Moderately distributed. Pánuco Mine.

DURANGO

Municipio de Cuencamé

Velardeña: Moderately distributed. Most noted from the Terneras, Tunel, and La Choua mines.

Municipio de Mapimí

Mapimí: Limited distribution. Ojuela Mine.

HIDALGO

Municipio de Hausca de Ocampo

San Miguel Regla: Widely distributed in the Barranca de Regla.

Municipio de Pachuca

Pachuca de Soto: Moderately distributed.

OAXACA

Municipio de San Francisco Telixtlahuaca

San Francisco Telixtlahuaca: Moderately distributed. Associated with beryl, albite, orthoclase, and spodumene.

BIRNESSITE

$Na_4Mn_{14}O_{27} \cdot 9H_2O$. An uncommon mineral of secondary origin forming by the breakdown of primary manganese minerals.

CHIHUAHUA

Municipio de Ahumada

Sierra Gellege: Limited distribution.

ZACATECAS

Municipio de Zacatecas

Zacatecas: Moderately distributed.

BISMITE

Bi_2O_3. A secondary mineral formed by the oxidation of other bismuth minerals.

DURANGO

Municipio del Oro

Santa María del Oro: Limited distribution. El Carmen[San Pedro] Mine. Associated with bismutite and psudomalachite.

SAN LUIS POTOSI

Municipio de San Luis Potosí

San Luis Potosí: Limited distribution.

BISMUTH

Bi. A rare mineral (element) found in hydrothermal veins or in pegmatites.

CHIHUAHUA

Municipio de Batopilas

Batopilas: Limited distribution. Los Trajos Mine.

Municipio de Morelos

Morelos: Limited distribution. Associated with bismuthinite.

DURANGO

Municipio del Oro

Santa María del Oro: Limited distribution. El Carmen[San Pedro] Mine. Massive.

Municipio de Mapimí

Mapimí: Limited distribution. Monterrey Mine. Massive. Associated with galena, sphalerite, pyrrhotite, magnetite, stannite, chalcopyrite, emplectite, umangite, and marcasite.

GUANAJUATO

Municipio de Guanajuato

Guanajuato: Limited distribution. Massive. Associated with bismuthinite and bismutite.

BISMUTHINITE

Bi_2S_3. A rare mineral formed in moderately high temperature hydrothermal veins and pegmatites.

CHIHUAHUA

Municipio de Camargo

Location unknown.

Municipio de Morelos

Morelos: Limited distribution. Associated with native bismuth.

DURANGO

Municipio de Mapimí

Mapimí: Monterrey Mine. Crystals to 4 mm. Associated with bismuth, umangite, emplectite, stannite, pyrrhotite, marcasite, pyrite, galena, and sphalerite.

GUANAJUATO

Municipio de Guanajuato

Guanajuato: Moderately distributed. Associated with native bismuth and bismutite.

Sierra de Santa Rosa: Limited distribution. Noted from the La Industria and Santa Catarina mines. [This location is often listed as Rancho Calvillo.]

Municipio de San Luis de la Paz

San Luis de la Paz: Limited distribution.

SAN LUIS POTOSI

Municipio de San Luis Potosí

Rancho del Santuario: Limited distribution.

SINALOA

Municipio de Cosalá

Cosalá: Las Canas Mine.

Municipio del Rosario

La Rastra: Moderately distributed. Most noted from the Plomosas and Mariposa mines. Associated with chalcopyrite and quartz.

SONORA

Municipio de Nacozari de García

Nacozari de García: Paulina Mine. Associated with stibiobismuthinite.

Municipio de Navajoa

Navajoa: Limited distribution. La Gloria Mine.

Municipio Unknown

Cabenzos: Limited distribution. Associated with chalcopyrite.

ZACATECAS

Municipio de Concepción del Oro

Concepción del Oro: Limited distribution. Associated with sphalerite and pyrite.

Municipio de Ojo Caliente

Cerro de Ganzules: Limited distribution.

BISMUTITE

$Bi_2(CO_3)O_2$. A rare secondary mineral formed by the alteration of other bismuth minerals.

CHIHUAHUA

Municipio de Santa Bárbara

Santa Bárbara: Limited distribution.

DURANGO

Municipio de Durango

Victoria de Durango: Limited distribution. [This location is better known as Durango.]

Municipio del Oro

Santa María del Oro: Limited distribution. El Carmen [San Pedro] Mine. Associated with native bismuth.

GUANAJUATO

Municipio de Guanajuato

Guanajuato: Limited distribution. Associated with native bismuth and bismuthinite.

Sierra de Santa Rosa: Limited distribution. Santa Catarina Mine. [This location has been also listed as Rancho Calvillo, which is the ranch where the mine is located.]

QUERETARO

Municipio de Cadereyta

Cerro de Marol: Limited distribution.

SAN LUIS POTOSI

Municipio de Cerro de San Pedro

Cerro de San Pedro: Moderately distributed. Noted from the Buenavista, La Cruz, and Santuario mines.

BIXBYITE

$(Mn,Fe)_2O_3$. A rare mineral formed in rhyolitic rocks and occasionally in metamorphosed manganese ores.

CHIHUAHUA

Location data not available.

DURANGO

Municipio de Durango

Victoria de Durango: Limited distribution. Cerro de Mecardo Mine. Crystals to 5 mm. [This location is better known simply as Durango.]

SAN LUIS POTOSI

Municipio de Tepetate

Tepetate: Moderately distributed. Crystals to 8 mm. Associated with topaz and opal.

BOHDANOWICZITE

$AGBiSe_2$. A rare mineral found in hydrothermal deposits.

CHIHUAHUA

Municipio de Janos

Janos: Limited distribution. Associated with sphalerite and calcite.

BOLEITE

$Pb_{26}Ag_9Cu_{24}Cl_{62}(OH)_{48}$. A rare secondary mineral found in the oxidized zones of copper/lead copper deposits.

BAJA CALIFORNIA SUR

Municipio de Mulegé

Santa Rosalía: Limited distribution in area around Cañada de Curuglu. Noted from the Amelia, Cumenge, and Curuglú mines. Bright blue cubic and modified crystals to 4 cm. Associated with pseudoboleite, cumengite, atacamite, anglesite, cerussite, phosgenite, pyromorphite, and gypsum. [This is the type location for this mineral. It was found here in the early 1890s in the area around the Cañada de Curuglú in the Boleo mining district. The mineral was named after the mining district. The crystals are usually found in the warm moist volcanic tuff ore bed number 3. This several meter thick ore bed dries quickly when mined, thus allowing the boleite crystals to become free of their matrix. The volcanic matrix has to be stabilized upon being mined to remain intact if matrix specimens are desired. Boleite occurs in and upon gypsum (selenite) and anglesite that provides excellent matrix for specimens. The Boleo Mining District is most remarkable and considered by some economic geologists as still being in the process of formation. This area of Mexico is one of the world's most active geological regions.]

SONORA

Municipio de Arizpe

Arizpe: Las Chipas Mine. [This mine has also been referred to in some literature as the Pedrazzini Mine.]

BORAX

$Na_2B_4O_5(OH)_4 \cdot 8H_2O$. A rare mineral formed in the evaporation of dry lakes and as efflorescence in arid regions.

HIDALGO

Municipio de Pachuca

Pachuca de Soto: El Cristo Mine. Massive.

BORNITE

Cu_5FeS_4. An important ore mineral of copper, usually forms in hypogene deposits, but can also be found in mafic rocks, in contact metamorphic, replacement deposits and in pegmatites.

AGUASCALIENTES

Municipio de Asientos

Asientos: Widespread. Noted from the El Corbre and San Bartolo mines. Massive.

Municipio de Tepezalá

Tepezalá: Widespread.

BAJA CALIFORNIA

Municipio de Ensenada

San Fernando: San Fernando Mine.

BAJA CALIFORNIA SUR

Municipio de Mulegé

Santa Rosalía: Widespread. Massive.

CHIAPAS

Municipio de Pichucalco

Pichucalco: Santa Fe Mine.

CHIHUAHUA

Municipio de Aldama

Las Vigas: Moderately distributed.

Municipio de Aquiles Serdán

Francisco Portillo: Moderately distributed. Noted from the El Potosí and Buena Tierra mines. Associated with pyrite, galena, and sphalerite.

San Antonio el Grande: San Antonio Mine. Associated with pyrite, sphalerite, and galena.

Municipio de Cusihuiriáchic

Cusihuiriáchic: Moderately distributed. La Reina Mine. [This mine is located on the old Hacienda de Huisache.]

Municipio de Hidalgo del Parral

Hidalgo del Parral: Moderately distributed. Noted from the La Inmensidad, Remedios, and Tecolte mines.

Municipio de Jiménez

Los Reyes: Limited distribution. Noted from the Josefina and Anexas mines.

Municipio de Manuel Benavides

San Carlos: Widespread. San Carlos Mine.

Municipio de Urique

Piedras Verdes: Moderate distribution. El Carmen and San Miguel mines.

COAHUILA

Municipio de Candela

Pánuco: Moderate distribution. Pánuco Mine.

DURANGO

Municipio de Cuencamé

Velardeña: Moderately distributed. Noted from the Terneras and Tiro Samuel mines.

Municipio del Oro

Santa María del Oro: Moderate distribution. La Buena Mine.

Municipio de Pánuco de Coronado

Pánuco de Coronado: Moderately distributed. Avino Mine.

Municipio de Guanaceví

Guanaceví: Moderately distributed. Most noted from the Barredon, San Gil, and Sirena mines.

Municipio de Indé

Indé: Limited distribution. La Union Mine.

Municipio de Mapimí

Mapimí: Limited distribution. Ojuela Mine. Associated with chalcopyrite and enargite.

Municipio de Poanas

Poanas: Moderately distributed. Most noted from the La Candelaria, La Concha, and La Purísima mines. [Mines are located on the Cerro Sacrificio.]

Municipio de San Dimas

Tayoltita: Moderately distributed. Most noted from the El Cobre, Guadalupe, La Montanesa, La Verde, and Candelaria mines.

GUANAJUATO

Municipio de Guanajuato

Cedro: Limited distribution. Noted from the Fe, Esperanza, and El Guapillo mines.

Municipio de San Luis de la Paz

Mineral de Pozo: Moderate distribution. Santa Brígida Mine.

GUERRERO

Municipio de Taxco

Taxco de Alcarón: Moderately distributed. Guerrero Mine.

MICHOACAN

Municipio de Angangueo

Mineral de Angangueo: Moderately distributed. Noted from the Catingon and San Cristóbal mines. Massive.

Inguaran: Widely distributed.

Municipio de Santa Clara

Oropeo: Limited distribution. Aztec Mine.

NAYARIT

Municipio de Santa María del Oro

Acuitapilco: Moderate distribution. La Parla and La Verde mines.

SAN LUIS POTOSI

Municipio de Charcas

Charcas: Moderately distributed. Veta Rica Mine.

Municipio de Villa La Paz

La Paz: Moderately distributed. Noted from the La Paz and Trinidad mines.

Municipio de Villa de Ramos

Ramos: Moderately distributed.

SINALOA

Municipio de Cosalá

Cosalá: Moderately distributed. Noted from the Nuestra Senora and El Cobre mines.

Municipio del Fuerte

Hacienda de Picachos: Moderately distributed.

SONORA

Municipio de Arizpe

Arizpe: Moderately distributed. Elenita Mine.

Municipio de Cananea

Cananea: Widespread. Crystals to 2 cm.

Municipio de Soyopa

San Antonio de las Huertas: Limited distribution. La Libertad Mine.

VERACRUZ

Municipio de Las Minas

Zomelahuacan: Widespeard. Noted from the Buena Suerte, Dorado, El Alto, Espíritu Santo, Jesús María, La Cruz, Miqueta, Nueva Zaragosa, Nuevo Descubrimiento, Pinos, Preciosa Sangre, Providencia, Rosario, Sabanilla, San Anselmo, San José de García, San Miguel, Soledad, Tequezquite, Vieja Zaragoza, and Vigilancia mines.

ZACATECAS

Municipio de Concepción del Oro

Aranzazú: Moderate distribution. Aranzazu Mine.

Bonanza: Moderate distribution. Bonanza Mine.

Municipio de Mazapil

Mazapil: Moderately distributed. Noted from the San Carlos and Jesús María mines.

Municipio de Noria de Angeles

Noria de Angeles: Moderately distributed. Noted from the Grande and Pachuca mines.

Municipio de Sombrerete

San Martín: Moderately distributed. San Martín Mine. Massive. Associated with chalcopyrite, covellite and native silver.

Sombrerete: Moderately distributed. San Antonio Mine.

Municipio de Zacatecas

Zacatecas: Widespread. Massive.

BOTRYOGEN

$MgFe(SO_4)_2(OH)\cdot 7H_2O$. An uncommon mineral found associated with secondary sulfates in sulfide ore deposits.

ZACATECAS

Municipio de Fresnillo

Cerro de San Marcos: Limited distribution.

BOULANGERITE

$Pb_5Sb_4S_{11}$. An uncommon mineral formed in low to moderate temperature veins and found associated with sulfosalts and sulfides.

BALA CALIFORNIA SUR

Municipio de San Antonio

Triunfo: Moderately distributed. Triunfo and Marronena mines.

CHIHUAHUA

Municipio de Hidalgo del Parral

Hidalgo del Parral: Moderately distributed.

DURANGO

Municipio de Cuencamé

Velardeña: Limited distribution. Espondio Mine.

Municipio de Guanaceví

Guanaceví: Limited distribution. El Verde Mine.

Municipio de Mapimí

Mapimí: Moderately distributed. Ojuela Mine. Associated with galena, pyrite, stibnite, calcite, and jamesonite. Crystals to 7 mm.

Municipio de San Dimas

Tayoltita: Limited distribution. La Rosa Mine.

SONORA

Municipio de Hermosillo

Hermosillo: Moderately distributed. Noted from the San Javier and Las Animas mines.

ZACATECAS

Municipio de Mazapil

Noche Buena: Moderately distributed. Noche Buena Mine. Needle-like crystals to 2 cm. Associated with quartz and pyrite. [This mineral was found as mats of crystals and was first sold as jamesonite, but later was found to be bourlangerite.]

BOURNONITE

$PbCuSbS_3$. A common mineral formed in hydrothermal veins of moderate temperatures.

AGUASCALIENTES

Municipio de Rincón de Ramos

Rincón de Ramos: Limited distribution. Most noted from the Nueva Velardena Mine.

BAJA CALIFORNIA SUR

Municipio de San Antonio

El Triunfo: Moderately distributed within this old district especially at the El Triunfo Mine.

CHIHUAHUA

Municipio de Aquiles Serdán

Francisco Portillo: Moderately abundant within St. Eulalia District's west camp. Associated with pyrite and sphalerite. Crystals to 1 cm.

Municipio de Saucillo

Naica: Moderately distributed especially at the Naica Mine. Associated with pyrite, galena, sphalerite, fluorite, and calcite. Cog wheel shaped crystals to 2 cm in diameter. [In the past several years, many fine crystal specimens have been found in the new mine on the Naica fault on the west side of the Sierra de Naica.]

DURANGO

Municipio de Indé

Indé: Limited distribution.

GUANAJUATO

Municipio de Guanajuato

La Luz: Limited distribution

SONORA

Municipio de Onavas

La Barranca: Moderately distributed especially at the Las Animas, Noche Buena, and Tarahumara mines.

ZACATECAS

Municipio de Mazapil

Noche Buena: Moderately distributed. Noche Buena Mine. Associated with tetrahedrite, pyrite, galena, sphalerite, and arsenopyrite. Cog wheel shaped crystals to 5 cm in diameter.

BRANNERITE

$(U,Ca,Y,Ce)(Ti,Fe)_2O_6$.
A rare mineral found in pegmatites and in sediments.

SONORA

Municipio de Aconchi

Sierra de Aconchi: Limited distribution within the Sierra de Aconchi.

BRAUNITE

Mn,Mn_6SiO_{12}. Occurs in veins as a secondary product of the metamorphism of other manganese minerals.

DURANGO

Municipio de Mapimí

Mapimí: Limited distribution especially at the San Pedro Mine. Micro grains within the manganese ores.

HIDALGO

Municipio de Mineral del Monte

Mineral del Monte: Limited distribution within district. Most noted from the Santa Brídgida, Carretera, and Jesús María mines.

JALISCO

Municipio de Autlán

Autlán de Navarro: Braunite has been found near here at the San Francisco Mine.

BREWSTERITE

$(Sr,Ba,Ca)Al_2Si_6O_{16} \cdot 5H_2O$. An uncommon member of the zeolite group.

BAJA CALIFORNIA SUR

Municipio de Mulegé

Santa Rosalía: Limited distribution especially at the Santa Rita Mine. Associated with native copper.

MEXICO

Municipio de Zacualpan

Zacualpan: Limited distribution within area.

BROCHANTITE

$Cu_4(SO_4)(OH)_6$. An uncommon secondary mineral found in oxidized copper deposits in arid regions.

AGUASCALIENTES

Municipio de Asientos

Asientos: Moderately distributed throughout district. Crystal tuffs to 5 mm. Associated with goethite.

Municipio de Tepezalá

Tepezalá: Moderately distributed. Crystal sprays. Associated with goethite.

CHIHUAHUA

Municipio de Ahumada

Los Lamentos: Moderately distributed at the Erupción/Ahumada Mine. Associated with calcite and goethite. Crystals sprays to 8 mm.

Municipio de Aquiles Serdán

Francisco Portillo: Moderately distributed throughout west side of the Santa Eulalia District. Associated with goethite and calcite.

San Antonio el Grande: San Antonio Mine. Associated with goethite.

Municipio de Saucillo

Naica: Moderately distributed in the older upper oxidized zone. Associated with goethite. Small crystal sprays.

DURANGO

Municipio de Mapimí

Mapimí: Moderately distributed throughout district's upper oxidized zone especially at the Ojuela and San Juan mines. Associated with cerussite malachite, and goethite. Acicular tuffs of crystals to 1 cm.

HIDALGO

Municipio Zimapán

Zimapán: Moderately distributed throughout district especially at the Santa Eleonora and Flojanales mines. Associated with goethite. Small crystal sprays.

SAN LUIS POTOSI

Municipio de Villa de Ramos

Ramos: Moderately distributed. Most noted from the Cocinera Mine.

SONORA

Municipio de Caborica

Heróica Caborica: Otoña Mine.

Municipio de Ures

Ures: La Loba Mine.

BRONZITE. See **ENSTATITE**
var. ferroan

BROMARGYRITE

AgBr. A rare secondary mineral
that forms from the surface
oxidation of silver ores where
there is an abundance of chlorine
and bromine.

BAJA CALIFORNIA SUR

   Municipio de La Paz

Cacachilas: Moderately distrib-
uted especially at the Anima
Sola, Animas, Bebelama, Casuali-
dad, Chivato, Jesús María,
Peruana, Rosario, San Gregorio,
Santa Lucía, Santa Teresa,
Soledad, Tesorito, and Trinidad
mines.

   Municipio de Mulegé

Santa Rosalía: Moderately
distributed throughout district.

CHIHUAHUA

   Municipio de Aquiles Serdán

San Antonio el Grande: Limited
distribution San Antonio Mine.

   Municipio de Batopilas

Batopilas: Moderately distributed
throughout district. Associated
with chlorargyrite.

   Municipio de Cusihuiriáchic

Cusihuiriáchic: Limited distribution. Princesa Mine.

   Municipio de Guadalupe y
   Calvo

Guadalupe y Calvo: Limited
distribution. El Rosario Mine. As
chlorine rich bromargyrite.

   Municipio de Julimes

Julimes: Carrizo Mine.

COAHUILA

   Municipio de Sierra Mojada

Sierra Mojada: Moderately
distributed especially at the
Guadalupe, Providencia, and, Tiro
Juárez mines.

DURANGO

   Municipio de Mapimí

La Cadena: Limited distribution
in small prospects on the Rancho
de Cadena.

   Municipio de Nombre de Dios

San Juan de la Parrilla: Limited
distribution.

   Municipio de San Pedro del
   Gallo

Peñoles: Limited distribution.
Most noted from the Guadalupe and
San Rafael mines.

GUANAJUATO

   Municipio de San Luis de La
   Paz

Mineral de Pozos: Moderately
distributed. Most noted at the
Cinco Señores Mine.

MEXICO

   Municipio del Oro

El Oro: Moderately distributed.
Noted from the San Rafael,
Carmen, and San Antonio mines.

   Municipio de Temascaltepec

Temascaltepec de Gonzalez:
Moderately distributed within
district.

MICHOACAN

   Municipio de Acuitzio del
   Canje

Curucopaseo: Moderately distributed within district. Most noted

from the El Angel, Dulces Nombres, and Refugio mines.

Municipio de Tlalpujahua

Tlalpujahua: Moderately distributed within this old district. Most noted from the Dos Estrellas Mine.

PUEBLA

Municipio de Tetla de Ocampo

Tetla de Ocampo: Limited distribution.

QUERETARO

Municipio de Cadereyta

Cadereyta de Montes: Moderately distributed. Especially noted at the La Luz and Santa Inés mines.

El Doctor: Moderately distributed on the San Onofre Hacienda. San Juan Nepomuceno Mine.

SAN LUIS POTOSI

Municipio de Catorce

Catorce: Moderately distributed. Most noted from the La Purísima Concepción, Medellín, and Refugio mines. Associated with silver, gold, and chlorargyrite. [Silver was first discovered in 1773 and was originally known as Mineral de la Purísima Concepción de Alamos de Catorce. The camp name was later shortened to la Purísima Concepción de Alamos de Catorce and now known as Catorce.]

Municipio de Cerro de San Pedro

Cerro de San Pedro: Limited distribution.

Municipio de Charcas

Charcas: Limited distribution. [This old mining district was founded as Santa María de las

Charcas with mining first being done on the Rancho de San Onofre near Charcas.]

Municipio de Villa de La Paz

La Paz: Limited distribution. Most noted from the La Paz and Santa Fe mines.

SONORA

Municipio de Caborca

Heróica Caborca: Moderately distributed. Associated with chlorargyrite.

Municipio de Hermosillo

Hermosillo: Limited distribution. Noted from the California Mine.

Municipio de Ures

Ures: Limited distribution. Most noted from the El Gavilán and La Blanca mines.

ZACATECAS

Municipio de Concepción del Oro

Albarradón: Moderately distributed.

Municipio de Fresnillo

Fresnillo de Gonzalez Echeverria: Moderately distributed as a chlorine rich bromargyrite.

Plateros: Moderately distributed. Associated with native silver. Crystals to 5 mm.

Municipio de Ojo Caliente

Ojo Caliente: Moderately distributed. Most noted from the Palestina and San Antonio mines.

Municipio de Pinos

Pinos: Moderately distributed Especially noted at the Candelaria, Grande, La Palma,

Matatuza, Porvenir, Refugio, San Bartolo, San José, Santa Rita, Tajo de Ibarra, and Trinidad Number one and Number two mines. Crystals to 4 mm.

Municipio de Sombrerete

Sombrerete: Moderately distributed. Most noted in the upper mine workings of the Veta Negra and Tocayos mines. Crystals to 4 mm.

Municipio de Zacatecas

Zacatecas: Moderately distributed. Most noted from the La Luz and San Vicente mines.

BROOKITE

TiO_2. Found in igneous and metamorphic rocks, also as detrital mineral derived from hydrothermal leaching of gneisses and schists.

SAN LUIS POTOSI

Municipio de Guadalcázar

Guadalcázar: Moderately distributed.

BRUCITE

$Mg(OH)_2$. An alteration product of periclase in contact metamorphic limestones and dolomites and in low temperature hydrothermal veins.

ZACATECAS

Municipio de Sombrerete

Sombrerete: Moderately distributed.

BUERGERITE

$NaFe_3Al_6(BO_3)_3Si_6O_{21}F$. A rare member of the tourmaline group of minerals that forms in granites, pegmatites, and metamorphic rocks. It also can be found in sedimentary rocks as detritus.

SAN LUIS POTOSI

Municipio de Mexquitic

Mexquitic: Limited distribution. Brown crystals to 4 cm on matrix. [This is the type location for this mineral, but the exact location is not known. The man who discovered and mined all the buergerite that reached the mineral world has died and told no one the exact location.]

BURCKHARDTITE

$Pb_2(Fe,Mn)Te(AlSi_3)O_{12}(OH)_2 \cdot H_2O$. A rare mineral found in few metal deposits.

SONORA

Municipio de Moctezuma

Moctezuma: Limited distribution. Noted from the Moctezuma Mine.

BURKEITE

$Na_6(CO_3)(SO_4)_2$. A rare mineral found in the clay zones of desert dry lakes.

Location data is unavailable.

BUSTAMITE

$(Mn,Ca)_3Si_3O_9$. An uncommon mineral found with other manganese minerals.

CHIHUAHUA

Municipio de Saucillo

Naica: Moderately distributed.

PUEBLA

Municipio unknown

Tetla de Xonotla: Moderately distributed.

BYSTROMITE

$MgSb_2O_6$. A rare mineral found in a few manganese/antimony deposits.

SONORA

Municipio de Agua Prieta

Limited distribution. 27 km south of Agua Prieta.

Municipio de Heróica Caborica

El Antimonio: Limited distribution. Most noted from the La Fortuna Mine. Associated with quartz.

BYTOWNITE

$(Ca,Na)Al(Al,Si)Si_2O_8$. An uncommon member of the feldspar group, which forms as phenocrysts in some basalts, layered mafic and ultramafic rocks.

CHIHUAHUA

Municipio de Nuevo Casas Grandes

Nuevo Casas Grandes: Moderately distributed. East of Nuevo Casas Grandes. Gemmy light yellow masses with occasional crystal faces up to 8 cm. [This material is almost completely gemmy and has yielded cut stones up to 80 ct.]

Municipio de Madera

Madera: Moderately distributed. Pale yellow masses to 3 cm. Much of this material is gemmy.

C

CALAVERITE

AuTe$_2$. A rare mineral found in in metal deposits.

OAXACA

Municipio de San Pedro Taviche

San Pedro Taviche: Limited distribution.

CALCITE

CaCO$_3$. A common member of the calcite group, which forms in a wide variety of conditions; sedimentary and metamorphic rock environments. In hydrothermal veins it develops its best crystal development.

AGUASCALIENTES

Municipio de Asientos

Asientos: Widespread throughout district. White to cream colored crystals to 60 cm. [Several years ago a large watercourse produced many large crystals. Many were over 60 cm long but most were under 15 cms in length. Most of the crystals contained phantoms and had shiny faces. This material was labeled as being from Zacatecas, which is 100 km north.]

A 10 × 12 cm crystal cluster of calcite from Areponapuchic, Chihuahua, Robert Jones Collection. (Photographed by Robert W. Jones)

Municipio de Tepezalá

Tepezalá: Widespread throughout the district especially on the vetas Nopal and Amarilla. Crystals to 60 cm.

BAJA CALIFORNIA

Municipio de Ensenada

Arroyo de Tule: Moderately distributed. Massive. Referred to locally as "Mexican Onyx."

Calamahi: Moderately distributed. Most noted from the Rio Salomon Mine.

El Marmol: Moderately distributed. Most noted from the Pedrara
quarry. Massive. (Onyx) locally
called "Mexican Onyx."

El Marmolito: Moderately distributed. Massive (Onyx). [This
massive calcite has been referred
to locally as "Yaqui Onyx."]

Rosarito: Moderately distributed.
South and east of Rosarito at the
Rancho Ballistero or Viasteros.
Reddish brown rhombohedral
crystals to 25 cm. Associated
with gypsum, stilbite, and
analcime. [This material was
often gemmy and yielded cut
stones to over forty carats in
size. The calcite is found in
vugs in the volcanic rock of the
area.]

BAJA CALIFORNIA SUR

Municipio de Mulegé

Santa Rosalía: Widespread
throughout the El Boleo Mining
district. Associated with
chrysocolla. Crystals to 2 cm.
[The chrysocolla often stains the
crystals a beautiful blue-green
color.]

CHIAPAS

Municipio de Pichucalco

Pichucalco: Limited distribution
especially at the Sante Fe Mine.
Crystals to 2 cm.

CHIHUAHUA

Municipio de Ahumada

Constitución: Moderately distributed at the Mojina Mine. Located
on Rancho de Mojina, south of
Constitución. Associated with
celestite. White masses and
crystals to 2 cm. [In the 1970s,
a large watercourse produced
masses of white knobby calcite
with bright blue celestite
crystals.]

Los Lamentos: Widespread throughout the area noted from the

Erupción/Ahumada Mine. White to
colorless crystals to 3 cm.
Associated with wulfenite,
descloizite, and goethite.

Municipio de Aquiles Serdán

Francisco Portillo: Widespread
throughout the west camp of the
Santa Eulalia mining district.
Most noted from the El Potosí and
Buena Tierra mines. Colorless to
white and shades of red and brown
crystals to 40 cm. Crystal plates
to 1 m. Associated with goethite, cerussite, acanthite,
pyrite, galena, hemimorphite,
sphalerite, pyrrhotite, and
rhodochrosite. [The large reddish
-brown calcite, with a two stage
crystal growth, which the
district is famous for, came from
the 19th level of the Buena
Tierra Mine. The water clear
and highly twinned calcite
crystals, which were usually on
a dark goethite matrix are from
the 10th level of the El Potosí
Mine. Many large caverns lined
with calcite have been found
within the district's west
camp.]

San Antonio el Grande: Widespread
throughout the Santa Eulalia
mining district's east camp
especially at the San Antonio
Mine. Colorless, through shades
of white, cream, apple green, tan
pink, and red crystals to 60 cm.
Associated with fluorite,
galena, sphalerite, pyrite,
chalcopyrite, and arsenopyrite.
[In the early 1900s a crystal
line cave was uncovered with
crystals up to 60 cms in length.
The miners joked about using
the crystals with the flattened
tops as stools to sit on!]

Municipio de Batopilas

Batopilas: Widespread throughout
the district. Most noted from the
San Rafael and San Nestor mines.
Crystals to 5 cm. Associated with
native silver and acanthite.

La Bufa: Widespread. Most noted
from the La Bufa Mine. Crystals
to 10 cm. Associated with native
copper. [In the early 1960s

several pockets were uncovered of
large crystals of calcite with
fine crystals of native copper.]

Municipio de Bocoyna

Areponapuchic: Widespread
throughout district. Most noted
from the El Porvenir, La Parla,
Flor de Esperanza, La Aurora, and
El Carman mines. Crystals to 20
cm. [This district is located in
the remote Barrranca de Urique,
is one of the Mexico's largest
calcite districts. The mines
are found on the steep sides of
the canyon wall, where the
calcite occurs in the faults and
fissures of the volcanic rocks.
The Porvenir Mine, the largest
within the district, is located
335 m below the rim of the Urique
Canyon and has produced many
fine crystallized specimens. The
crystal habits take two forms; a
modified or distorted scaleno-
hedron with flat basal terminations and arrowhead twins
or as the Mexican miner call
them, "corazon" or heart shaped.
A few of these heart or arrowhead
crystals found in the mid the
1940s weighed over 11 kg, but the
average was less than 450 g! The
crystal surfaces were etched, but
their interiors were for the
mostly water clear. The crystals
did have areas within them that
contained some minor color
zoning, a pale yellow or brown.
Most of the better crystals came
from a large cavity near the main
adit of the mine. A few cavities
where found with crystals that
did not exhibit the usual etched
surface, but most of these were
sold for optical calcite during
the later stages of W W II. The
La Aurora Mine, located 3.2 km.
northeast of El Porvenir Mine and
600 m below the canyon rim has
also produced good calcite
crystal. The crystals were shaped
like a flat shaft with a barb on
the end rather than like an
arrowhead. The Flor de Esperanza
Mine, located next to El Porvenir
Mine produced small but exceptionally fine crystals of
calcite. Most were of the
arrowhead twin type and well
formed!]

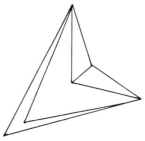

A large 8 cm "V" twin crystal of calcite from
Areponapuchic, Chihuahua. (Modified by
Shawna Panczner from Fries, 1948)

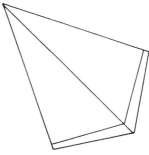

A 4 cm arrowhead twin crystal of calcite from
Areponapuchic, Chihuahua. (Modified by
Shawna Panczner from Fries, 1948)

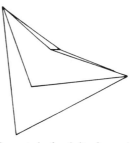

A twinned crystal of calcite from Areponapuchic, Chihuahua. (Modified by Shawna
Panczner from Fries, 1948)

Barranca del Cobre: Moderately
distributed. Crystals to 2 cm.
Associated with native copper.

Municipio de Chínipas

Chínipas: Moderately distributed.
Crystals to 4 cm.

Municipio de Ciudad Camargo

La Perla: Moderately distributed
especially at the La Perla and La

Negra mines. Crystals to 4 cm.
Associated with hematite and
magnetite.

Municipio de Guadalupe y Calvo

Guadalupe y Calvo: Moderately
distributed. Most noted from the
Refugio and Rosario mines.
Crystals to 4 cm.

Municipio de Guazapares

Monterde: Moderately distributed.
Most noted from the Arechuchique
and Las Trojas mines. Crystals to
6 cm. [The mines are found on
the side of the Oteros Canyon.
The small mine of Arechuchique
is 560 m below the top of the
canyon and the Las Trojas Mine
is 400 meters below the canyon
rim.]

Municipio de Madera

Chihuichupa: Widespread. Crystals
to 60 cm. [Crystals have been
found in some of the local caves
up to 1,000 kg in size!]

Municipio de Manuel Benavides

Rancho del Murcielago: Widespread. Fifty kilometers southeast of Manuel Benavides.
Crude crystals to 8 cm. Associated with pectolite. [This
location is known as the El
Murcielago mining district and is
composed of 300 shafts and
prospect pits in the faults and
fissures of the local basaltic
rock. In the Arroyo de la
Monilla, "boxwork" calcite
crystals, like those of Durango,
have been found. They were a
pleasing pale yellow color and
reached up to single crystal
plates of 45 kg.]

Rancho del Parran: Moderately
distributed. Crude crystals to 4
cm. [This location is next to
Rancho del Murcielago.]

Municipio de Nuevo Casas Grandes

San Pedro Corralitos: Widespread
throughout district. Most noted

from the León Mine. Crystals to 3
cm. Associated with goethite and
mimetite.

Municipio de Saucillo

Naica: Widespread. Colorless and
through shades of white, pink,
gray, and tan crystals to 15 cm.
Associated with pyrite, galena,
sphalerite, quartz, fluorite.
[The crystals from here are
usually sharp rhombs, but their
forms are also seen but are not
as common.]

Municipio de Urique

Agua Caliente: Widespread.
Waterclear crystals to 2 cm.
[This location is on the Río
Oteros near Pito real.]

Guagueybo: Moderately distributed. Most noted from the La
Esmeralda Mine. Crystals to 12
cm. [Calcite crystals from here
have been found up to 11 kg.
Most of the crystals were not
twinned, but on occasion fine
pale yellow "V" twin crystals
were found.]

DISTRITO FEDERAL

Sierra de Guadalupe: Moderately
distributed within local volcanic
rocks.

DURANGO

Municipio de Coloma

Palo Blancar: Moderately distributed within this area.

San Isidro: Moderately distributed within the canyon of the Río
de Las Viborillas. Scalenohedral
crystals to 20 cm and to over 14
kg.

Municipio de Cuencamé

Cuencamé de Ceniceros: Moderately
distributed. Tan to pink scalahe-
dral crystals to 5 cm. Associated
with stibiconite.

Municipio de Mapimí

Buruguilla: Widespread. Most
noted from the Descubridora Mine.

Waterclear to white stalatitic crystals to 4 cm. Associated with goethite. [The Descubridora district is located 35 km northwest of Mapimí and 2 km west northwest of Buruguilla. Much of what is found here is often mislabled as being from Mapimí.]

Mapimí: Widespread throughout district. Most noted from the La China, La Reina, and Ojuela mines. Colorless, white, pink, and gray crystals to 5 cm. Associated with wulfenite, aurichalcite, smithsonite, goethite, adamite, and hemimorphite. [The La China and La Reina mines are located west of Mapimí, but are within the district. Bright blood red calcite crystals were found at the Ojuela Mine in the 1960s and in the mid 1980s bright blue calcite crystals were found. These bright colors were caused by other minerals included within the calcite crystals.]

Municipio de San Pedro del Gallo

Peñoles: Widespread noted from the Veta Peñoles. Common vein mineral.

Municipio del Oro

Santa María del Oro: Moderately distributed. Most noted from the La Cacinera Mine.

Municipio de Topia

Topia: Moderately distributed. Most noted from the El Salto Mine.

Municipio de Rodeo

San Antonio: Widespread throughout area. Most noted from the La Amparo (La Fe) Mine. Colorless, white, pink, and tan crystals plates to 60 cm. [This location is Mexico's largest producer of optical calcite. The mine was originally called the La Fe Mine, but several years ago was changed to the La Amparo Mine. The present operating mines are located 3.2 km north and south of

the original La Amparo. The new mines are extensions of older operations and still carry the same mine name, La Amparo. The calcite is found in faults and fissures in the rhyolitic rock. The crystals occur as large plates or "boxwork" called "angelwing" calcite. The mine is operated for the fine grade of optical calcite found at the base of the "Angle wing" crystals. These crystals are broken off to expose the small zone of optical (iceland spar) beneath. In 1984 several small pockets were uncovered that produced clear, pink cleavage. This location has been listed as Rodeo, but the closest village is the small pueblo of San Antonio.]

Municipio de Poanas

Poanas: Widepread on the Cerro del Sacrificio. Most noted from La Candelaria and La Purísima mines.

GUANAJUATO

Municipio de Guanajuato

Guanajuato: Widespread throuhout district. Most noted from the Valanciana, San Juan de Rayas, and the Peregrina mines. Crystals to 40 cm. Associated with quartz (amethyst and rock), orthoclase, dolomite, acanthite, pyrargyrite, polybasite, and native silver. [Many years ago the Valanciana Mine produced, large, tan colored crystals to over 30 cm in length! The San Juan de Rayas mine is still producing fine specimens with some of the white to water-clear crystals reaching almost 15 cm in length. The Peregrina Mine is producing fine white to tan colored crystals. There is a new mine just east of the San Juan de Rayas, which in the mid 1980s produced large water-clear to white wheel like crystals up to 15 cm in diameter!]

La Luz: Widespread. Most noted from Jesús María Mine. Associated with quartz (amethyst and rock), orthoclase, dolomite, acanthite, pyrargyrite, polybasite, and native silver.

Municipio de San Luis de La Paz

Mineral de Pozos: Widespread. Most noted from the Escondida and Garibaldi mines.

Municipio de San Miguel de Allende

Mineral de Puerto de Nieto: Moderately distributed. Most noted from the San Antonio, Guadalupe, and La Argentina mines. Crystals to 4 cm. Associated with pyrite, galena, quartz, chalcopyrite, chalcocite, malachite, and azurite.

GUERRERO

Municipio de Huitzuco

Huitzuco de Los Figueroa: Moderately distributed. Most noted from the Tumbaga, Trinidad and La Cruz mines.

Municipio de Taxco

Taxco de Alcarón: Widespread throughout district. Most noted from the Acatitlan Mine. Crystals to 6 cm. Associated with quartz, silver, and acanthite.

Tehuilotepec: Widespread. Most noted from the San Agustin Mine.

HIDALGO

Municipio de Mineral del Chico

Mineral del Chico: Widerspread. Especially at the Arevalo Mine.

Municipio de Mineral del Monte

Mineral del Monte: Widespread. Most noted at the Carretera Mine.

Crystals to 4 cm. Associated with quartz (amethyst) and chlorargyrite.

Municipio de Pachuca

Pachuca de Soto: Widespread. Most noted from the San Nicanor, Rosario, San Cayetano, Guadalupe, El Encino, Jacal, San Rafael, and Barron mines. Crystals to 5 cm. Associated quartz (amethyst) and chlorargyrite. [This location is better known simply as Pachuca.]

Municipio de Zimapán

Bonanza: Moderately distributed. Most noted from the San Judas Mine.

Zimapán: Widespread. Crystals to 5 cm.

JALISCO

Municipio de Bolaños

Bolaños: Moderately distributed. Most noted from the Santa Fe Mine.

Municipio de Venustiano Garranza

Jiquilpan: Moderately distributed. Crystals to 2 cm.

MEXICO

Municipio de Zacualpan

Zacualpan: Moderately distributed. Most noted from the Alacrán and La Marsellesa mines. Crystals to 4 cm.

OAXACA

Municipio de San Pedro Taviche

San Pedro Taviche: Moderately distributed. Most noted from the San Pedro Taviche Mine. Crystals to 4 cm. Associated with native silver. [This location has produced large course wires of native silver in and on rhombs of calcite.]

PUEBLA

Municipio de Puebla

Puebla: Widespread.

Municipio de Tecali de Herrera

Tecali de Herrera: Widespread in mountains to the south. Massive. (Onyx).

Municipio de Zapotitlán:

La Sopresa: Widespread. Massive. (Onyx).

San Antonio: Widespread. Massive. (Onyx). [The onyx is mined here in both underground mines and small pits. Some of the onyx is carved here, but most is sent to Tehuacán or Puebla to be carved and polished.]

QUERETARO

Municipio de Cadereyta

Cedereyta: Moderately distributed. Most noted from the Santa Inés and La Luz mines. Crystals to 4 cm.

Municipio de Tolimán

Soyatal: Widespread throughout district. Most noted from the Claveo Negro and Santo Nino mines. Apple green to white scalenohedral crystals to 10 cm. [They have been found both as singles and as clusters. In the early 1980s fine cleaveges of green calcite where uncovered.]

SAN LUIS POTOSI

Municipio de Catorce

Mineral de La Puríisima Concepción de Alamos de Catorce: Widespread.

Most noted from the Santa Inés, La Puríisima Concepción, and San Agustin mines. Common vein mineral. [This location is usually referred to simply as Catorce.]

Wadley: Moderately distributed. Most noted from the San Jose Mine.

Municipio de Cedral

Limited distribution on the Hacienda de Espíritu Santo.

Municipio de Charcas

Charcas: Widespread throughout the district. Most noted from the San Sebastián and San Bartolo mines. Colorless, gray, white to cream crystals to 25 cm. [In the early 1960s a remarkable watercourse was uncovered in the mining operations at the San Sebastián Mine. It was full of calcite crystals that were in the shape of poker chips. These white to grayish crystals ranged up to almost 25 cm in diameter and 5 cm thick and were found in clusters of stacked crystal groups. This area of the mine has been totally mined out and officals now are saying the chances of finding more of this tpye of crystal at the San Sebastián Mine are slim. Calcite does occur in other crystal forms, the San Bartolo Mine over the past several years has produced nice calcite scalenhedrals up to 15 cm long. Charcas has also been called Santa María de Las Charcas.]

SINALOA

Municipio de Badiraguato

Los Terreros: Limited distribution.

Pitayita: Limited distribution.

San Javier: Limited distribution the Cerro del Yeso. Pale yellow crystals to 4 cm.

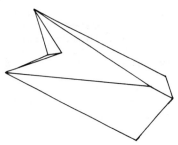

A 6 cm calcite crystal twin from Bacadehuachic, Sonora. (Modified by Shawna Panczner from Fries, 1948)

Municipio del Fuerte

El Fuerte: Moderately distributed.

SONORA

Municipio de Arizpe:

Arizpe: Moderately distributed. Most noted from the Las Chipas Mine. Associated with acanthite, polybasite, and fluorite.

Municipio de Bacadéhuachi

Bacadéhuachi: Moderately distributed on the Rancho del Hueringo. Twin "V" crystals to 15 cm and 20 kg in size.[The larger crystals from here were fine butterfly twins.]

Municipio de Baviacora

Mazocahui: Limited distribution in the Arroyo de la Junta, Rancho de la Noria de Lares. Most noted from the El Caliche mine. Rhombic crystals to 8 cm. [This mine has produced specimens of large rhombohedrons, both single and twin.]

Tarahuacachic: Limited distribution.

Municipio de Bavispe

Esqueda: Limited distributed, west side of the Río de Bavispe. Crystals to 25 cm and 28 kg in size.

Municipio de Cumpas

Cumpas: Moderately distributed. Most noted from the El Cristal mine. Pale yellow crystals to 5 cm.

Municipio de Nacozari de García

Oputo: Limited distribution. Twin "V" crystals to 4 cm.

Municipio de Sahuaripa

Sahuaripa: Moderately distributed.

TAMAULIPAS

Municipio de Reynosa

Rancho de Santa Ana: Limited distribution.

TLAXCALA

Municipio de Panotla

Cerro Anacoreta: Moderately distributed.

ZACATECAS

Municipio de Concepción del Oro

Aranzazú: Moderately distributed. Most noted from the Aranzazú Mine. Colorless, white, and green crystals to 4 cm. Associated with quartz. [Green scalenohedral crystals have been found up to 4 cm in length on 3 to 5 cm long quartz crystals.]

Concepción del Oro: Widespread througout district. Crystals to 5 cm. Associated with pyrite and quartz.

El Cobre: Widespread. Crystals to 3 cm.

Municipio de Fresnillo

Fresnillo de Gonzalez Encheverria: Widespread. Crystals

to 4 cm. Associated with pyrite, marcasite, quartz, native silver, pyrargyrite, and polybasite.

Municipio de General Francisco Murguía

Nieves: Moderately distributed. Most noted from the Santa Rita, Rosario, and San Francisco mines. Crystals to 4 cm. Associated with jamesonite, pyrite, stibnite, and barite.

Municipio de Sombrerete

Sombrerete: Moderately distributed. Crystals to 4 cm. Associated with pyrite, proustite, pyrargyrite, and native silver.

Municipio de Zacatecas

Zacatecas: Widespread throughout district. Crystals to 5 cm. Associated with chalcopyrite, native silver, and native copper.

CALCIUM TELLURITE. See **CARLFRIESITE**

CALOMEL

Hg_2Cl_2. A rare secondary mineral forming from the alteration of other mercury minerals.

GUANAJUATO

Municipio de San José Iturbide

San José Iturbide: Limited distribution. Associated with cinnabar.

Municipio de San Luis de La Paz

Mineral de Pozos: Limited distribution. Associated with cinnabar.

GUERRERO

Municipio de Taxco

Hauhauxtla: Limited distribution. Most noted from the San Luis Mine. Crystals to 4 mm. Associated with metacinnabar, eglestonite, and montroydite.

QUERETARO

Municipio de Cedereyta

El Doctor: Limited distribution. Most noted from the San Juan Nepomuceno Mine.

SAN LUIS POTOSI

Municipio de Guadalcázar

Guadalcazár: Limited distribution.

Municipio de Moctezuma

Moctezuma: Limited distribution. Most noted from the Los Dulces Nombres Mine. Associated with cinnabar.

ZACATECAS

Municipio de Fresnillo

Plateros: Limited distribution. Crystals to 1 cm. Associated with cinnabar.

CANFIELDITE

Ag_8SnS_6. A rare sulfosalt mineral found in silver/tin deposits.

GUANAJUATO

Municipio de Guanajuato

Guanajuato: Limited distribution on the Veta Madre.

CARBONATE-HYDROXYLAPATITE

$Ca_5(PO_4,CO_3)_3(OH)$. An uncommon mineral and member of the apatite mineral group found in few mineral deposits.

DURANGO

Municipio de Durango

Victoria de Durango: Limited distribution. Cerro de Mercado Mine.

CARLFRIESITE

$CaTe_2TeO_8$. A rare mineral found in tellurium deposits.

SONORA

Municipio de Moctezuma

Cumobabi: Limited distribution. San Judas Mine. Crystals to 2 mm. Associated with poughite, ourayite, mendozavilite, and paramendozavilite. [This location is in the old La Verde Mining district.]

Moctezuma: Limited distribution. Most noted from the Bambollita, Moctezuma, and La Oriental mines. Associated with hessite and choloaite. [This is the type location for this mineral.]

CARMINITE

$PbFe_2(AsO_4)_2(OH)_2$. A rare mineral found in arsenic-rich lead deposits.

DURANGO

Municipio de Mapimí

Mapimí: Moderate distribution. Most noted from the Ojuela and San Juan mines. Red crystals to 1 cm. Associated with scorodite, dussertite, mimetite, cerussite, anglesite, wulfenite, and plumbojarosite. [In the early 1980s, a pocket of of crystallized calcite, which was covered with crystals of carminite.]

SONORA

Municipio de Trincheras

La Mur: Limited distribution. Most noted from the Los Animas Mine. Southwest of Benjamin Hill. Red crystals to 2cm. Associated with arseniosiderite, beudantite, and, akaganeite. [The crystals were laying flat on the matrix and were not free standing.]

Municipio de Caborca

El Antimoneo: Limited distribution. Most noted from the San Felix and El Antimoneo mines. Red crystals to 4 mm. Associated with chlorargyrite, quartz, corkite, and beudantite.

CARNOTITE

$K_2(UO_2)_2(VO_4)_2 \cdot 3H_2O$. A secondary mineral formed by the actions of meteoric waters on other uranium and vanadium minerals.

CHIHUAHUA

Municipio de Aldama

Sierra Peña Blanca: Moderately distributed. Most noted from the Nopal one, two, and three and Margaritas one and two mines. Associated with margaritasite, weeksite, and uraninite.

Municipio de Manuel Benavides

Lajitas: Limited distribution. El Sotolar Mine.

CARPHOSIDERITE. See
HYDRONIUM JAROSITE

CASSITERITE

SnO_2. An uncommon mineral found
in high temperature veins and
pyrometasomatic deposits in
igneous rocks.

AGUASCALIENTES

Municipio de Asientos

Cerro del Chiquihuitillo: Limited
distribution.

Municipio de Calvillo

Calvillo: Limited distribution.

Municipio de Tepezalá

Cerro San Juan: Moderately
distributed on Cerros San Juan,
San Miguel, and La Lega.

CHIHUAHUA

Municipio de Aquiles Serdán

San Antonio el Grande: Limited
distribution. San Antonio Mine's
tin chimneys and Dolores fissures. Associated with topaz,
quartz, hematite, and fluorite.

DURANGO

Municipio de Coneto de
Comonfort

América: Widespread throughout
the Sierra de San Francisco,
Arroyo de Grant. Most noted from
the Grant and Varocitas mines.
Massive (wood tin) nodules to 12
cm. [This is the América-Sapioris
(Potrillos) mining district.]

Amerila: Widespread throughout
the Sierra de San Francisco. Most
noted from the Twentyninth of
June Mine. Massive (wood tin).
Associated with hematite. [This
location is in the América-Sap-

ioris (Potrillos) mining district.]

Potrillos: Widespread throughout
the Sierra de San Francisco.
Noted from the Orozco Mine.
Massive (wood tin). [This
location is in the América-Sapioris (Potrillos) mining district.]

Sapioris: Widespread. Most noted
from the El Pipiano Mine. Massive
(wood tin). [This location is
in the América-Sapioris mining
district.]

Municipio de Durango

Cacaria: Moderately distributed
within the Sierra de Cacaria.
Most noted from the La Martao
mine. Massive (wood tin).
Associated with hematite.
[This location is northwest of
the capitol, Victoria de Durango.]

Victoria de Durango: Widespread throughout the Cerro de
Los Remedios. [This mining
district is now buried beneath
the southwest part of city of
Victoria Durango.]

Municipio de Mapimí

Mapimí: Limited distribution.
Most noted from the Monterrey
Mine. Acicular crystals to 4 mm.
Associated with magnetite and
pyrrhotite.

Municipio de Mezquital

Mezquital: Moderately distributed
in the Sierra de Michu. Massive
(wood tin). [This location is
east of Mezquital.]

Municipio de Topia

Topia: Moderately distributed.

Municipio de Poanas

Cerro de Sacrificios: Moderately
distributed.

GUANAJUATO

Municipio de Dolores
Hidalgo

Dolores Hidalgo: Limited distributed in the Arroyo de Salta Pena

Colorado. Massive (wood tin) nodules. [This location is 24 km from Dolores Hidalgo.]

Municipio de Guanajuato

Santa Rosa: Moderately distributed in the Sierra de Santa Rosa. Massive. (Wood tin).

Campuzano: Moderately distributed within the Arroyo del Aguilar. (Rancho de Campuzano)

Municipio de San Felipe

San Felipe: Moderately distributed.

Municipio de San Miguel Allende

Cerro de la Mina: Moderately distributed. Massive (wood tin) nodules. Associated with hematite.

Municipio de Santa Catarina

Santa Catarina: Moderately distributed on the Cerro de Las Fajas. Most noted from the Santin Mine.

Municipio de Tierra Blanca

Penital: Moderately distributed. Santin Mine.

Tierra Blanca: Moderately distributed in the Arroyo de Carrizal. Massive (wood tin).

GUERRERO

Municipio de Taxco

Taxco de Alcarón: Limited distribution.

JALISCO

Municipio de Teocaltiche

Paso de Sotos: Moderate distribution.

MICHOACAN

Municipio unknown

Tepuxtepec: Limited distribution. Cubires Mine. Massive.

QUERETARO

Municipio de La Cañada

Hacienda de Chichimequillas: Moderate distribution.

Municipio de Tolimán

Panales: Moderate distribution.

Municipio unknown

Tlachiquera: Moderately distributed. Massive (wood tin). Associated with hematite.

SAN LUIS POTOSI

Municipio de Guadalcázar

Guadalcázar: Moderately distributed on the San Nicolás claim. Noted from the Santa Elena Mine.

Municipio de Santa María del Oro

Santa María del Oro: Widespread. Massive (wood tin) nodules to 5 cm, white, tan, brown, and black.

Municipio unknown

Villa Arriago: Moderate distribution. Noted from the El Tocho Mine. Associated with tridymite.

ZACATECAS

Municipio de Mazapil

El Aguila: Widespread on the Sierra Chapultepec. Most noted from the Leona and Tulita mines.

Municipio de Pinos

Pinos: Moderately distributed in the Arroyo de San Juan de los Herreros.

CELADONITE

$K(Mg,Fe)(Fe,Al)Si_4O_{10}(OH)_2$. A member of the mica mineral group which forms in basaltic rocks.

BAJA CALIFORNIA SUR

Municipio de Mulegé

Santa Rosalía: Moderately distributed throughout the Boleo mining district.

CHIHUAHUA

Municipio de Aquiles Serdán

Francisco Portillo: Moderately distributed throughout the west camp of the Santa Eulalia Mining district.

San Antonio el Grande: Moderately distributed throughout the east camp of the Santa Eulalia Mining district. Most noted from the San Antonio Mine. Associated with fluorite, hedenbergite, and epidote.

Municipio de Meoqui

Moderately distributed in the Cerro del Chanate.

CELESTITE

$SrSO_4$. An uncommon mineral and member of the barite group forming in sedimentary rocks and as a gangue mineral in lead-zinc ores.

BAJA CALIFORNIA SUR

Municipio de Mulegé

Santa Rosalía: Moderately distributed in the Boleo Mining district.

A 25 cm × 15 cm crystal cluster of celestite from Ramos Arispe, Coahuila from the Romero Mineralogical Museum, Tehuacan, Puebla.

CHIHUAHUA

Municipio de Ahumada

Constitución: Moderately distributed. Mojina Mine. Light blue crystals to 2 cm. Associated with knobby masses of calcite.

Municipio de Aquiles Serdán

Francisco Portillo: Limited distribution within the west camp of the Santa Eulalia mining district.

Municipio de Saucillo

Naica: Limited distribution. Light blue crystals to 2 cm.

COAHUILLA

Municipio de Matamoras

Laguna Matamoras: Moderate distribution. Blue crystals to 6 cm. [Crystals from here have been found in large groups.]

Municipio de Melchor Muzquiz

Melchor Muzquiz: Moderately distributed. Most noted from the El Tule Mine. Colorless to white crystals to 30 cm. Associated with fluorite. [The crystals found in the clay zones in the mine are often doubly terminated.]

Municipio de Ramos Arispe

Ramos Arispe: Moderately distributed. Most noted from the San Augustín Mine. Pale blue crystals to 12 cm. [Crystals have been found in large groups up to 60 cm in diameter.]

Municipio de Sierra Mojada

Sierra Mojada: Moderate distribution.

DURANGO

Municipio de Gómez Palacio

Dinamita: Moderately distributed in the El Vergel stock. Pale blue crystals to 15 cm. Associated with calcite, gypsum, and sulfur. [Specimens from here have made their way to the collector's marketplace labeled as Mapimi, which is 19 km northwest.]

Municipio de Mapimí

Mapimí: Moderately distributed. Ojuela upper levels and the El Padre (La Esperanza) mines . Colorless to light blue crystals to 2 cm. Associated with fluorite, barite, and calcite. [At the El Padre or La Esperanza mine it is found as water clear crystals on limestone.]

GUANAJUATO

Municipio de Guanajuato

Guanajuato: Moderately distributed. Associated with calcite.

SAN LUIS POTOSI

Municipio de Villa de La Paz

La Paz: Moderately distributed. Most noted from the La Luz Mine. Light blue blocky crystals to 15 cm. [The large blocky crystals were found lining watercourses and were removed in large plates. The crystals averaged about 15 cm in length, but the miners talk about finding crystals over 30 cm

long. This location has been listed as Matehuala, which is 10 km east of La Paz.]

TLAXCALA

Municipio de Zacatelco

Inmediaciones de Tepeyanco: Moderately distributed.

CERIANITE

$(Ce,Th)O_2$. A rare mineral found in few mineral deposits.

DURANGO

Municipio de Mapimí

Sierra de Bermjillo: Limited distribution in the Canon de Colorado. Associated with kaolinite.

CERUSSITE

$PbCO_3$. A common secondary mineral and member of the aragonite group, which is formed by the reaction of carbonated waters on lead minerals in the upper ore zone.

AGUASCALIENTES

Municipio de Asientos

Asientos: Widespread. Most noted from the El Orito Mine. Associated with goethite, malachite, and azurite.

BAJA CALIFORNIA SUR

Municipio de Mulegé

Santa Rosalía: Widespread throughout the Boleo Mining district. Most noted from the Amelia Mine. Associated with boleite.

Municipio de San Antonio

El Triunfo: Widespread. Most noted from the Gobernadore, Mina Rica, San Joaquín, and San Antonio mines.

San Antonio: Widespread. Most noted from the Mexicana, Rosario, Enriqueta, Piacho, Comstock, Rancheria, and Guadalupe mines.

CHIHUAHUA

Municipio de Ahumada

Los Lamentos: Moderately distributed. Erupción/Ahumada Mine. Crystals to 1 cm. Associated with goethite and calcite. [This mineral is often found here as psudomorphs after calcite.]

Municipio de Aldama

Aldama: Moderately distributed. Most noted from the Aurora Mine.

Municipio de Aquiles Serdán

Francisco Portillo: Moderately distributed in the west camp of the Santa Eulalia mining district. Most noted from the oxidized zones of the Velardeña, El Potosí, Santo Domingo, and Buena Tierra mines. White crystals to 1 cm. Associated with calcite, hemimorphite, and goethite.

San Antonio el Grande: Moderately distributed in the east camp of the Santa Eulalia mining district. Most noted from the San Antonio mine's oxide zone. Associated with hemimorphite, massicot, calcite, and goethite.

Municipio de Camargo

Sierra de las Encinillas: Widespread. Most noted from the San Francisquito, El Carmen, San Antonio, Santa Rita, San Vicente, San Felipe, El Convento, and Pajaritos mines. Associated with goethite.

Municipio de Coyame

Cuchillo Parado: Widespread.

Municipio de Cusihuiriáchic

Cusihuiriáchic: Moderately distributed. Most noted from the Soledad and La Reina mines. Associated with galena.

Municipio de San Francisco del Oro

San Francisco del Oro: Widespread.

Municipio de Santa Bárbara

Santa Bárbara: Widespread.

Municipio de Saucillo

Naica: Widespread throughout the mines of the Sierra de Naica.

COAHUILA

Municipio de Matamores

Jimulco: Moderately distributed.

Municipio de Múzquiz

La Entantada: Moderately distributed. Most noted from the La Encantada Mine.

Melchor Múzquiz: Moderately distributed. Most noted from the La Salvadora Mine.

Municipio de Sierra Mojada

Sierra Mojada: Moderately distributed. Most noted from the La Encantada and Esmeralda mines.

DURANGO

Municipio de Indé

Indé: Widespread. Most noted from the Caballo, Desmasia, and Portillo mines.

Municipio de Mapimí

Buruguilla: Moderately distributed. Most noted from the Descubridora and La Candelaria mines. [This location is often referred to as Hornillas because of its location on the Rancho de Hornillas.]

Mapimí: Moderately distributed. Most noted from at the Ojuela Mine's oxidized ore zone. Crystals to 2 cm. Associated with goethite, mimetite, plumbojarosite, wulfenite, massicot, rosasite, plattnerite, nantockite, malachite, and aurichalcite. [This mineral is more common massive, but twinned crystals both as reticulated clusters and as individual twinned crystals have been found. Cerussite pseudomorphs after anglesite and what appears to be hemimorphite have been found at the Ojuela Mine.]

Municipio de San Luis del Cordero

San Luis del Coddero: Moderately distributed. Noted from the San Antonio Mine.

Municipio de San Perdo del Gallo

Peñoles: Moderately distributed. Most noted from the San Antonio and San Salvador mines.

GUNAJUATO

Municipio de Guanajuato

Guanajuato: Moderately distributed within the Veta Madre.

Municipio de San Luis de la Paz

Mineral de Pozos: Widespread. Most noted from the Santa Brígida and San Pedro mines.

Municipio de San Miguel de Allende

Mineral de Puerto de Nieto: Moderately distributed. Noted from the La Argentina Mine. Associated with goethite.

HIDALGO

Municipio de Jacala

Jacala: Widespread. Noted from the El Oro, La Prieta, Plomosa,

Los Ingleses, Jesús, Estaca, Presa, and La Milpa mines. Associated with goethite.

Municipio de Zimapán

Zimapán: Widespread. Noted from the Miguel Hidalgo, Lomo del Toro, Santa Rita, and Guadalupe mines. Crystals to 4 cm. Associated with goethite and plumbojarosite. [At the Lomo del Toro Mine the crystals occurred as 4 and 6 ray stars stellate twins to 4 cm in length. The Miguel Hidalgo mine produced cerussite was found often as pseudomorphs after plumbojarosite, but more commonly as fine needle crystals.]

JALISCO

Municipio de Bolaños

Bolaños: Moderately distributed. Noted from the San José, Pichardo, Santa Fe, and Iguana Mines. Associated with native silver and calcite.

MEXICO

Municipio de Tlatlaya

Rancho de los Ocotes: Moderately distributed. Noted from the La Fama Mine.

Municipio de Zacualpan

Zacualpan: Moderately distributed. Noted from the Socavón de Dios nos Guie Mine. Associated with goethite.

QUERETARO

Municipio de Cadereyta

El Doctor: Moderately distributed. Associated with goethite.

SAN LUIS POTOSI

Municipio de Catorce

Catorce: Widespread throughout district. Noted from the Concepción and San Pedro mines. Associated with goethite.

Municipio de Cerro de San Pedro

Cerro de San Pedro: Moderately distributed. Noted from the Juárez and Begona mines. Associated with goethite.

Municipio de Guadalcázar

Guadalcázar: Moderately distributed.

Municipio de Villa de La Paz

La Paz: Moderately distributed throughout district. Noted from the La Paz, Providencia, and Plomosa mines. Associated with goethite.

SINOLOA

Municipio de Cosalá
Cosalá: Moderately distributed. Noted from the Nuestra Senora Mine.

SONORA

Municipio de Alamos

Alamos: Moderately distributed throughout district. Noted from the Los Chinos and Caballo Muerto mines.

Municipio de Cucurpe

Cucurpe: Limited distribution in Cerro Prieto. Noted from the San Francisco Mine. White to clear crystals to 4 cm. Associated with goethite, wulfenite, and mimetite. [Crystals are found here as reticulated masses and as single crystals.]

Municipio de Hermosillo

Hermosillo: Widespread throughout district. Noted at the San Javier and Santa Ana mines.

VERACRUZ

Municipio de Las Minas

Zomelahuacán: Moderately distributed. Noted from the Zomelahuacán Mine.

ZACATECAS

Municipio de Concepción del Oro

Aranzazú: Moderately distributed. Noted from the Aranzazú Mine.

Municipio de Fresnillo

Fresnillo de Gonzalez Echeverria: Moderately distributed. Noted from the San Ricardo mine.

Municipio de Mazapil

Mazapil: Widespread. Most noted from the San Carlos, Dulces Nombres, and Santa Rosa mines. Associated with goethite, azurite, and malachite.

Municipio Ojo Caliente

Ojo Caliente: Moderately distributed. Most noted at the Santa Rita, San Francisco, and Trinidad mines.

Municipio de Zacatecas

Zacatecas: Moderately distributed. Associated with galena.

CERVANTITE

$SbSbO_4$. A secondary mineral formed by the oxidation of other antimony minerals.

BAJA CALIFORNIA

Municipio de Tecate

Rosa de Castillo: Limited distribution. Noted from the San Felipe mine.

BAJA CALIFORNIA SUR

Municipio de San Antonio

El Truifo: Moderately distributed. Noted from the San Antonio and Gobernadora mines.

DURANGO

Municipio de Cuencamé

Cuencamé de Ceniceros: Limited distribution within the Berrendos mining district. Associated with stibnite and stibiconite.

Municipio de Mapimí

Mapimí: Limited distribution. Noted from the San Ilario, América Dos, and Ojuela mines. Massive. Associated with stibiconite and bindheimite.

Municipio de Pánuco de Coronado

Pánuco de Coronado: Limited distribution. Casualidad Mine.

GUERRERO

Municipio de Huitzuco

Huitzuco de los Figueroa: Limited distribution. note from the La Cruz and Tumbasa mines. Massive. Associated with stibnite and stibiconite.

Municipio de Tetipac

Pipichahuasco: Limited distribution.

MEXICO

Municipio de Zacualpan

Zacualpan: Limited distribution. Noted from the El Carmen and Coronas mines. Massive.

OAXACA

Municiapio de San Juan Mixtepec-Juxtlahuaca

San Jaun Mixtepec: Limited distribution. Noted from the

Matutina mine. Associated with stibnite and stibiconite.

QUERETARO

Municipio de Tolimán

Soyatal: Moderate distribution on Cerro Pinguicas. Noted from the San Antonio and Santa María de Miera mines. Associated with stibiconite, semarmontite, valentinite, gypsum, cinnabar, alunite, and varisite. [This antimony district is located 10 km north of Motoxi. This mineral has been found as pseudomorphs after stibnite at the San Antonio Mine. At the Santa María de Miera Mine cervantite has been found associated with cinnabar, alunite, gypsum, and varisite.]

SAN LUIS POTOSI

Municipio de Catorce

Wadley: Moderately distributed. Noted from the San José Mine. Associated with cinnabar and calcite. [The cervantite is usually pseudomorphic after stibnite.]

Municipio de Gaudalcázar

Guadalcázar: Limited distribution. Noted from the La Providencia Mine.

SONORA

Municipio de Caborca

El Antimonio: Moderately distributed. Noted from the La Fortuna Mine.

Municipio de Naco

La Morita: Limited distribution.

Municipio de Moctezuma

Moctezuma: Limited distribution. Associated with stibiconite and allemontite.

Municipio de Trincheras

El Mur: Limited distribution. Los Animas Mine. Associated with carminite, argenosiderite, and beudantite. [This location is often referred to as Benjamín Hill, which is 40 km east of the mine.]

ZACATECAS

Municipio de Mazapil

Sierra de Santa Rosa: Limited distribution. Los Gallos Mine.

CESBRONITE

$Cu_5(TeO_3)_2(OH)_6 \cdot 2H_2O$. A rare mineral found in few mineral deposits.

SONORA

Municipio de Moctezuma

Moctezuma: Limited distribution. La Oriental Mine. [This is the type location for this mineral.]

CHABAZITE

$CaAl_2Si_4O_{12} \cdot 6H_2O$. A member of the zeolite mineral group which forms from the alteration of volcanic glass in tuffs.

GUANAJUATO

Municipio de Guanajuato

Guanajuato: Moderately distributed throughout the Veta Madre. Associated with huelandite and laumontite.

HIDALGO

Municipio de Huasca de Ocampo

San Miguel Regla: Moderately distributed in the Barranca de Regla.

Municipio de Mineral del Chico

Mineral del Chico: Limited distribution. Noted from the Tetitlán mine.

OAXACA

Municipio de la Pe

La Planchita: Limited distribution. La Planchita Mine. Associated with huelandite.

CHALCANTHITE

$CuSO_4 \cdot 5H_2O$. A water soluble secondary mineral found in the oxidized zone of copper deposits. decomposes in dry atmosphere.

BAJA CALIFORNIA

Municipio de Ensenada

San Fernando: Limited distribution. Noted from the Chalmita Mine.

CHIHUAHUA

Municipio de Aquiles Serdán

San Antonio el Grande: Limited distribution. San Antonio mine's 8th level. Associated with chalcopyrite, sphalerite, and pyrite. [This is the transition zone between the oxide and sulfide ore zones.]

Municipio de Batopilas

Batopilas: Limited distribution. Noted from the Santo Niño and Nopalera mines.

Municipio de Rosario

Valle Los Olivos: Limited distribution. Tuffs to 15 mm.

COAHUILA

Municipio de Candela

Panuco: Moderately distributed. Noted from the Pánuco Mine.

DURANGO

Municipio de Mapimí

Mapimí: Moderately distributed. Noted from the Ojuela Mine. Massive.

GUANAJUATO

Municipio de San Miguel de Allende

Mineral de Puerto de Nieto: Limited distribution. Noted from the La Argentina Mine. Massive. Associated with goslarite and melanterite.

HIDALGO

Municipio de Zimapán

Zimapán: Limited distribution. Noted from the Lomo del Toro Mine's Los Bronces winze.

SONORA

Municipio de Nacozari de García

La Caridad: Moderate distribution. La Caridad Mine. Tuffs to 1 cm.

Nacozari de García: Moderate distribution throughout district. Stalactites to 10cm.

ZACATECAS

Municipio de Concepción del Oro

Concepción del Oro: Limited distribution within the Arroya de Cata.

Municipio de Pinos

Pinos: Moderate distribution. Massive.

CHALCOCITE

Cu_2S. An uncommon primary mineral, but also can be found as a secondary mineral in copper deposits.

AGUASCALIENTES

Municipio de Asientos

Asientos: Widespread throughout district. Massive.

Municipio de Tepezalá

Tepezalá: Widespread. Massive.

BAJA CALIFORNIA

Municipio de Ensenada

San Fernando: Widespread throughout district. Noted from the Chalcocita, San Fernando, and Adela mines. Massive. Associated with chalcopyrite, cuprite, and bornite.

BAJA CALIFORNIA SUR

Municipio de Mulegé

Santa Rosalía: Widespread throughout the Boleo copper mining district. Massive. Associated with chalcopyrite.

CHIHUAHUA

Municipio de Aquiles Serdán

Francisco Portillo: Moderately distributed throughout the west camp of the Santa Eulalia mining district. Massive.

San Antonio el Grande: Moderately distributed throughout the east camp of the Santa Eulalia mining district. Noted from the San

Antonio Mine. Associated with chalcopyrite and pyrite.

Municipio de Coyame

Cuchillo Parado: Moderately distributed.

Municipio de Cusihuiriáchic

Cusihuiriáchic: Moderately distributed. Noted from the Princesa Mine. Associated with chalcopyrite.

Municipio de Hidalgo del Parral

Hidalgo del Parral: Moderately distributed throughout the district.

COAHUILA

Municipio de Candela

Pánuco: Moderately distributed. Noted from the Panuco Mine. Massive.

Municipio de Sierra Mojada

Sierra Mojada: Moderately distributed. Noted from Porvenir and San Juan mines. Massive.

DURANGO

Municipio de Cuencamé

Velardeña: Widespread. Noted from the Velardeña and Santa María mines. Massive.

Municipio de Mapimí

Buruguilla: Widespread throughout the Descubridora mining district. Most noted from the Descubridora Mine. Massive.

Mapimí: Limited distribution. Noted from the America Dos and Ojuela mines. Massive.

Municipio del Oro

Magistral: Moderately distributed. Noted from the Porvenir mine. Massive. Associated with chalcopyrite, gold, and silver.

Municipio de Poanas

Cerro del Sacrificio: Moderately distributed. Noted from the La Candelaria, La Purísima, and El Desengano mines. Massive.

GUANAJUATO

Municipio de Guanajuato

Guanajuato: Moderately distributed throughout the Veta Madre. Massive.

Municipio de San Luis de la Paz

Mineral de Pozos: Moderately distributed. Noted from the Santa Brídgida Mine. Massive.

Municipio de San Miguel de Allende

Mineral de Puerto de Nieto: Moderately distributed. Noted from the La Argentina Mine. Massive. Associated with chalcopyrite, calcite, pyrite, and galena.

GUERRERO

Municipio de Huitzuco

Huitzuco de Figueroa: Limited distribution.

Municipio de Taxco

Taxco de Alcarón: Moderately distributed. Massive.

HIDALGO

Municipio de Zimapán

Bonanza: Limited distribution. Noted from the Bonanza Mine.

Zimapán: Limited distribution.
Guadalupe Mine.

JALISCO

Municipio de Bolaños

Bolaños: Moderate distribution.
Noted from the Zuloaga Mine.

MEXICO

Municipio de Tlatlaya

Rancho de los Ocotes: Moderate
distribution. Noted from the La
Fama Mine.

Municipio de Zacualpan

Zacualpan: Moderately distributed. Noted from the Coetzillos
Mine.

MICHOACAN

Municipio de Angangueo

Mineral de Angangueo: Moderately
distributed.

Municipio de La Huacana

Inguaran: Moderate distribution.
Massive.

Municipio de Santa Clara

Opopeo: Moderate distribution.
Noted from the La China Mine.

MORELOS

Municipio de Tlaquiltenango

Huautla: Noted from the Santa Ana
Mine.

NUEVO LEON

Municipio de Cerralvo

Sierra de Cerralvo: Moderately
distributed. El Carmen Mine.

PUEBLA

Municipio de Chietla

Rancho de la Cofradia: Moderately
distributed.

Municipio de Chinantla

Cerro del Guajolote: Moderately
distributed.

Municipio de Zautla

Santiago Zautla: Moderately
distributed. Noted from the
Zautla Mine.

SAN LUIS POTOSI

Municipio de Charcas

Charcas: Widespread throughout
district. Massive.

Municipio de Villa de la Paz

La Paz: Moderately distributed.
Noted from the Dolores Mine.
Massive.

Municipio de Villa de Ramos

Ramos: Moderately distributed.
Noted from the Cocinera Mine.
Massive.

SINALOA

Municipio de Badiraguato

Soyatita: Widespread. Most noted
from the Dulces Nombres, El
Salto, El Tabor, San Antonio, San
José, San José de Ledesma, and
Tres Reyes mines. Massive.

Municipio de Cosalá

Cosalá: Moderately distributed.
Most noted from the Nuestra
Señora Mine. Massive.

Municipio de San Ignacio

San Ignacio: Moderately distributed in the Veta Hondo on the
Cerro del Camaron. Massive.

SONORA

Municipio de Alamos

Alamos: Moderately distributed.
Noted from the Veta Grande and
Espiritu Santo mines. Massive.
Associated with native silver and
chlorargyrite.

Municipio de Cananea

Cananea: Widely distributed.
Crystals to 1 cm.

Municipio de Nacozari
de García

La Caridad: Widespread. Noted
from the La Caridad Mine.
Massive. Associated with chalcopyrite, covellite, digenite,
and pyrite.

TAMPAULIPAS

Municipio de Villagrán

San José: Limited distribution on
the Rancho de San Pedro.

VERACRUZ

Municipio de Las MInas

Zomelahuacan: Moderate distribution. Most noted from the Santa
Cruz Mine.

ZACATECAS

Municipio de Concepción del
Oro

Concepción del Oro: Widespread.
Massive.

Municipio de Fresnillo

Fresnillo de Gonzalez Echeverria:
Limited distribution. Noted from
the La Purísima Mine.

Municipio de Jalpa

Jalpa: Limited distribution.
Noted from the La Leona Mine.

Municipio de Mazapil

Mazapil: Widely distributed
throughout the district. Massive.

Municipio de Noria de Los
Angeles

Noria de Los Angeles: Moderately
distributed. Noted from the San
Antonio and San Francisco
mines.

Municipio de Pinos

Pinos: Moderately distributed
throughout district. Massive.

Municipio de Sombrerete

Sombrerete: Moderately distributed. Massive.

CHALCOMENITE

$CuSeO_3 \cdot 2H_2O$. A rare mineral found
in few metal deposits.

SONORA

Municipio de Moctezuma

Moctezuma: Limited distribution.
Noted from the Moctezuma Mine.
Associated with tellurium,
selenium, klockamnnite, and
bambollaite. [This is the type
location for this mineral.]

CHALCOPHANITE

$(Zn,Fe,Mn)Mn_3O_7 \cdot 3H_2O$. A rare
mineral formed as an alteration
product of other zinc and
manganese mineral.

DURANGO

Municipio de Mapimí

Mapimí: Limited distribution.
Most noted from the Ojuela Mine's

Esperanza stope. Thin velvetlike grayish black crust. Associated with adamite, calcite, goethite, and smithsonite.

CHALCOPYRITE

$CuFeS_2$. A common mineral forming in late magmatic environments either as disseminated or metalliferous veins of hydrothermal nature. Also formed in metamorphic or sedimentary environments.

AGUASCALIENTES

Municipio de Asientos

Asientos: Widespread throughout district. Noted from the El Orito, Merced, and San Francisco mines. Massive.

El Cobre: Moderately distributed. Noted from the San Bartolo Mine. Crystals to 1 cm.

Municipio de Tepezalá

Tepezalá: Widely distributed.

BAJA CALIFORNIA

Municipio de Ensenada

Real del Castillo: Moderately distributed. Most noted from the Chalcocita and San Fernando mines. Massive.

Municipio de Tijuana

Tijuana: Moderately distributed. El Swino Mine. Massive. [This location is 28 km south of Tijuana.]

BAJA CALIFORNIA SUR

Municipio de San Antonio

El Triunfo: Widespread. Noted from the San Antonio, Altar, San José, and Humboldt mines. Massive.

Municipio de Mulegé

Santa Rosalía: Widespread throughout the Boleo Mining district. Massive. Associated with chalcocite.

CHIAPAS

Municipio de Pichucalco

Pichucalco: Moderately distributed. Most noted from the Santa Fe Mine.

CHIHUAHUA

Municipio de Aquiles Serdán

Aquiles Serdán: Moderately distributed throughout the west camp of the Santa Eulalia district. Noted from the El Potosí and Buena Tierra mines. Crystals to 1 cm. Associated with pyrite, calcite, galena, and sphalerite.

San Antonio el Grande: Moderately distributed throughout the east camp of the Santa Eulalia Mining district. Noted from the San Antonio Mine. Crystals to 15 mm. Associated with pyrite, galena, calcite, and sphalerite.

Municipio de Batopilas

La Bufa: Moderately distributed. Massive.

Municipio de Chihuahua

Labor de Terrazas: Widespread. Most noted from the Rio Tinto Mine. Massive.

Municipio de Camargo

Sierra de Encinillas: Moderately distributed. Most noted from the Dolores, El Carmen, and Sirena mines.

Municipio de Cusihuiriáchic

Cusihuiriáchic: Moderately distributed. Noted from the Le Reina and San Nicolasito mines. Massive.

Municipio de Manuel Bena-
vides

San Carlos: Widespread. Noted
from the San Carlos Mine.
Massive.

Municipio de Nuevo Casas
Grandes

San Pedro Corralitos: Moderately
distributed. Noted from the
Candelaria Mine.

Municipio de Santa Bárbara

Santa Bárbara: Widespread. Noted
from the Guadalupe and Los
Canarios mine. Massive.

Municipio de Saucillo

Naica: Widespread throughout the
district. Crystals to 1 cm.
Associated with calcite, sphalerite, galena, and fluorite.

Municipio de Urique

Cerocahui: Widespread. Noted from
the Cerocahui Mine. Massive.
Associated with galena, sphalerite, and pyrite.

Piedras Verdes: Widespread. Most
noted from the Piedras Verdes, El
Truyo, and Trillo mines. Massive.

COAHUILA

Municipio de Candela

Pánuco: Moderately distributed.
Noted from the Pánuco Mine.
Massive.

DURANGO

Municipio de Cuencamé

Velardeña: Widespread throughout
district. Noted from the La
Adelfa, La Chona, Velardeña, and
Reina del Cobre mines. Crystals
to 1 cm. Associated with pyrite,
chalcocite, and quartz. Moderately abundant.

Municipio de Guanaceví

Guanaceví: Widespread throughout
district. Massive.

Municipio de Indé

Indé: Moderaetly distributed.
Most noted from the Union,
Constancia, and Matracal mines.
Massive.

Municipio de Mapimí

Buruguilla: Moderately distributed. Noted from the La Descubridora Mine. Massive.

Mapimí: Widespread. Most noted
from the Ojuela Mine. Massive.
Associated with pyrite, galena,
sphalerite, enargite, arsenopyrite, tetrahedrite, and marcasite. [Chalcopyrite is the most
important primary copper mineral
at Mapimí.]

Municipio del Oro

Magistral: Moderately distributed. Noted from the Porvenir
Mine. Massive. Associated with
native silver, gold, and chalcocite.

Municipio de Pánuco de
Coronado

Pánuco de Coronado: Moderately
distributed. Noted from the Avino
Mine. Massive.

Municipio de San Dimas

Tayoltita: Widespread. Noted from
the Candelaria, La Montañesa,
Concepción, Castellana, San
Vincente, and Tayoltita mines.
Massive.

Municipio de San Pedro del
Gallo

Peñoles: Moderately distributed.
Noted from the Jesús María Mine.
Massive.

Municipio de Trinidad

Trinidad: Moderately distributed.

Municipio de Topia

Topia: Widespread throughout the district. Noted from the El Toro, Palmira, and the San José del Tigre mines. Massive.

Municipio de Poanas

Poanas: Moderate distribution. Noted at the La Condelaria and La Purísima mines. Massive.

GUANAJUATO

Municipio de Guanajuato

Guanajuato: Moderately distributed. Noted from the Rayas and Valenciana mines. Massive. Associated with pyrite, quartz, acanthite, and native silver.

La Luz: Moderately distributed. Most noted from the La Luz and San Pedro mines. Massive. Associated with pyrite, acanthite, and quartz.

Municipio de San Luis de la Paz

Mineral de Pozos: Moderately distributed. Noted from the Santa Brídgida, Trinidad, and La Joya mines. Massive.

Municipio de San Miguel de Allende

Mineral de Puerto Nieto:

Moderately distributed. Noted from the San Antonio, Guadalupe, and La Argentina mines. Massive. Associated with calcite, pyrite, galena, and chalcocite.

GUERRERO

Municipio de Taxco

Taxco de Alcarón: Moderately distributed. Most noted from the Agua Bendita and Espiritu Santo mines. Massive.

HIDALGO

Municipio de Mineral del Chico

Mineral del Chico: Moderate distribution. Noted from the San Marcial, San Nicolas, and Arevalo mines. Massive.

Municipio de Mineral de Monte

Mineral del Monte: Widely distributed throughout district. Massive. [This location is also known as Real del Monte.]

Municipio de Pachuca

Pachuca de Soto: Widely distributed throughout the district. Note from the Barron and El Cristo mines. Massive.

Municipio de Zacualtipan

Zacualtipan: Moderate distribution. Noted from the Chacuaco Mine. Massive.

Municipio de Zimapán

Zimapán: Moderately distributed througout district. Noted from the Concordia Mine. Massive.

JALISCO

Municipio de Exatlán

Exatlán: Moderate distribution. Noted from the Regeneración Mine.

Municipio de Tonila

Pihuamo: Moderately distributed. Noted from the El Agostadero Mine. Massive.

MEXICO

Municipio de Tlatlaya

Rancho Los Ocotes: Widely distributed. Noted from the La

Fama and La Pinta mines.
Massive.

Municipio de Zacualpan

Zacualpan: Widespread. Noted from the Carboncillo, Coronas, Cuchara, Guadalupe y San Juan Bautista, and El Toro mines. Massive.

MICHOACAN

Municipio de Angangueo

Mineral de Angangueo: Widespread. Noted from the Carrillos, El Carmen, and San Cristobal mines. Massive. Common.

Municipio de Huacana

Inguaran: Widely distributed. Crystals to 5 cm.

Oropeo: Moderate distribution. Noted from the Puerto de Mayapito Mine. Massive.

Municipio de Tlalpujahua

Tlalpujahua: Moderate distribution. Most noted from the La Borda, Coronas, and Dos Estrellas mines. Massive.[The Dos Estrellas Mine does not appear in any records in the district. It was probably renamed, however, records do not show its former name.]

Municipio de Zitacuaro

Zitacuaro: Moderately distributed. Note from the Chiranganguco Mine. Massive.

MORELOS

Municipio de Ayala

Cerro de Galván: Moderately distributed.

NAYARITT

Municipio de Ixtlán

Ixtlán: Moderate distribution. Noted from the Dolores Mine. Massive.

Municipio de Santa María del Oro

Acuitapilco: Moderately distributed. Noted from the La Noria, Paredes, and Todos Santo mines. Massive.

NUEVO LEON

Municipio de Montemorelos

Ríncon de Toros: Moderately distributed.

OAXACA

Municipio de Ejutla de Crespo

Eujutla de Crespo: Moderately distributed. Noted from the Los Ocotes Mine.

PUEBLA

Municipio de Cuyoaco:

Temextla: Moderate distribution.

QUERETARO

Municipio de Cadereyta

El Doctor: Moderately distributed within the Cerro de San Nicolás. Noted from the San Juan Nepomuceno, Las Aguas, and La Dificultad mines. Massive.

SAN LUIS POTOSI

Municipio de Catorce

Catorce: Widespread. Noted from the San Carlos, San Agustín, and Pobres mines. Massive.

Municipio de Cerro de San Pedro

Cerro de San Pedro: Moderately distributed. Noted from the Victoria, El Rey, San Pedro mines. Massive. Associated with pyrite, galena, and sphalerite.

Municipio de Charcas

Charcas: Moderately abundant.
Noted from the Tiro General and
La Recompensa mines. Crystals to
1 cm. Associated with pyrite,
galena, and sphalerite.

Municipio de Guadalcázar

Guadalcázar: Moderately distributed. Noted from the La Trinidad
Mine. Massive. [This location
started its life known as San
Francisco de Guadálcazar, but
later was changed to Gaudalcázar.
It was an important mining center
in the 1620s as a producer of
silver and gold. In the 1700s it
also produced mercury. In the
1800s copper became the district's important metal. At the
present time the mines of the
district are closed.]

Municipio de Villa de La Paz

La Paz: Moderately distributed.
Noted from the Santa Fe and La
Paz mines. Crystals to 1 cm.

SINALOA

Municipio de Badiraguato

Soyatita: Moderately distributed.
Noted from the San José, Ledesma,
and San Antonio mines.

Municipio de Cosalá

Cosalá: Moderately distributed.
Noted from the El Refugio,
Clarina, and Culebra mines.
Crystals to 1 cm. [The El Refugio
Mine has produced brilliant
clusters of crystals up to 10 to
15 cm in diameter.]

Municipio del Fuerte:

Picachos: Moderately distributed.
Noted from the Aquincuari Mine.

Municipio de San Ignacio

Arroyo de Tecuilapan: Limited
distribution within pegmatites.

San Ignacio: Moderately distributed. Noted from the Las Canas,
Tecolotes, and Tecomates mines.

SONORA

Municipio de Alamos

Alamos: Moderately distributed.
Noted from the Dr. Brown, Los
Chinos, Tesacho, and Quiteria
mines. Massive.

Municipio de Arizpe

Arizpe: Moderate distribution.
Noted from the Las Chipas Mine.
Associated with pyrargyrite.

Municipio de Bacanora

Bacanora: Moderate distribution.
Noted from the El Sacramento
Mine.

Municipio de Cananea

Cananea: Widespread. Crystals to
1 cm. Associated with rhodochrosite pyrite and quartz.

Municipio de Cumpas

Cumpas: Moderately distributed.
Noted from the Gran Republic and
Sonora mines.

Municipio de Nacozari de García

La Caridad: Widespread. Noted
from the La Caridad Mine.
Associated with chalcocite,
covellite, digenite, and pyrite.

Nacozari de García: Moderately
distributed. Noted from the
Pilares mine. Massive.

Municipio de Sahuaripa

El Encinal: Moderately distributed. Noted from the Lydia Mine.
Associated with pyrite.

Municipio de San Javier

San Javier: Moderately distributed. Noted from the Las Animas
Mine. Massive.

TAMPAULIPAS

Municipio de Villagrán

San José: Moderate distribution. Noted from the Santa Cruz and Los Reyes mines.

VERACRUZ

Municipio de Los Minas

Zomelahuacan: Widespread. Noted from the Asunción, Jesús María, and La Cruz mines. Massive.

ZACATECAS

Municipio de Concepción del Oro

Aranzazú: Widespread. Noted from the Sacavón and San Antonio mines. Crystals to 10 cm. Associated with quartz and pyrite.[The San Antonio Mine has produced probably the largest crystals of chalcopyrite from Mexico. These etched crystals are almost 10 cm in length and are found in clusters of crystals associated with quartz.]

Concepción del Oro: Widespread. Noted from the Arroyo de Cata mine. Crystals to 1 cm. Associated with quartz and pyrite.

El Cobre: Widespread. Noted from the El Cobre Mine. Crystals to 2 cm. Associated with quartz and calcite.

Municipio de Fresnillo

Fresnillo de Gonzalez Echeverria: Moderate distribution. Noted from the Amarilla Mine.

Municipio General Francisco Murguía

Nieves: Moderately distributed. Noted from the Santa Catarina Mine. Massive.

Municipio de Mazapil

Mazapil: Moderately distributed. Noted from the Jesus Maria and Santa Rosa mines. Associated with azurite and malachite.

Municipio de Noria de Angeles

Noria de Angeles: Moderately distributed. Noted from the Dios nos Guie, El Pinto, La Verde, and Cubierta mines. Massive.

Municipio de Sombrerete

San Martín: Widespread. Noted from the San Martin and Sabino mines. Crystals to 2 cm. Associated with covellite, bornite, and sheets of native silver. [Chalcopyrite is found at the San Martín Mine as both veins and as limestone replacements. It has been found in large masses 100s of meters thick.]

Sombrerete: Moderately distributed. Noted from the Los Tocayos, Trinidad, and Bartolo mines. Crystals to 1 cm.

Municipio de Vetagrande

Vetagrande: Moderately distributed. Crystals to 3 cm. Associated with native silver. [Vetagrande is part of the Zacatecas mining district.]

Municipio de Zacatecas

Zacatecas: Moderately distributed. Crystals to 1 cm.

CHALCOSTIBITE

$Cu_6Tl_2SbS_4$. A rare mineral found in few metal deposits.

SONORA

Municipio de Moctezuma

Moctezuma: Limited distribution. Noted from the Guadalupe Mine.

CHENEVIXITE

$Cu_2Fe_2(AsO_4)_2(OH)_4 \cdot H_2O$. An uncommon secondary mineral forming in the oxidized zone of copper deposits.

DURANGO

Municipio de Mapimí

Mapimí: Limited distribution. Noted from the San Juan Mine. Botryoiadal olive-green crust. Associated with mimetite and goethite.

SONORA

Municipio de Naco: La

Morita: Limited distribution.

CHLORALUMINITE

$AlCl_3 \cdot 6H_2O$. A common secondary mineral found in volcanic rocks.

MICHOACAN

Municipio de Nuevo San Juan Parangaricuto

Volcan Paricutín: Moderate distribution.

CHLORARGYRITE

AgCl. An uncommon secondary mineral that forms a series with bromargyrite, the chlorine being replaced by bromine. This mineral is formed in the oxidized zones of silver deposits from the alteration of primary silver minerals.

BAJA CALIFORNIA SUR

Municipio de La Paz

Cacachilas: Limited distribution. Noted from the La Luz Mine.

Municipio de San Antonio

El Triunfo: Moderate distribution. Noted from the Jesús María, Rosario, and San Cayetano mines.

San Antonio: Limited distribution. Noted from the San Antonio Mine.

CHIHUAHUA

Municipio de Ahumada

Los Lamentos: Limited distribution. Ahumada/Erupción Mine.

Municipio de Aquiles Serdán

Francisco Portillo: Limited distribution within the west camp of the Santa Eulalia mining district. Noted from the Viejo and Parcionera mines. Stalactites to 20 cm. [This mineral was found in the early years of mining. It has been recorded that the upper enriched ore caves of the Mina Viejo contained stalactites of chlorargyrite over 20 cms long!]

San Antonio el Grande: Limited distribution within the east camp of the Santa Eulalia mining district. Noted from the San Antonio Mine's oxide ore zone. Bromian enriched chlorargyrite has been found here.

Municipio de Batopilas

Batopilas: Limited distribution. Associated with bromargyrite.

Municipio de Cusihuiriáchic

Cusihuiriáchic: Moderate distribution. Noted from the La Reina Mine. Associated with stephanite.

Municipio de Julimes

Julimes: Limited distribution. El Barril Mine.

Municipio de Guazapares

Monterde: Moderately distributed.

Municipio de Saucillo

Naica: Moderate distribution throughout the Sierra de Naica. Associated with native silver.

COAHUILA

Municipio de Sierra Mojada

Sierra Mojada: Modrately distributed throughout district. Noted from the Fortuna, Guadalupe, and Esmeralda mines.

DURANGO

Municipio de Indé

Indé: Moderate distribution. Noted from the Caballo, Demasia, and Portillo mines.

Municipio de Mapimí

Mapimí: Moderately distributed throughout district. Noted from the Ojuela Mine. Massive. Associated mimetite, cerussite, anglesite, and wulfenite.[During the early years of mining within the district, this mineral was one of the important silver minerals. The chlorargyrite from the Ojuela Mine is rich in bromine.]

Municipio de Mezquital

Mezquital: Limited distribution.

Municipio de San Dimas

Tayolita: Moderately distributed. Massive. Associated with native silver, acanthite, and dioptase. [This old silver/gold location has also been referred to in the older records as Guarisemeny or San Dimas in older records.]

GUANAJUATO

Municipio de Guanajuato

Guanajuato: Moderate distribution throughout Veta Madre. Noted from the Mellado and Cata mines. [Bromian rich chlorargyrite has been found throughout the district.]

Municipio de la Paz

Mineral de Pozos: Moderately distributed.

GUERRERO

Municipio de Taxco

Taxco de Alcarón: Moderate distribution. Noted from the Apaga Candela mine.

HIDALGO

Municipio de Mineral de Monte

Mineral de Monte: Moderate distribution. Massive. Associated with acanthite, pyrargyrite, and calcite.

Municipio de Pachuca

Pachuca de Soto: Moderate distribution. Massive. Associated with native silver, acanthite, pyrargyrite, and calcite.

MEXICO

Municipio del Oro

El Oro: Moderately distributed. Noted form the Carmen, San Antonio, and San Rafael mines.

Municipio de Temascaltepec

Temascaltepec de Gonzalez: Moderate distribution.

Municipio de Zacualpan

Zacualpan: Moderate distribution. Massive.

MICHOACAN

Municipio de Angangueo

Mineral de Angangueo: Moderate distribution. Massive.

Municipio de Tlalpujahua

Tlalpujahua: Moderate distribution.

QUERETARO

Municipio de Tolimán

Río Blanco: Limited distribution. Nueva California Mine. Massive.

SAN LUIS POTOSI

Municipio de Catorce

Catorce: Widespread throughout district. Noted from the Zavala (Padre Flores) and Concepción Mine. Crystals to 4 mm. Associated with gold, native silver, phosgenite, wulfenite. [This famous mining location, officially named, La Purísima Concepción de Alamos de Catorce, was discovered in 1773. Miners in 1778, sinking the first shaft, discovered chlorargyrite and gold just beneath the surface. In one month they took out $546,915.00 in gold and silver. The early mines were extremely rich in chlorargyrite and it was not uncommon to find large "bovedas" or "vaulted chambers" of silver chloride. The Zavala or Padre Flores mine yielded two such chambers lined with chlorargyrite crystals. The mine owners split the handsome sum of $7 million over a three year period! Before the War of Independence of 1810, Catorce was the third leading producer of silver in Mexico.]

Municipio de Cerro de San Pedro

Cerro de San Pedro: Moderately distributed. Noted from the El Rufugio Mine.

Municipio de Charcas

Charcas: Moderately distributed within the Cerro de Santa Inés.

Municipio de Guadalcázar

Guadalcázar: Moderate distribution. Noted from the El Rufugio Mine.

Municipio de Villa de Ramos

Ramos: Limited distribution. Noted from the La Purísima Mine. Associated with azurite, malachite, and pyrargyrite. [This mineral from here was bromian rich.]

SONORA

Municipio de Alamos

Alamos: Moderately distributed. Noted from the Vetagrande and Vallorecas mines. Associated with native silver and chalcocite. [The Vallorecas Mine produced a bromian rich chlorargyrite.]

Municipio de Altar

Altar: Moderately distributed. Noted from the Marenena Mine.

Municipio de Arizpe

Arizpe: Moderately distributed. Las Chispas Mine. Crystals to 6 mm. Associated with acanthite. [A bromian rich chlorargyrite associated with chlorargyrite was found on the second level of the mine in the Las Chispas Vein.]

Municipio de Caborca

Heroica Caborca: Moderately distributed. Noted from the Felix Mine. Associated with bromargyrite.

ZACATECAS

Municipio de Concepción del Oro

Albarradón: Moderately distributed. Noted from the Albarradón Mine.

Municipio de Fresnillo

Fresnillo de Gonzalez Escherria: Moderately distributed within the Cerro de Proano. Massive. Associated with bromargyrite.

Municipio de Ojo Caliente

Ojo Caliente: Limited distribution. El Cebezón Mine.

Municipio de Sombrerete

Sombererete: Limited distribution. Veta Negra Mine.

Municipio de Vetagrande

Vetagrande: Moderate distribution.

Municipio de Zacatecas

Zacatecas: Moderately distributed. Noted from the Quebradillas

Mine. [Bromine was commonly contained within the chlorargyrite from this district.]

CHLORITE is a group name, not a specific mineral.

CHOLOALITE

$PbCu(TeO_3)_2 \cdot H_2O$. A rare mineral found in few base metal deposits.

SONORA

Municipio de Moctezuma

Moctezuma: Limited distribution. Noted from the La Oriental Mine. Associated with hessite and carlfriesite.

CHROMITE

$FeCr_2O_4$. A member of the spinel mineral group which forms in ultramafic rocks, some anorthositic rocks, and basalts.

OAXACA

Municipio de Ejutla de Crespo

Ejutla de Crespo: Limited distribution.

PUEBLA

Municipio de Tehuitzingo

Cerro de Chinantla: Limited distribution.

Cerro de Tehuytla: Limited distribution.

CHRYSOBERYL

$BeAl_2O_4$. An uncommon mineral found in granitic pegmatites, mica schists and alluvium.

GUERRERO

Municipio de Chilpancingo

Chilpancingo de los Bravos: Limited distribution.

HIDALGO

Municipio Tulancingo

Tulancingo: Limited distribution.

CHRYSOCOLLA

$(Cu,Al)_2H_2Si_2O_5(OH)_4 \cdot nH_2O$. A common secondary mineral formed in the oxidized zones of copper deposits.

AGUASCALIENTES

Municipio de Asientos

Asientos: Moderately distributed. Massive.

Municipio de Tepezalá

Tepezalá: Moderately distributed. Massive.

BAJA CALIFORNIA

Municipio de Ensenada

Chapala: Moderately distributed. 17 km east of. Noted from the Rey Salomón and San Antonio mines. Massive. Associated with malachite.

Las Palomas: Moderately distributed. 25 km north of El Arco. Noted from the Alejandra Mine.

San Fernando: Moderately distributed. Noted from the Calumet and Rosario mines. Massive.

BAJA CALIFORNIA SUR

Municipio de Mulegé

Santa Rosalía: Moderately distributed. Massive.

Municipio de San Antonio

Rancho Naranjo: Moderate distribution. Noted from the Altar and Amelia mines. Massive.

CHIHUAHUA

Municipio de Chihuahua

Labor de Terrazas: Moderately distributed. Río Tinto Mine. Massive.[Has been found at times as replacement of hemimorphite.]

Municipio de Coyame

Cuchillo Parado: Moderately distributed. Massive.

Municipio de Cuishuiriáchic

Cuishuiriáchic: Moderately distributed. Noted from the La Reina Mine.

Municipio de Julimes

Julimes: Moderately distributed. noted from the El Barril Mine.

Municipio de Hidalgo del Parral

Hidalgo del Parral: Moderately distributed throughout the Cerro la Cruz. Noted from the El Tajo and La Alfarena mines.

Municipio de Saucillo

Naica: Moderately distributed.

COAHUILA

Municipio de Candela

Pánuco: Moderately distributed. Pánuco Mine.

Municipio de Sierra Mojada

Sierra Mojada: Moderate distribution. Noted from the El Porvenir Mine.

DURANGO

Municipio de Mapimí

Mapimí: Moderately distributed. Massive. [It has been found as pseudomorphs after malachite, aurichalcite, and rosasite. Large masses of a hard compact chrysocolla were mined within the district in the past. It has been reported that some of it was cut into fine gemstones of good blue color.]

Municipio de San Dimas

Tayolita: Moderate distribution. Noted from the San Nicolás, Candelaria, and Sacramento mines.

GUANAJUATO

Municipio de San Luis de la Paz

Mineral de Pozos: Moderate distribution. Noted from the Santa Brígida Mine. Massive.

GUERRERO

Municipio de Huitzuco

San Antonio: Moderate distribution.

HIDALGO

Municipio de Pachuca

Pachuca de Soto: Moderate distribution. Noted from the San Antonio, and Entrometida mines. Massive.

Municipio de Zimapán

Zimapán: Moderate distribution.

MEXICO

Municipio de Zacualpan

Zacualpan: Moderate distribution. Noted from the Coetzillos Mine.

MICHOCAN

Municipio de La Huacana

Inguaran: Moderately distributed. Massive.

Oropeo: Moderate distribution. Noted from the San Cristóbal Mine.

SAN LUIS POTOSI

Municipio de Guadalcázar

Guadalcázar: Limited distribution. Noted from the La Trinidad Mine.

SINALOA

Municipio de Mazatlán

La Calera: Moderately distributed within the Mazatlán Mining district. Massive.

SONORA

Municipio de Alamos

Alamos: Moderately abundant. Noted from the Adriana, Quinteros, and La Paloma mines. Massive.

Municipio de Cananea

Cananea: Widespread. Massive. Associated with malachite.

Municipio de Nacozari de García

Nacozari de García: Widespread. Massive.

Municipio de Oquioa

Oquioa: Moderate distribution. Noted from the Tony Mine.

Municipio de Trincheras

La Mur: Limited distribution. Noted from the Las Animas Mine. Massive. Associated with argentosiderite, beudantite, chenivixite, and goethite. [This location is west of Benjamín Hill and south of La mur.]

TAMAULIPAS

Municipio de Villagrán

San Carlos: Moderately distributed. Noted from the San Antonio and Santa Ana mines. Massive.

VERACRUZ

Municipio de La Minas

Zomelahuacán: Moderate distribution. Noted from the Zomelahuacán Mine. Massive.

ZACATECAS

Municipio de Concepción del Oro

Aranzazú: Moderately distributed. Noted from the San Juan and El Pando mines. Massive.

Municipio de Mazapil

Mazapil: Moderately distributed. Noted from the San Carlos Mine. Massive.

CHRYSOTILE

$Mg_3Si_2O_5(OH)_4$. A member of the serpentine group and forms in metamorphic conditions.

TAMAUILIPAS

Municipio de Ciudad Victoria

Canyón del Novillo: Moderately distributed. Massive.

CINNABAR

HgS. An uncommon mineral forming in veins and replacement or impregnation deposits near the surface, found in rocks and hotsprings of recent volcanic envirnoments.

CHIHUAHUA

Municipio de Aquiles Serdán

Francisco Portillo: Limited distribution. Las Cocineras Mine. [This is within the west camp of the Santa Eulalia mining district.]

San Antonio el Grande: Limited distribution. San Antonio Mine. Associated with quartz and calcite. [This is within the east camp of the Santa Eulalia mining district.]

Municipio de Batopilas

Batopilas: Limited distribution.

Municipio de Camargo

Sierra de Encinilla: Limited distribution. Noted from the San Miguel Mine.

Municipio de Jiménez

Escalón: Limited distribution. Associated with calcite.

Municipio de Saucillo

Naica: Limited distribution.

Municipio de Uruáchic

Uruáchic: Limited distribution.

COAHUILA

Municipio de Candéla

Pánuco: Limited distribution. Pánuco Mine.

DURANGO

Municipio de Cuencamé

Santa María: Limited distribution. Noted from the Macarena Mine.

Municipio de Guananceví

Guanaceví: Limited distribution. Noted from the Barradon Mine.

Municipio de Indé

Indé: Limited distribution. Noted from the Gran Tenoxtitlan Mine. Massive. Associated with calcite.

Municicpio del Oro

Magistral: Limited distribution. Noted from the Porvenir Mine. Massive. Associated native mercury.

Santa María del Oro: Limited distribution. Massive. Associated with native mercury and galena.

Municipio de Nombre de Dios

San José de Parrilla: Moderate distribution. Noted from the Encarnación and San José de Cantuna mines. Masive. Associated with calcite and native mercury.

Municipio de San Bernardo

Villa Cinabrio: Moderately distributed throughout the Cuarenta mining district. Noted from the Porvenir, Faro, El Sol, and Foco mines. Massive. Associated with native mercury and calcite.

GUANAJUATO

Municipio de Guanajuato

Guanajuato: Moderate distribution. Noted from the Almadén and Chico mines. Massive. Associated with calcite and native mercury. [Cinnabar was found at many of the mines of the Veta Madre in the early years of mining.]

La Luz: Moderately distributed. Noted from the Purísima and San Pedro mines. Massive. Associated with native mercury.

Municipio de San Felipe

San Felipe: Widely distributed. Most noted from the Buen Sucesu and Castellanos mines. Massive. Associated with calcite and native mercury.

Municipio de San luis de la
Paz

Mineral de Pozos: Moderately
distributed. Noted from the Santa
Brígida, Sierra Gorda, Cinco
Señores, and El Progreso mines.
Massive. Associated with calcite.

San Luis de la Paz: Moderate
distribution on the Rancho El
Rodeo at the Cerro de Las
Beatas. Noted from the Animas,
Encarnación, and Haminón mines.
Associated with calcite, native
mercury.

Municipio de Xichú

Xichú: Moderately distributed,
Noted from the Almadenes,
Esperanza, and Victoria mines.

GUERRERO

Municipio de Huitzuco

Huitzuco de los Figueroa:
Widespread. Noted from the La
Cruz, San Agustín, Agua Salada,
and La Unión mines. Massive.
Associated with metacinnabar,
terlinguarite, selenium rich
terlinguarite, (formerly called
onofrite), tiemannite, livingstonite, stibnite, gypsum,
sulfur, and anhydrite.

Municipio de Taxco

Huahuaxtla: Widespread. Noted
from the San Luis, Aurora, and
Esperanza mines. Massive.
Associated native mercury,
metacinnabar, calomel,
eglestonite, terlinguaite, and
montroydite.

HIDALGO

Municipio de Atotonilco el
Grande

Guadalupe de la Atarjetea:
Limited distribution.

Municipio de Zacualtipán

Zacualtipán: Limited distribution.

Municipio de Zimapán

Zimapán: Limited distribution.

JALISCO

Municipio de Mascota

El Moral: Limited distribution.

MICHOACAN

Municipio de La Huacana

Inguaran: Limited distribution.
Noted from the La Reina Mine.

NAYARIT

Municipio de San Blas
Rancho de Navarrete: Limited
distribution. Noted from the La
Costerna Mine.

Municipio de San Pedro
Lagunillas

Huaynamota: Limited distribution.

QUERETARO

Municipio de Cadereyta

El Doctor: Moderate distribution.
Noted from the El Durazno and Las
Cabras mines. Massive.

Municipio de Tolimán

Soyatal: Widespread within the
Cerro Pinguicas. 10 km north of
Motoxi. Noted from the Santa
María de Miera Mine. Massive.
Associated with stibiconite,
cervantite, and semarmontite.
[In the El Raizal stope of the
Santa María de Miera Mine,
cinnabar was found as a coating
or "paint" on pseudomorphs of
stibiconite after stibnite.]

SAN LUIS POTOSI

Municipio de Catorce

Catorce: Limited distribution.
Noted from the Concepción and
Guadalpana mines. Massive.

Wadley: Moderately distributed.
Noted from the San José Mine.
Massive. Associated with stibi-
conite and calcite.

Muicipio de Cerro de San Pedro

Cerro de San Pedro: Limited
distribution. Noted from the San
Cristóbal Mine.

Municipio de Charcas

Charcas: Limited distribution.
Noted from the San Bartlo Mine.
Crystals to 1 cm. Associated with
calcite. [This is Mexico's best
producing location for crystallized cinnabar.]

Municipio de Guadalcázar

Guadalcázar: Widespread. Massive.
Associated with metacinnabar,
native mercury, and calcite.

Municipio de Moctezuma

Hacienda de Santa Antonio de Rul:
Limited distribution. Noted from
the Dulces Nombrers Mine.
Massive. Associated with native
mercury and calcite.

TLAXCALA

Municipio de Tlaxco

Cerro de la Mesa: Limited
distribution on the Ranchos del
Convento and de Teteles.

ZACATECAS

Municipio de General Francisco Murguía

Nieves: Moderate distribution.
Noted from the El Tequezquite
mine. Massive.

Municipio de Guadalupe

Santa Teresa: Limited distribution.

Municipio de Mazapil

Mazapil: Moderate distribution.
Noted from the El Cinabrio and
Todos Santos mines. Massive.
Associated with native mercury.

Municipio de Morelos

Morelos: Limited distribution.
Noted from the Santa Rosa Mine.

Municipio de Pinos

San Miguel: Moderately distributed. Noted from the San Acacio.

Municipio de Sáin Alto

Sáin Alto: Moderately distributed. Noted from the Santa Lucía
Mine. Massive.

Municipio de Sombrerete

Sombererete: Limited distribution.

CLAUSTHALITE

PbSe. A rare mineral found in
complex lead/selenium ores.

DURANGO

Municipio de Mapimí

Mapimí: Limited distribution.
Monterrey Mine. Massive. Associated with umangite, pyrrhotite,
bismuth, sphalerite, and enargite.

CLIFFORDITE

UTe_3O_9. A rare mineral found in
few metal deposits.

SONORA

Municipio de Moctezuma

Moctezuma: Limited distribution.
Discovered at the San Miguel

Mine. Associated with Mackayite. [This is the type location for this mineral.]

CLINOCHLORE

$(Mg,Fe)_5Al(Si_3Al)O_{10}(OH)_8$. A member of the chlorite mineral group which formed by hydrothermal and thermal metamorphism.

CHIHUAHUA

Municipio de Aquiles Serdán

Francisco Portillo: Limited distribution. Noted from the El Potosí Mine's silicate ore body. Associated with rhodochrosite and fluorite.

HIDALGO

Municipio de Pachuca

Pachuca de Soto: Limited distribution.

CLINOZOISITE

$Ca_2Al_3(SiO_4)_3(OH)$. A member of the epidote group, which forms by local and regional metamorphism.

BAJA CALIFORNIA

Municipio de Tecate

Laguna Hansen: Moderately distributed. Noted from the Olivia Mine, 15 km east northeast of Rose de Castillo. Associated with diopside and axinite.

Rosa de Castilla: Widely distributed. Noted from the El Fenomeno mine. Greenish brown crystals to 10 cm. Poor crystal faces or terminations.

DURANGO

Municipio de Mapimí

Mapimí: Limited distribution.

HIDALGO

Municipio de Pachuca

Pachuca de Soto: Limited distribution.

OAXACA

Municipio de la Pe

La Planchita: Limited distribution. Noted from the La Planchita mine. Crystals to 2 cm.

SONORA

Municipio de Alamos

Rancho de Tepueste: Moderately distributed. 24 km northwest of Alamos. Pinkish red Fibrous masses.

CLINTONITE

$Ca(Mg,Al)_3(Al_3Si)_{10}(OH)_2$. An uncommon mineral and member of the mica mineral group.

MICHOACAN

Municipio de Zitácuaro

Susupuato: Moderately distributed within the Cerro Mazahua skarn.

COBALTITE

CoAsS. An uncommon mineral forming in high temperature deposits, in metamorphic rocks, and vein deposits.

DURANGO

Municipio de Guanaceví

Guanaceví: Limited distribution.

JALISCO

Municipio de Pihuamo

Pihuamo: Limited distribution. Noted from the Esmeralda Mine.

SINALOA

Municipio de Cosalá

Cosalá: Moderately distributed. Noted from the Nuestra Mine.

SONORA

Municipio de Alamos

Alamos: Limited distribution.

San Bernardo: Limited distribution. Noted from the Sara Alicia Mine. Associated with erytherite.

Municipio de Cananea

Cananea: Limited distribution.

COBALTOCALCITE. See **SPHALAEROCOBALTITE**

COHENITE

Found in meteorites at a few locations within Mexico.

COLORADOITE

HgTe. A rare mineral found in few telluride deposits.

SONORA

Municipio de Moctezuma

Moctezuma: Limited distribution. Noted from the La Bambollita Mine.

COLUMBITE. See **FERRO-COLUMBITE**

CONICHALCITE

$CaCu(AsO_4)(OH)$. An uncommon secondary mineral and member of the adelite mineral group found in the oxide zone of few base metal deposits.

CHIHUAHUA

Municipio de Saucillo

Naica: Limited distribution. Noted from the Naica Mine. Associated with goethite.

DURANGO

Municipio de Mapimí

Buruguilla: Moderately distributed. Noted from the Descubridora Mine. Associated with goethite.

Mapimí: Moderately distributed. Noted from the Ojuela Mine. Light olive green to bright green botryoidal crust. Associated with malachite, calcite, austinite, mimetite, and goethite.

SAN LUIS POTOSI

Municipio de Villa de La Paz

La Paz: Limited distribution. Noted from the Dolores Mine. Associated with goethite.

ZACATECAS

Municipio de Concepción del Oro

Concepción del Oro: Limited distribution. Noted from the Cata Arroyo Mine. Associated with goethite.

COPIAPITE

$FeFe_4(SO_4)_6(OH)_2 \cdot 20H_2O$. An uncommon mineral forms from the

oxidation of pyrite and precipi-
tates from acidic mine waters.

CHIHUAHUA

Municipio de Aquiles Serdán

Francisco Portillo: Moderately
distributed. Massive.

San Antonio el Grande: Moderately
distributed. Noted from the San
Antonio Mine. Massive.

Municipio de Saucillo

Naica: Moderately distributed.
Massive.

DURANGO

Municipio de Cuencamé

Velardeña: Limited distribution.
Noted from the Santa Maria Mine.
Massive.

Municipio de Mapimí

Mapimí: Moderately distributed. A
massive yellowish white botryoidal coating. Associated with
pyrite.

COPPER

Cu. An uncommon mineral (element), found in oxide zones of
base metal deposits.

AGUASCALIENTES

Municipio de Asientos

Asientos: Moderately distributed.

Municipio de Tepezalá

Tepezalá: Moderately distributed.
Noted from the San Bartolo Mine.

BAJA CALIFORNIA

Municipio de Ensenada

San Fernando: Moderately distributed. Noted from the San Fernando
and Adela mines.

BAJA CALIFORNIA SUR

Municipio de Mulegé

Mulegé: Limited distribution.
Crystals to 4 mm. Associated with
volcanic glass (obsidian).

Santa Rosalía: Moderate distribution. Noted from the Santa Rita
Mine. Associated with cuprite and
gypsum.

CHIAPAS

Municipio de Motozintla de Mendoza

Ojo de Agua: Limited distribution.

CHIHUAHUA

Municipio de Aldama
El Pastor: Limited distribution.
Nos Plus Ultra Mine.

Municipio de Aquiles Serdán

Francisco Portillo: Limited
distribution. Associated with
cuprite and gypsum.

Municipio de Ascensión

Sabinal: Limited distribution.

Municipio de Batoplis

La Bufa: Moderately distributed.
Noted from the La Bufa Mine.
Crystals to 2 cm. Associated with
calcite. [This location produced
in the 1960s fine spray of
arborescent crystals.]

Batopilas: Moderate distribution.
Noted from the El Carmen mine's
15th level. Arborescent crystals
to 4 cm. Associated with calcite.

Municipio de Beunaventura

San Lorenzo: Moderate distrib-
ution. Noted from the Carlos
Pacheco, Libertdad, and San
Rafael mines.

Municipio de Chihuahua

Labor de Terrazas: Moderately
distributed. Noted from the Río

Tinto Mine. Associated with cuprite and malachite.

Municipio de Coyame

Cuchillo Parado: Moderately distributed.

Municipio de San Francisco del Oro

San Francisco del Oro: Moderately distributed. Noted from the El Tajo and Guadalupe mines.

Municipio de Saucillo

Naica: Moderately distributed. Noted from the Naica Mine. Associated with cuprite and malachite.

COAHUILA

Municipio de Sierra Majada

Sierra Majada: Moderate distribution. Noted from the Veta Rica Mine. Crystals to 5 mm. Associated with gypsum (selenite).

DURANGO

Municipio de Guanaceví

Guanaceví: Limited distribution.

Municipio de Indé

Indé: Limited distribution.

Municipio de Mapimí

Buruguilla: Moderately distributed. Noted from the Descubridora Mine. Associated with goethite and cuprite.

Mapimí: Moderately distributed. Noted from the Ojuela Mine. Crystals to 6 mm. Associated with tenorite, goethite, cuprite, native silver, gypsum (selenite),and delafossite.

Municipio del Oro

Santa María del Oro: Limited distribution.

GUANAJUATO

Municipio de San Luis de la Paz

Mineral de Pozos: Limited distribution. Noted from the Santa Brígida Mine.

HIDALGO

Municipio de Pachuca

Pachuca de Soto: Limited distribution. Noted from the San Rafael Mine.

Municipio de Zimapán

Bonanza: Limited distribution. Noted from the Bonanza Mine.

Zimapán: Limited distribution. Noted from the La Luz Mine. Massive.

MEXICO

Municipio de Zacualpan

Teocalcingo: Limited distribution. Noted from the El Cobre Mine. Massive.

MICHOACAN

Municipio de Angangueo

Mineral de Angangueo: Moderately distributed.

Municipio de La Huacana

Inguaran: Moderately distributed. Associated with cuprite.

Municipio de Tuzantla

Tuzantla: Moderately distributed. Noted from the La Cabra Mine. Crystals to 6 cm. [In the early years of mining at the La Cabra mine, large masses up to 1 m in size of native copper crystals were encountered. Some of these crystal masses contained crystals "as big as fingers."]

SAN LUIS POTOSI

Municipio de Guadalcázar

Guadalcázar: Limited distribution.

SONORA

Municipio de Cananea

Cananea: Moderately distributed. Crystals to 2 cm.

Municipio de Nacozari de García

Nacozari de García: Moderately distributed. Crystals to 2 cm.

TAMAULIPAS

Municipio de Villagrán

San José: Moderately distributed.

VERACRUZ

Municipio de Las Minas

Zomelahuacan: Limited distribution.

ZACATECAS

Municipio de Concepción del Oro

Aranzazú: Moderate distribution. Noted from the Aranzazú Mine.

Concepción del Oro: Moderately distributed. Noted from the Arroyo de Cata Mine. Associated with cuprite.

Municipio de Sombrerete

San Martín: Limited distribution. Noted from the Parroquia Mine. Arborescent crystals to 1 cm. [The copper was also found as casts of crystals of calcite up to 2 cm in length.]

Municipio de Zacatecas

Zacatecas: Limited distribution. Noted from the El Bote Mine. Crystal wires to 1 cm. Associated with calcite and cuprite.

COQUIMBITE

$Fe_2(SO_4)_3 \cdot 9H_2O$. An uncommon mineral found in the oxide zone of many base metal deposits.

HIDALGO

Municipio de Pachuca

Pachuca de Soto: Limited distribution.

CORDIERITE

$Mg_2Al_4Si_5O_{18}$. An uncommon mineral formed by the metamorphism of argilliaceous sediments.

DURANGO

Municipio de Mapimí

Mapimí: Limited distribution.

CORKITE

$PbFe_3(PO_4)(SO_4)(OH)_6$. An uncommom mineral found in the upper zones of few base metal deposits.

CHIHUAHUA

Municipio de Aquiles Serdán

Francisco Portillo: Limited distribution. Noted from the El Potosí Mine.

SONORA

Municipio de Caborca

Heróica Caborca: Limited distribution. Noted from the San Felix Mine. Associated with carminite.

CORNWALLITE

$Cu_5(AsO_4)_2(OH)_4 \cdot H_2O$. A rare secondary mineral found in the oxidized zone of copper deposits.

SONORA

Municipio de Naco

La Morita: Limited distribution.

CORONADITE

$Pb(Mn,Mn)_8O_{16}$. A rare mineral and member of the cryptomelane group, which is found in oxidized ore zone of metal deposits.

CHIHUAHUA

Municipio de Allende

Villa de Allende: Limited distribution.

CORUNDUM

Al_2O_3. A fairly common mineral and member of the hematite group found in contact metamorphic and pegmatitic environments.

SAN LUIS POTOSI

Municipio de Guadalcázar

Guadalcázar: Limited distribution.

COSALITE

$Pb_2Bi_2S_5$. A rare sulfosalt which forms in moderate temperature veins of contact metamorphic and pegmatitic deposits.

CHIHUAHUA

Municipio de Chihuahua

Candamene: Limited distribution. Noted from the Loreto and Candamene mines. [This location is near Rayon.]

Municipio de Saucillo

Naica: Limited distribution.

SINALOA

Municipio de Cosalá

Cosalá: Limited distribution. Noted from the Nuestra Señora Mine. [This is the type location for this mineral named after the location.]

COVELLITE

CuS. An uncommon mineral usually found as a secondary mineral in the oxide zones of copper deposits, but it also can be found as a primary mineral.

AGUASCALIENTES

Municipio de Asientos

Asientos: Moderately distributed.

BAJA CALIFORNIA

Municipio de Ensenada

San Fernando: Moderately distributed. Noted from the Chalcocita Mine. Massive.

BAJA CALIFORNIA SUR

Municipio de Mulegé

Santa Rosalía: Widespread. Crystals to 4 mm. Massive.

CHIHUAHUA

Municipio de Saucillo

Naica: Moderately distributed. Noted at the Naica Mine.

DURANGO

Municipio de Mapimí

Mapimí: Widespread. Noted from the Ojuela Mine. Crystals to 6 mm. Associated with galena, copper, tenorite, cuprite, delafossite, chalcopyrite, and goethite. [Covellite has been found as a replacement of arborescent crystal masses of native copper on goethite from the Ojuela Mine.]

GUANAJUATO

Municipio de Guanajuato

Guanajuato: Moderately distributed within the Veta Madre.

Municipio de San Luis de la Paz

Mineral de Pozos: Moderately distributed. Santa Brígida Mine. Massive.

OAXACA

Municipio de San Miguel Peras

San Miguel Peras: Moderate distribution. Noted from the Zavaleta Mine.

SONORA

Municipio de Cananea

Cananea: Moderate distribution. Crystals to 3 cm. Associated with quartz.

ZACATECAS

Municipio de Sombrerete

San Martín: Widely distributed. San Martín Mine. Massive. Associated with bornite, chalcopyrite, and native silver.

CRANDALLITE

$CaAl_3(PO_4)_2(OH)_5 \cdot H_2O$. A rare mineral of the crandallite group and found in few metal deposits.

DURANGO

Municipio de Mapimí

Mapimí: Limited distribution. San Ilario Mine.

CREASEYITE

$Pb_2Cu_2Fe_2Si_5O_{17} \cdot 6H_2O$. A rare mineral found in the oxidization zone of lead/copper deposits.

SONORA

Municipio de Arizpe

Sinoquipe: Limited distribution.

CREDNERITE

$CuMnO_2$. A rare mineral found in a few metal deposits.

BAJA CALIFORNIA SUR

Municipio de Mulegé

Santa Rosalía: Limited distribution within the Boleo mining district. Massive.

CREEDITE

$Ca_3Al_2(SO_4)(F,OH)_{10} \cdot 2H_2O$. A rare mineral found in hydrothermal veins.

CHIHUAHUA

Municipio de Aquiles Serdán

Francisco Portillo: Limited distribution. Noted from the El Potosí's 5th level-silicate ore body. Inglaterra, and Condesa mines. Clear to deep violet crystals to 8 cm. Associated with gypsum (selenite), calcite, mimetite, and rhodochrosite. [In 1984, a water course was uncovered within the El Potosí mine that was lined with creedite. Crystals up to 8 cm were found long and were of the deepest violet color ever found within the district.]

CRISTOBALITE

SiO_2. An uncommon mineral which is a high temperature polymorph of quartz and found in volcanic rocks in metastable conditions.

DURANGO

Municipio de San Dimas

Saplores: Limited distribution within the Arroyo Liendres in the Sierra de San Francisco.

HIDALGO

Municipio de Pachuca

Cerro San Cristóbal: Limited distribution. [This is the type location for this mineral and was named for the location.]

MICHOACAN

Municipio unknown

Tepoxtepec: Limited distribution. Noted from the Cabires Mine.

Municipio de Tzintzuntzan

Lake Pátzcuaro: Moderately distributed along east shore.

SAN LUIS POTOSI

Municipio unknown

San Lorenzo: Limited distribution.

CROCOITE

$PbCrO_4$. A rare mineral found in the oxidized zones of lead deposits, where the country rocks contain chromite.

JALISCO

Municipio de Bolaños

Bolaños: Limited distribution. Noted from the Pichardo, Santa Fe, Iguana, and Verde mines.

MEXICO

Municipio de Zacualpan

Zacualpan: Limited distribution.

ZACATECAS

Municipio de Mazapil

Mazapil: Limited distribution.

CRONSTEDTITE

$Fe_2Fe(Si,Fe)O_5(OH)_4$. An uncommon member of the kaolinite-serpentine group found in metal deposits.

CHIHUAHUA

Municipio de Aquiles Serdán

Francisco Portillo: Limited distribution. Noted form the El Potosí mine's silicate ore body. Associated with rodochrosite, pyrite, huebnerite, and pyrrhotite.

San Antonio el Grande: Limited
distribution. Noted from the San
Antonio Mine's silicate ore body.

CRYPTOMELANE

$K(Mn,Mn)_8O_{16}$. An uncommon mineral
probably of secondary origin.

BAJA CALIFORNIA SUR

Municipio de Mulegé

Santa Rosalía: Moderately
distributed. Massive.

Lucifer: Moderately distributed.
Noted from the Lucifer Mine.
Massive.

CHIHUAHUA

Municipio de Ahumada

Constutición: Limited distribution. Massive.

DURANGO

Municipio de Mapimí

Mapimí: Moderately distributed on
the western edge of the Sarnoso
stock. Noted from the Ojuela
Mine. Massive.

CSIKLOVAITE

$Bi_2Te(S,Se)_2(?)$. A rare mineral
found in few metal deposits.

CHIHUAHUA

Municipio de Janos

Janos: Limited distribution. Associated with sphalerite and
calcite.

CUMENGITE

$Pb_4Cu_4Cl_8(OH)_8 \cdot H_2O$. A rare
mineral found in the oxidized
zones of copper/lead deposits.

BAJA CALIFORNIA SUR

Municipio de Mulegé

Santa Rosalía: Limited distribution within the Cañada de
Curuglú. Noted from the Amelia
and Curuglú mines. Crystals to 8
cm. Associated with boleite,
atacamite, and gypsum (selenite).
[Cumengite was discovered in the
late 1880s at the Amelia
Mine. The discovery occurred when
the Chiflón Cumenge (Cumenge
shaft) was being sunk and had
just struck the top of the #3 ore
bed. A pocket of crystals,
different from those that had
been seen before was discovered.
The new mineral was brought to
the attention of the superintendent of the French operation,
Eduard Cumenge, who sent them to
his friend and associate Professor Mallard in France for
identification. Mallard found
the new mineral to be different
from the mineral that he and
Cumenge had worked on before and
jointly identified in 1891 as
boleite. In 1893, Mallard named
the new mineral after his friend,
Cumengite. Several more pockets
were uncovered while drifting a
few meters from the Cumenge
Shaft toward the Amelia Mine
workings. The crystals from
these pockets were extremely
large in size, several over 6 cm
in length and in four specimens,
the crystals are almost 8 cm
in size and on matrix! Most of
the crystals from these 1890s
pockets were either on or in a
gypsum (selenite) matrix.]

CUPRITE

Cu_2O. An uncommon mineral found
in the upper oxidation zone of
copper deposits.

AGUASCALIENTES

Municipio de Asientos

Asientos: Moderately distributed.
Noted from the Santa Elena, No
Pensada, and San Bartolo mines.
Associated with native copper.

BAJA CALIFORNIA

Municipio de Ensenada

Real del Castillo: Limited distribution.

San Fernando: Moderately distributed. Noted from the Calumet, Rosario, and Adela mines.

BAJA CALIFORNIA SUR

Municipio de Mulegé

Santa Rosalía: Moderately distributed. Associated with native copper.

Municipio de San Antonio

Rancho del Naranjo: Limited distribution.

CHIAPAS

Municipio de Arriaga

San Luis: Limited distribution.

CHIHUAHUA

Municipio de Chihuahua

Labor de Terrazas: Moderately distributed. Noted from the Río Tinto Mine. Associated with native copper, malachite, and azurite.

Municipio de Coyame

Cuchillo Parado: Moderately distributed within the Cerro de La Espumosa. Associated with azurite and malachite.

Las Vigas: Moderately distributed within the Sierra de la Boquilla. Associated with malachite and azurite.

Municipio de Cusihuiriáchic

Cusihuiriáchic: Limited distribution on the Hacienda Huisochic. Noted from the La Reina Mine.

Municipio de Hidalgo del Parral

Hidalgo del Parral: Moderately distributed. Noted from the El Tajo and Solodad mines.

Municipio de Saucillo

Naica: Limited distribution. Associated with native copper and malachite.

COAHUILA

Municipio de Candela

Pánuco: Moderate distribution. Noted from the Pánuco Mine. Associated with native copper.

COLIMA

Municipio de Manzanillo

Miraflores: Limited distribution.

DURANGO

Municipio de Indé

Indé: Limited distribution. Buenavista Mine.

Municipio de Mapimí

Buruguilla: Moderately distributed. Noted from the Descubridora Mine. Associated with native copper.

Mapimí: Moderately distributed. Noted from the Ojuela Mine. Crystals to 4 mm. Associated with goethite, tenorite, native copper, malachite, and azurite.

Municipio del Oro

Santa María del Oro: Moderately distributed. Noted from the La Candela and La Reina mines.

HIDALGO

Municipio de Mineral del Chico

Mineral del Chico: Moderately distributed. Cuprite (chalcotrichite) has also been found.

Municipio de Zimapán

Flojonales: Limited distribution.

MEXICO

Municipio de Zacualpan

Zacualpan: Moderately distributed. Noted from the El Cobre and Coetzillos mines. Cuprite (chalcotrichite) has also been found.

MICHOACAN

Municipio de Angangueo

Mineral de Angangueo: Moderately distributed.

Municipio de la Huacana

Inguaran: Moderate distribution. Associated with native copper.

Oropeo: Limited distribution. Noted from the China Mine.

PUEBLA

Municipio de Piaxtla

Cerro Yucuyuxi: Limited distribution.

SAN LUIS POTOSI

Municipio de Villa de Ramos

Ramos: Moderate distribution. Noted from the Cocinera Mine. Associated with native copper.

SINALOA

Municipio de Mocorito

Ranchos: Limited distribution. Associated with native copper.

SONORA

Municipio de Alamos

Alamos: Moderate distribution. Espíritu Santo Mine. Crystals to 4 mm.

Municipio de Cananea

Cananea: Moderate distribution. Crystals to 8 mm. Associated with native copper.

Municipio de Imuris

Imuris: Limited distribution. Bonanza Mine.

Municipio de Onavas

La Barranca: Moderate distribution. Noted from the Creton and Los Cristos mines.

TAMAULIPAS

Municipio de Villagrán

San José: Limited distribution.

VERACRUZ

Municipio de Las Minas

Zomelahuacan: Moderate distribution. Associated with native copper.

ZACATECAS

Municipio de Concepción del Oro

Aranzazú: Moderate distribution. Noted from the Aranzazú Mine.

Concepción del Oro: Moderate distribution. Noted from the Arroyo de Cata Mine. Associated with native copper.

Municipio de Mazapil

Mazapil: Moderate distribution. Noted from the San Carlos and Jesús María mines.

Municipio de Zacatecas

Zacatecas: Moderately distributed. Noted from the El Bote, San Salvador, La Asturiana, and La Capilla mines. Crystals to 2 cm. Associated with calcite and native copper.

CUPRODESCLOIZITE. See
MOTTRAMITE

CUPROTUNGSTITE

$Cu_2(WO_4)(OH)_2$. A rare secondary mineral formed by the alteration of scheelite.

BAJA CALIFORNIA SUR

Municipio de L Paz

La Paz: Limited distribution. Associated with scheelite.

CHIHUAHUA

Municipio de Buenaventura

San Lorenzo: Limited distribution. Noted from the San Lorenzo Mine. Associated with scheelite.

SINALOA

Municipio de Choix

Choix: Limited distribution.

Municipio del Fuerte

El Fuerte: Limited distribution. Associated with scheelite.

SONORA

Municipio de Huépac

Huépac: Moderate distribution. San Lorenzo Mine. Associated with scheelite.

Municipio de Nacozari de García

Pilares de Nacozari: Limited distribution. Associated with scheelite.

Municipio unknown

Milopilas: Limited distribution.

Municipio de Yécora

San Nicolás: Limited distribution. Noted from the La Cruz Mine. Associated with scheelite.

CUSPIDINE

$Ca_4Si_2O_7(F,OH)_2$. An uncommon mineral found in few deposits.

MICHOACAN

Municipio de Zitácuaro

Susupuato: Moderately distributed throughout the Cerro Mazahua skarn.

CUZTICITE

$Fe_2TeO_6 \cdot 3H_2O$. A rare mineral found in few metal deposits.

SONORA

Municipio de Moctezuma

Moctezuma: Limited distribution. Noted from the Bambolla Mine. Associated with emmosite, schmitterite, kuranaknite, and eztlite.

D

DAHLLITE. See **CARBONATE-HYDROXYLAPATITE**

DANBURITE

$CaB_2(SiO_4)_2$. An uncommon mineral found in low tempature hydrothermal veins and in marbles.

BAJA CALIFORNIA

Municipio de Enseneda

El Alamo: Limited distribution. [This location is 65 km southeast of Ensenada.]

La Huerta: Limited distribution. Noted from the La Verde Mine. Crystals to 11 cm. Associated with elbaite, microcline, stilbite, smoky quartz, and muscovite. [This location is in a weathered pegmatite 6 km south of the village of La Huerta and 13 km east of Ojoa Negros. It is interesting to note that granitic pegmatites are not the normal environments for danburite to develop. Also the associations are most unusual for this mineral. The location was discovered by the local Indians in 1964 and the La Verde Mine quickly developed. In the early 1970s it was operated for the gem tourmaline crystals that were found associated with the danburite. Several pockets of danburite were mined and produced about 24 large crystal specimens and many smaller ones. The crystals ranged up to 11 cm long and about 9 cm in diameter and were white to a faint yellow color! The larger crystals were mostly opaque, but the small ones had areas that were very gemmy and yielded several fine cut stones. The crystals were terminmated in a most unusual fashion, they had two faces that looked like a blunt wedge.]

Rosa de Castilla: Moderate distribution. Noted from the Chuqui Mine (2.5 km, north of the El Fenomeno Mine). Crystals to 3 cm. [The crystals from here were in a cuboid habit.]

SAN LUIS POTOSI

Municipio de Charcas

Charcas: Moderately distributed. Noted from the La Bufa, San Bartolo, and San Sebastián mines. Crystals to 16 cm. Associated with datolite, apophyllite, stilbite, bakerite, calcite, sphalerite, cinnabar, chalcopyrite, and quartz (rock crystal and chalcedony). [The San Sebastián Mine has been the best producer of specimens with some of crystals reaching a length of 16 cm and 7 cm in diameter! The crystals are usually opaque white, but do have tips that occassionally are water clear. The smaller crystals, less than 5 cm, have brilliant crystal faces. The larger crystals usually have crystal faces that are dull. The quartz is of interest because on the larger danburite crystals, it is the high temperature form of crystals and on the smaller, it is the lower temperature crystal morphology.]

Municipio de Guadalcázar

Guadalcázar: Limited distribution.

HIDALGO

Municipio de Zimapán

Zimapán: Limited distribution within the La Sirena ore body. Associated with diopside and hedenbergite.

SONORA

Municipio unknown

Jalisco: Limited distribution. Associated with axinite and tourmaline.

DATOLITE

$CaBSiO_4(OH)$. An uncommon secondary mineral found in basalts and metamorphosed limestones.

GUANAJUATO

Municipio de Guanajuato

Guanajuato: Limited distribution. Noted from the San Carlos, and Caliche mines. Crystals to 2 cm.

SAN LUIS POTOSI

Municipio de Charcas

Charcas: Moderate distribution. Noted from the La Bufa Mine. Pale green crystals to 8 cm. Associated with danburite, pyrite, bakerite, sphalerite, and chalcopyrite. [Usually the crystals are in clusters, but at times single crystals have been found.]

DAUBREEITE

Found in meteorites at a few locations within Mexico.

DAVIDITE

$(La,Ce)(Y,U,Fe)(Ti,Fe)_{20}(O,OH)_{38}$. A rare mineral of the crichtonite mineral group. Found mostly in pegmatites and hydrothermal deposits.

SONORA

Municipio de Nacozari de García

Nacozari de García: Limited distribution.

DELAFOSSITE

$CuFeO2$. A rare secondary mineral found in the oxide zone of copper deposits.

DURANGO

Municipio de Mapimí

Mapimí: Limited distribution. Noted from the Ojuela Mine. Crystals to 4 mm. Associated with goethite, cuprite, and tenorite. [This mineral has been found as small brown to black botryoidal coatings, rosettes, and elongated crystals.]

DENNINGITE

$(Mn,Zn)Te_2O_5$. A rare mineral found in tellurium deposits.

SONORA:

Municipio de Moctezuma

Moctezuma: Limited distribution. Noted from the Moctezuma Mine. Associated with zemannite, mroseite, tellurite, and spiroffite.

DESCLOIZITE

$PbZn(VO_4)(OH)$. An uncommon secondary mineral and member of the descoloizite group found in the oxide zone of lead/zinc/copper deposits.

CHIHUAHUA

Municipio de Ahumada

Los Lomentos: Moderate distribution. Noted from the Erupción/Ahumada Mine. Crystals to 4 mm. Associated with wulfenite, vanadinite, goethite, calcite, and willemite. [It is usually found around the ore pipe as spongy crystal masses, but may be found as irregular shaped nodules and as coatings on other minerals.]

Municipio de Aquiles Serdán

Francisco Portillo: Limited distribution within the Santa Eulalia mining district west camp. Noted from the Buena Tierra and Vieja mines. Crystals to 4 mm. Associated with calcite and goethite.

San Antonio el Grande: Moderately distributed within the Santa Eulalia district east camp. Noted from the San Antonio Mine's tin chimney and oxide zone. Crystals to 4 mm. Associated with vanadinite, plattnerite, wulfenite, mimetite, calcite, and goethite.

Municipio de Coyame

Cuchilla Parado: Limited distribution. Noted from the Oruro and La Aurora mines. Crystals to 4 mm. Associated with vanadinite and calcite.

COAHUILA

Municipio de Sierra Mojada

Sierra Mojada: Limited distribution.

DURANGO

Municipio de Mapimí

Mapimí: Moderate distribution. Noted from the first level of the Ojuela, La China and La Reina mines. Brown tabular crystals to 4 mm. Associated with wulfenite, fluorite, barite, hydrozincite, murdockite, wulfenite, aurichalcite, hemimorphite, rosasite, and goethite. [This mineral is the most common vanadium mineral in the Mapimi district. The descloizite associated with fluorite was found at the La China and La Reina mines west of Mapimí.]

GUANAJUATO

Municipio de San Luis de la Paz

Mineral de Pozos: Limited distribution. Noted from the Santa Brígada Mine.

SAN LUIS POTOSI

Municipio de Catorce

Catorce: Limited distribution. Noted from the Concepción Mine.

ZACATECAS

Municipio de Zacatecas

Zacatecas: Limited distribution.

DIAMOND

C. An uncommon mineral found in kimberlite pipes and dikes or in alluvial conglomerates and gravels.

SONORA

Municipio unknown

Unknown canyon: Limited distribution. Greenish octahedral crystals to 4 mm. [The local Yaqui Indians have found on occasions small industrial grade diamond crystals in the aulluvium of a small canyon. The canyon is located in one of the three small northeastern Sonoron municipios of Bavispe, Bacerac, or Huachinera. The Indians will not tell the location of their find. The Yaquis, it must be remembered, were the last Indians in Mexico to be controlled by the Mexican government and the last to practice cannibalism (1940s).]

DIAPHORITE

$Pb_2Ag_3Sb_3S_8$. An uncommon mineral found in the primary ore zones of base metal deposits.

SAN LUIS POTOSI

Municipio de Catorce

Catorce: Limited distribution.

DICKITE

$Al_2Si_2O_5(OH)_4$. A member of the kaolinite mineral group which forms in hydrothermal environments.

CHIHUAHUA

Municipio de Cuishuiriáchic

Cusihuiriáchic: Limited distribution.

DIGENITE

Cu_9S_5. An uncommon mineral and a member of the argentite mineral group, found both in hypogene and supergene environments.

SONORA

Municipio de Cananea

Cananea: Limited distribution. Massive.

DIOPSIDE

$CaMgSi_2O_6$. An uncommon mineral and a member of the pyroxene group formed in metamorphic environments.

AGUASCALIENTES

Municipio de Tepezalá

Tepeazalá: Moderately distributed.

BAJA CALIFORNIA

Municipio de Tecate

Los Gavilanes: Moderately distributed. Noted from the Raza Mine.

Rose de Castilla: Moderate distribution. Noted from the El Fenómeno Mine. Crystals to 2 cm in length. Associated with grossular, scheelite, vesuvianite, and calcite.

Laguna Hason: Moderately distributed. Noted from the Oliva Mine. Crystals to 1 cm.

CHIHUAHUA

Municipio de Aquiles Serdán

San Antonio el Grande: Limited distribution. San Antonio Mine's skarn zone. Associated with grossular, hedenbergite, epidote, and actinolite.

Municipio de Carmargo

Santa Elena: Moderately distributed. Noted from the Charvira Mine. Associated with fosterite and enstatite.

Municipio de Saucillo
Naica: Moderately distributed. Skarns of the Sierra de Naica.

DURANGO

Municipio de Mapimí

Mapimí: Limited distribution. Noted from the Ojuela Mine-Ojuela Stope, 15th level.

MICHOACAN

Municipio de Zitácuaro

Susupuato: Moderately distributed throughout the Cerro Mazahua skarn.

SAN LUIS POTOSI

Municipio de Guadalcázar

Guadalcázar: Moderately distributed.

ZACATECAS

Municipio de Fresnillo

Fresnillo de Gonzalez Echeverria: Moderate distribution.

DIOPTASE

$CuSiO_2(OH)_2$. A rare secondary mineral found in the oxide zones of copper deposits.

BAJA CALIFORNIA

Municipio de Ensenada

Las Palomas: Limited distribution. Noted from the Alejandra

Mine. Crystals to 6 mm. Associated with malachite, chrysocolla, and goethite. [The mine is located about 38 km north of El Arco.]

CHIHUAHUA

Municipio de Chínipas

Chínipas: Limited distribution. Crystal needles to 1 cm. Associated with goethite.

DURANGO

Municipio de San Dimas

Tayolita: Moderate distribution. Noted from the Nuestra Señora de Candelaria Mine. Crystals to 5 mm. Associated with chlorargyrite.

SONORA

Municipio de Alamos

Alamos: Limited distribution. Noted from the Malaquita Mine. Crystals to 6 mm. Associated with malachite, quartz, chrysocolla, and goethite. [The dioptase occurs as a bright druze and as very small crystals on quartz.]

Municipio de Hermosillo

Hermosillo: Limited distribution. Noted from the La Verde Mine. Crystals to 4 mm.

Municipio Unknown

Portales: Limited distribution.

Potrerillos: Limited distribution. Crystals to 4 mm.

DJURLEITE

$Cu_{31}S_{16}$. A common mineral found as a supergene mineral in many copper deposits.

CHIHUAHUA

Municipio de Urique

Barranca del Cobre: Limited distribution (undisclosed location).

Municipio de Chínipas

Milpillas: Moderate distribution. Noted from the Salvadora Mine. Massive.

DOLOMITE

CaMg(CO$_3$)$_2$. A common mineral which forms in sedimentary environments and within hydrothermal mineral deposits.

CHIHUAHUA

Municipio de Aquiles Serdán

Francisco Portillo: Moderately distributed. Crystals to 3 cm. Associated with goethite, calcite, pyrite, sphalerite, and galena.

San Antonio el Grande: Widespread. Noted from the San Antonio Mine. Crystals to 5 cm. Associated with goethite, calcite, hematite, pyrite, sphalerite, and galena. [Dolomite

A 10 cm × 15 cm crystal cluster of dolomite from the San Antonio Mine, San Antonio el Grande, Chihuahua, from the Romero Mineralogical Museum, Tehuacan, Puebla.

is often found replacing calcite. The San Antonio Mine several years ago produced several pockets of brilliant snow white crystals of dolomite on goethite.]

Municipio de Batopilas

Batopilas: Limited distribution.

Municipio de Saucillo

Naica: Moderate distribution. Noted from the Naica Mine. White crystals to 6 cm. Associated with calcite, pyrite, galena, and fluorite. [Dolomite is often found as a replacement of calcite.]

DURANGO

Municipio de Mapimí

Mapimí: Moderately distributed throughout the district. Noted from the Ojuela Mine. Crystals to 4 cm. Associated with quartz and goethite.

GUANAJUATO

Municipio de Guanajuato

Guanajuato: Moderately distributed. Noted from the Cata and San Juan de Rayas mines. Crystals to 4 cm. Associated with calcite and quartz. [Dolomite has been found here in large crystal masses with extremely high luster.]

La Luz: Moderate distribution. Crystals to 2 cm.

Municipio de San Luis de la Paz

Mineral de Pozos: Limited distribution. Noted from the Trinidad and Angustias mines.

GUERRERO

Municipio de Taxco

Taxco de Alcarón: Moderately distributed.

HIDALGO

Municipio de Pachuca

Pachuca de Soto: Widespread.

Municipio de Mineral del Monte

Mineral del Monte: Widespread. [This old mining area is also known as Real del Monte.]

MICHOACAN

Municipio de Angangueo

Mineral de Angangueo: Moderately distributed. Crystals to 2 cm.

Municipio de Tlalpujahua

Cerro del Compo del Gallo: Limited distribution.

PUEBLA

Municipio de Tetela del Oro

Tetla del Oro: Moderate distribution. Noted from the Covadonga Mine.

SONORA

Municipio de Cananea

Cananea: Moderate distribution.

Municipio de Hermosillo

Hermosillo: Moderate distribution. Noted from the Amarillas and San José de las Arenillas mines.

Municipio de Soyopa

Soyopa: Moderate distribution. Prietitla Mine.

VERACRUZ

Municipio de Las Minas

Zomelahuacan: Moderate distribution.

ZACATECAS

Municipio de Fresnillo

Fresnillo de González Echeverría: Moderately distributed throughout the Cerro de Proano. Crystals to 2 cm.

Municipio de Sombrerete

Sombrerete: Moderate distribution. Noted from the La Joya Mine.

DRAVITE

$NaMg_3Al_6(BO_3)_3Si_6O_{18}(OH)_4$. A member of the tourmaline mineral group forming in metamorphic sedimentary, and occasionally igneous rocks, especially those rich in lime.

BAJA CALIFORNIA

Municipio de Tecate

El Condor: Moderate distribution. Associated with microcline and quartz.

Municipio de Ensenada

Pino Solo: Moderately distributed. Crystals. Associated with quartz and epidote.

A 6 cm crystal cluster of dravite from Santa Cruz, Sonora, from the William Larson Collection, Fallbrook, California.

Rancho Viejo: Widespread through-out pegmatites. Crystals. Associated with microcline, muscovite, and quartz.

SONORA

Municipio de Santa Cruz

Santa Cruz: Widespread in the pegmatites on Cerro de Chevito. Noted from the Guadalapana Mine. Crystals to 30 cm. Associated with uvite, scheelite, quartz, microcline, and fluorapatite. [In the early 1970s large pockets of dravite had been found on the side of Cerro de Chevito. These pockets often exceeded 1 meter in diameter thus allowing for large crystal development.The bottoms of these pockets are lined with the larger blocky crystals and the pocket tops lined with sprays of iridescent needle crystals of dravite.]

DUFTITE

$PbCu(AsO_4)(OH)$. A rare secondary mineral found in the oxidized parts of arsenic rich lead/copper deposits.

DURANGO

Municipio de Mapimí

Mapimí: Limited distribution. Noted from the Ojuela Mine. Dark olive green botryoidal crust. Associated with wulfenite, goethite, mimetite, calcite, and various manganese oxides.

DUMORTIERITE

$Al_7(BO_3)(SiO_4)_3O_3$. An uncommon mineral found in metamorphic rocks and occasionally in igneous environments.

SAN LUIS POTOSI

Municipio de Guadalcázar

Guadalcázar: Limited distribu-tion.

DURANGITE

$NaAl(AsO_4)F$. A rare mineral found within a few arsenic rich mineral deposits.

DURANGO

Municipio de Coneto de Comonfort

Coneto de Comonfort: Light distribution. Noted from the Barranca Tin Mine. Associated with cassiterite. [This is the type locality for this mineral that was named after the state in which it was found.]

ZACATECAS

Municipio de Pinos

Villa García: Limited distribu-tion.

DUSSERTITE

$BaFe_3(AsO_4)_2(OH)_5$. A rare mineral and a member of the crandallite group.

DURANGO

Municipio de Mapimí

Mapimí: Limited distribution. Noted from the Ojuela Mine. Pale green crystals to 4 mm. Associated with scorodite, carminite, goethite, and arseniosiderite.

DYSCRASITE

Ag_3Sb. A rare mineral found in silver/antimony deposits.

HIDALGO

Municipio de Mineral del Chico

Mineral del Chico: Limited distribution. Arévalo Mine.

E

EGLESTONITE

$Hg_6Cl_3O_2H$. A rare secondary mineral found in mercury deposits.

GUERRERO

Municipio de Huahuaxtla

Huahuaxtla: Limited distribution. Noted from the San Luis Mine. Associated with terlinguaite, montroydite, and calomel. [Eglestonite was once fairly abundant in the older mine workings, above the 30 meter level.]

SAN LUIS POTOSI

Municipio de Guadalcázar

Guadalcázar: Limited distribution.

Municipio de Moctezuma

Moctezuma: Limited distribution. Noted from the Dulces Nombres Mine.

ELBAITE

$Na(Li,Al)_3Al_6(BO_3)_3Si_6O_{18}(OH)_4$. An uncommon member of the tourmaline group formed in pegmatic environments.

BAJA CALIFORNIA

Municipio de Ensenada

El Topo: Limited distribution. Noted from the La Olivia and Chucqui mines. Dark blue slender crystals to 10 cm.

La Huerta: Limited distribution. Noted at the El Verde Mine. Green crystals to 5 cm. Associated with danburite. [The mine was operated during the 1970s for gem elbaite.]

Rancho Agua Blanca: Moderately distributed. Green crystals to 4 cm. Associated with quartz and epidote. [The crystals were found as small needles and pencils.]

Rancho Viejo: Moderately distributed. Most noted from the El Socorro and Las Delicias mines. Bi-colored crystals to 10 cm. [The two important producers of this mineral are located in the pegmatites west of Rancho Viejo, the Las Delicias Mine to the northern end and the El Socorro Mine at the southern end. Both of these mines have produced very fine crystals of bi-color elbaite and "watermelon tourmaline," but the largest and finest crystals came from the Las Delicias Mine. The "watermelon" crystals have a dark green exterior and a bright pinkish red interior. The crystals of elbaite ranged in size from small "pencil" sized crystals to large crystals up to 10 cm. Not all the crystals were bicolored, some were solid green, pink, or dark blue. The Las Delicias Mine produced dark blue colored crystal that reached a length of just over 10 cm and a diameter of 10 cm and weighed almost 2.3 kg. All the crystals from these mines were well developed and had modification on the flat basil terminations.]

Municipio de Ensenda

Rosa de Castilla: Moderate distribution. Crystals to 12 cm in length. Noted from the Chuqui Mine. Associated with orthoclase, andesine, garnet, and danburite. [This location is made up of several small pegmatites and located throughout the area. The Chuqui Mine is 2.5 km north of the El Fernomeno Mine. The crystals from here are dark blue and often gemmy.]

Municipio de Tecate

El Condor: Moderately distributed. Green crystals to 5 cm.

ELECTRUM

AgAu. A rare alloy found in base metal deposits.

DURANGO

Municipio de San Dimas

Tayoltita: Moderately distributed. Noted from the Tayoltita Mine. Associated with calcite, pyrite, and rhodonite. [This location has also been called San Dimas and Guarisamey in the older listings. This was due to the fact that there were three small mining camps which were usually listed as separate districts, but are now listed as one, the San Dimas mining district with its center at

Tayoltita. These three older districts produced over $150 million in gold and silver since the 1700s. At the present time, electrum is one of the three principal ore minerals.]

GUANAJUATO

Municipio de Guanajuato

Guanajuato: Moderately distributed throughout the Veta Madre.

La Luz: Limited distribution. Noted from the Santa Rita Mine (formerly known as the Bolañitos Mine).

OAXACA

Municipio unknown

Unknown location: Limited distribution. Crystals to 6 mm.

EMBOLITE. See CHLORARGYRITE var. bromian or BROMARGYRITE var. chlorian

EMMONSITE

$Fe_2Te_3O_9 \cdot 2H_2O$. A rare mineral found in tellurium deposits.

SONORA

Municipio de Moctezuma

Moctezuma: Limited distribution. Noted from the Moctezuma and Bambolla mines. Associated with schmitterite, eztlite, and cuzticite.

EMPLECTITE

$CuBiS_2$. A rare mineral found in the sulfide zone of copper/bismuth deposits.

DURANGO

Municipio de Mapimí

Mapimí: Limited distribution. Noted from the Monterrey Mine. Massive. Associated with galena, pyrrhotite, sphalerite, bismuth, enargite, bismuthinite, and umangite.

ZACATECAS

Municipio de Concepción del Oro

El Cobre: Limited distribution. Noted from the El Cobre Mine.

ENARGITE

Cu_3AsS_4. A rare mineral found in veins and replacement deposits formed in moderate temperature environments.

CHIAPAS

Municipio de Pichucalco

Pichualco: Limited distribution. Noted from the Santa Fe Mine.

CHIHUAHUA

Municipio de Cusihuiriáchic

Cusihuiriáchic: Limited distribution.

Municipio de Ojinaga

Real de Milpillas: Limited distribution.

DURANGO

Municipio de Mapimí

Mapimí: Moderately distributed. Noted from the Ojuela Mine. Associated with arsenopyrite, marcasite, pyrrhotite, galena, and chalcopyrite. [It is one of the first sulfide ore minerals formed and is the primary ore mineral of the deeper sulfide ore zone.]

SONORA

Municipio de Cananea

Cananea: Moderately distributed. Crystals to 3 cm.

ENDELLITE

$Al_2Si_2O_5(OH)_4 \cdot 2H_2O$. A member of the kaolinite/serpentine group found in metamorphic environments as a hydrothermal replacement.

QUERETARO

Municipio de Tolimán

Soyatal: Moderately distributed. Noted from the Santa María de Miera Mine. Massive. [This mineral when exposed to the air quickly alters to the more stable halloysite.]

ENDLICHITE. See VANADINITE var. arsenatian

ENSTATITE

$Mg_2Si_2O_6$. A member of the pyroxene group found commonly in mafic igneous rocks.

CHIHUAHUA

Municipio de Carmargo

Santa Elena: Limited distribution. Noted from the Chavira Olivine-Peridot Mine. Massive. Associated with spinel and fosterite.

JALISCO

Municipio de Tomatlán

Chargantillo: Limited distribution.

EPIDOTE

$Ca_2(Al,Fe)_3(SiO_4)_3(OH)$. A common mineral forming in low to moderate thermal metamorphism of metamorphic and igneous rocks.

BAJA CALIFORNIA

Municipio de Ensenada

El Alamo: Moderate distribution. Noted from the Escorpión mine.

Isla Guadelupe: Moderate distribution. Crystals to 4 cm. [Crystals have been found both as singles, often doublly terminated, and in clusters.]

Pino Solo: Modrate distribution. Crystals to 3 cm.

Punta China: Moderate distribution. Crystals to 3 cm.

San Quintin: Widespread (48 km east) in the Arroyos of San Simon and Santa María. Crystals to 10 cm. Associated with quartz and titanite. [The crystals from here brillant faces and have been found both as clusters and single crystals.]

Municipio de Tecate

Los Gavilanes: Moderate distribution. Noted from the El Fenomeno Mine. Crystals to 6 cm. Associated with clinozoisite, axinite, prehnite, and calcite.

Municipio de Mexicali

San Felipe: Moderate distribution (8 km north). Crystals.

BAJA CALIFORNIA SUR

Municipio de Mulegé

Santa Rosalía: Moderate distribution within the Boleo mining district.

Municipio de San Antonio

El Triunfo: Moderate distribution.

San Antonio: Moderately distributed.

CHIHUAHUA

Municipio de Aqulies Serdán

San Antonio el Grande: Limited distribution. Noted from the San Antonio Mine's skarn zone. Associated with quartz, grossular, celadonite, and fluorite.

Municipio de Saucillo

Naica: Moderate distribution. Associated with fluorite, calcite, and grossular.

COLIMA

Municipio de Tecomán

Cerro de Ortega: Moderate distribution. Crystals.

DURANGO

Municipio de Mapimí

Mapimí: Limited distribution. Noted from the Ojuela Mine.

GUANAJUATO

Municipio de Guanajuato

Guanajuato: Limited distribution in the Arroyo de Santa Ana.

GUERRERO

Municipio de Acapulco

Xaltianguis: Limited distribution. Crystal sprays to 11 cm.

JALISCO

Municipio de Tonila

Pihuamo: Moderate distribution.

MICHOACAN

Municipio de Coahuayana

Coahuayana: Moderate distribution.

NAYARIT

Municipio de Santiago
Ixcuintla

Zopilote: Limited distribution.
Noted from the Restaurador Mine.

NUEVO LEON

Municipio de Lampazos de
Naranjo

Piacho del Carrizal: Moderate
distribution.

OAXACA

Municipio de San Francisco
Telixtlahuaca

San Francisco Telixtlahuaca:
Moderate distribution. Noted from
the El Muerto Mine. Crystals to
6 cm. Associated with allanite
and zircon.

SONORA

Municipio de Aconchi

Sierra de Aconchi: Moderately
distributed. Crystal sprays to 10
cm.

Municipio de Alamos

Alamos: Moderate distribution.

Municipio de Sahuaripa

Trinidad: Moderate distribution.
Noted from the Dios Pradre Mine.

Municipio de Santa Cruz

Santa Cruz: Moderately distributed on the Cerro Chivato. Noted
from the Guadalupana Mine. Crystal sprays. Associated with
quartz, dravite, and scheelite.

EPSOMITE

$MgSO_4 \cdot 7H_2O$. A common secondary
mineral found as efflorescent
coatings on old mine workings and
caves.

CHIHUAHUA

Municipio de Aquiles Serdán

San Antonio el Grande: Limited
distribution. Massive. Noted from
the San Antonio Mine.

DURANGO

Municipio de Mapimí

Mapimí: Limited distribution.
Noted from the Ojuela Mine.
Massive.

GUANAJUATO

Municipio de Guanajuato

Guanajuato: Limited distribution
on the Veta Madre. Massive.

MICHOACAN

Municipio de Angangueo

Mineral de Angangueo: Limited
distribution. Noted from the San
Cayetano Mine. Massive.

ERYTHRITE

$Co_3(AsO_4)_2 \cdot 8H_2O$. A rare secondary mineral and member of the
vivianite group. It forms from
the oxidation of cobalt and
nickel minerals and often
contains some amounts of nickel.

CHIHUAHUA

Municipio de Maguarichic

Maguarichic: Limited distribu-
tion. Noted from the La Luz Mine.
Tuffs of crystals to 8 mm.
Associated with goethite.

COAHUILA

Municipio de Sierra Mojada

Sierra Mojada: Limited distribution. Noted from the Veta Rica
mine. Crystals to 6 mm. Associated with native silver, acanthite, and proustite.

JALISCO

Municipio de Tonila

Pihuamo: Limited distribution. Noted from the Esperanza and Esmeralda mines. Crystal to 3 mm.

SINALOA

Municipio de Badiraguato

Alisitos: Moderate distribution. Noted from the Gloria Mine. Associated with gold, millerite, gersdorfffite, jarosite, goethite, pentlandite, violarite, and pyrite. Crystals.

SONORA

Municipio de Alamos

San Bernardo: Limited distribution. Noted from the Sara Alica Mine. Crystals to 1 cm. Associated with stainierite, cobaltian Mansfieldite, and goethite. [The erythrite was found as hairlike crystals coating the local country rock and at times masses of crystals were found up to almost 30 cm in diameter. This location is often referred to as Alamos, but San Bernardo, 44 km north of Alamos is much closer to the mine and even then to reach the mine one must walk or go on horseback for several hours. The mine is located on a high ridge line in the Sierra Madre's and allowed the mine operators in the early 1900s, with the aide of a tower, to watch for ships entering the harbor at Guaymas. The Germans, who were buying the cobalt and nickel ores from the mine would have their ships signal the mine as they approached the harbor. The miners would then load the ore on the backs of burros to be taken to Alamos where it was then loaded on wagons for the trip to Guaymas and the awaiting ship.]

Municipio de Cananea

Cananea: Limited distribution.

ETTRINGITE

$Ca_6Al_2(SO_4)_3(OH)_{12} \cdot 26H_2O$. An uncommon mineral found in metamorphic rocks.

MICHOACAN

Municipio de Zítacuaro

Susupuato: Moderately distributed throughout Cerro Mazahua skarn (7 km southwest).

EZTLITE

$Pb_2Fe_6(Te,O_3)_3(Te_6O_6)(OH)_{10} \cdot 8H_2O$. A rare mineral found in tellurium deposits.

SONORA

Municipio de Moctezuma

Moctezuma: Limited distribution. Noted from the Bambolla Mine. Associated with Emmosite, schmitterite, kuranakhite, and cuzticite.

F

FAMATINITE

Cu_3SbS_4. A rare mineral found in the sulfide ore zone of few base metal deposits.

Location data is unavailable.

FAYALITE

Fe_2SiO_4. An uncommon mineral of the olivine group found in alkaline and felsic volcanic and plutonic igneous rocks and thermally metamorphosed sediments that are iron rich.

CHIHUAHUA

Municipio de Aquiles Serdán

Francisco Portillo: Limited distribution. Noted from the El Potosí Mine-silicate ore body. Associated with rhodochrosite, llvaite, cronstradite, pyrrhotite, sphalerite, and galena. [A manganoan fayalite (formerly called knebelite) has been found on the 10th level of the El Potosí Mine. This is the west camp of the Santa Eulalia mining district.]

FEITKNECHTITE

$MnO(OH)$. An uncommon mineral, found in oxide zone of manganese deposits.

DURANGO

Municipio de Mapimí

Mapimí: Limited distribution. Noted from the Ojuela and San Pedro mines. Associated with other manganese oxides.

FERGUSONITE

$YNbO_4$. A rare mineral found in few pegmatites, skarns, and carbonate rocks.

CHIHUAHUA

Municipio de Aldama

Puerto del Aire: Limited distribution. Noted from the Puerto del Aire Mine.

OAXACA

Municipio de San Francisco Telixtlahuaca

San Francisco Telixtlahuaca: Limited distribution. Noted from the Santa Ana Mine. Associated with columbite, allanite, and ilmenite.

FERRIMOLYBDITE

$Fe_2(MoO_4)_3 \cdot 8H_2O$. An uncommon secondary mineral formed by the alteration of molybdenite.

DURANGO

Municipio de Mapimí

Mapimí: Limited distribution. Noted from the Ojuela Mine. Massive. Associated with molybdenite.

SONORA

Municipio de Nacozari de García

Nacozari de García: Limited distribution. Noted from the Canutillo Mine. Massive.

FERROAXINITE

$Ca_2FeAl_2BSi_4O_{15}(OH)$. An uncommon mineral of the axinite group formed in contact metamorphic conditions were the intrusive rocks have invaded sediments.

BAJA CALIFORNIA

Municipio de Ensenada

Trinidad: Limited distribution.

Municipio de Tecate

Laguna Hansen: Moderate distribution. (16km ENE of Rosa de Castillo) Noted from the Olivia

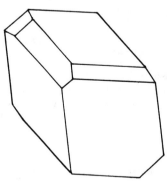

A 4 cm crystal of ferroaxinite from Los Gavilanes, Baja California. (Modified by Shawna Panczner from Sinkankas, 1964)

Mine. Crystals to 10 cm. Associated with epidote, clinozoisite, quartz, and diopside. Moderately distributed. [In 1966, a pocket was uncovered that produced over 550 kg of violet colored crystals of this mineral with several of the crystals reaching sizes of over 10 cm in length.]

Los Gavilanes: Limited distribution. Noted from the Raza Mine. Crystals to 6 cm. Associated with garnet, scheelite, epidote, clinozoisite, and diopside. [The crystals from the Raza Mine have been found up to 6 cm in length and almost 5 cm in diameter.]

Rosa de Castilla: Moderate distribution. Noted from the El Fenómeno Mine. Crystals to 8 cm. Associated with scheelite, grossular, epidote, prehnite, and diopside.

Tres Pinas: Moderate distribution. Crystals to 4 cm.

SAN LUIS POTOSI

Municipio de Guadalcázar

Guadalcázar: Limited distribution.

ZACATECAS

Municipio de Fresnillo

Fresnillo de Gonzáles Echeverría: Limited distribution. Associated with calcite.

FERROCOLUMBITE

$FeNb_2O_6$. An uncommon mineral found in lithium bearing pegmatites.

CHIHUAHUA

Municipio de Aquiles Serdán

Francisco Portillo: Limited distribution. Santa Eulalia mining district's west camp.

San Antonio el Grande: Limited distribution. Noted from the San Antonio Mine. Santa Eulalia mining district's east camp.

OAXACA

Municipio de San Francisco Telixtlahuaca

San Francisco Telixtlahuaca: Limited distribution. Noted from the Santa Ana Mine. Associated with fergusonite, orthoclase, muscovite, and biotite.

FERRO-HORNBLENDE

$Ca_2(Fe,Mg)_4Al(Si_7Al)O_{22}(OH,F)_2$. A common member of the amphoble mineral group, found in igneous and metamorphic rocks.

BAJA CALIFORNIA SUR

Municipio de Mulegé

Santa Rosalía: Moderately distributed.

DURANGO

Municipio de Mapimí

Mapimí: Moderate distribution. Noted from the Ojuela Mine.

HIDALGO

Municipio de Pachuca

Pachuca de Soto: Moderately distributed.

OAXACA

Municipio de Santo Domingo de Tehuantepec

Santo Domingo de Tehuantepec: Moderate distribution.

ZACATECAS

Municipio de Concepción del Oro

Concepción del Oro: Widespread.

FLUELLITE

$Al_2(PO_4)F_2(OH)\cdot 7H_2O)$. An uncommon mineral found in pegmatites.

HIDALGO

Municipio de Zimapán

Zimapán: Limited distribution

FLUOBORITE

$Mg_3(BO_3)(F,OH)_3$. An uncommon mineral found in hydrothermal veins.

CHIHUAHUA

Municipio de Aquiles Serdán

San Antonio el Grande: Limited distribution. Noted from the San Antonio Mine.

FLUORAPATITE

$Ca_5(PO_4)_3F$. A common member of the apatite group found in

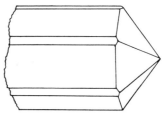

A typical 2 cm fluorapatite crystal from Cerro Mercado Mine, Victoria de Durango, Durango. (Modified by Shawna Panczner from Sinkankas, 1959)

igneous, metasomatic calc-silicate, impure carbonate, and
phosphorite rocks. [This mineral
was formerly known as apatite.]

BAJA CALIFORNIA

Municipio de Tecate

El Mesquite: Moderate distribution. Associated with quartz,
muscovite, albite, and beryl.

CHIHUAHUA

Municipio de Ciudad Camargo

La Negra: Limited distribution.
Noted from the La Negra Mine.
Crystals to 5 cm.[Fluorapatite
has been found here as replacements of magnetite crystals.]

DURANGO

Municipio de Cuencamé

Velardeña: Moderate distribution.

Municipio de Durango

Victoria de Durango: Moderate
distribution within the Cerro
Mercado. Noted from the Cerro
Mercado mine. Yellow crystals to
10 cm. Associated with calcite,
martite, amphibole, and opal
hyalite. [In 1552, Gines Vasquez
de Mercado, searching for gold
and silver discovered a hill of
solid iron. The Spanish not
needing iron at the time and
especially from this far remote
of an area, left the discovery
untouched. Mining did not start
until 1828, but it was not until
1888 that large scale mining
began. The fluorapatite is found
throughout the replacement ore
deposit and is mined along with
the iron ore. Until recently the
fluorapatite was left for the
miners to collect, but several
years ago the mining company
began processing the mineral for
its phosphate content. The
mineral is called by the miners
"amarillas" because of its yellow
color. Matrix specimens are

difficult to obtain, but are
found occasionally in the blasted
rubble.]

Municipio de Mapimí

Mapimí: Limited distribution.
Noted from the San Ilario Mine.
Massive.

GUANAJUATO

Municipio de Santa Catarina

Santa Catarina: Limited distribution. Noted from the El Santín
Mine.

MICHOACAN

Municipio de Tlalpujahua

Tlalpujahua: Limited distribution.

OAXACA

Municipio de La Pe

La Panchita: Limited distribution. Noted from the La Panchita
Mine. Dark green crystals to 12
cm. Associated with salmon
colored calcite. [The crystals
have been found up to about 12 cm
in length and about 5 cm in
diameter, but are usually much
smaller in size. The minerals
from this location looks like the
material from Canada. This small
mine is west of Ayoquezco and
north of La Panchita.]

SAN LUIS POTOSI

Municipio de Guadalcázar

Guadalcázar: Limited distribution.

SINALOA

Municipio de San Ignacio

Sindicatura San Juan: Limited
distribution.

SONORA

Municipio de Santa Cruz

Santa Cruz: Moderate distribution within the pegmatites of Cerro Chivato. Noted from the Guadalapana Mine. Cream colored crystals to 15 cm. Associated with quartz, scheelite, dravite, and epidote. [The crystals from here have been found up to 15 cm long and 8 cm in diameter. They are usually opaque, but at times some gemmy areas occur within the crystals.]

ZACATECAS

Municipio de Concepción del Oro

Concepción del Oro: Limited distribution.

FLUORITE

CaF_2. A common mineral, found in sedimentary and igneous rocks both as a primary and gangue mineral.

AGUASCALIENTES

Municipio de Tepezalá

Tepezalá: Moderate distribution. Noted from the No Pensada Mine.

CHIHUAHUA

Municipio de Ahumada

Los Lomentos: Limited distribution.

Municipio de Aquiles Serdán

Francisco Portillo: Moderate distribution. Noted from the El Potosí Mine's silicate ore body and the Buena Tierra mine. Crystals to 15 mm. Associated with huebnerite, galena, kutnahorite, rhodochrosite, sphalerite, arsenopyrite, manganite, cronstedtite, calcite, quartz, pyrite, and pyrrhotite. [The crystals are found colorless to shades of blue and purple and are found both as simple cubes or very highly modified cubes. The fluorite from the Buena Tierra Mine has a different association: calcite, quartz, sphalerite, pyrite, and galena, but other wise is similar in size and color to the El Potosí fluorite.]

San Antonio el Grande: Moderate distribution. Noted from the San Antonio Mine's skarn zone. Crystals to 2 cm. Associated with calcite, quartz, epidote, garnet, and celadonite. [It ranges in color from colorless through shades of blue to purple cubic crystals.]

Municipio de Chínipas

Palmarejo: Limited distribution. Noted from the Las Hundidas Mine.

Municipio de Hidalgo del Parral

Hidalgo del Parral: Moderate distribution. Noted from the El Tajo and Crestas de La Veta mines. Crystals to 2 cm.

Municipio de San Francisco del Oro

San Francisco del Oro: Moderate distribution. Noted from the San Francisco del Oro Mine.

Municipio de Saucillo

Naica: Moderate distribution. Noted from the Naica Mine. Crystals to 8 cm. Associated with calcite, sphalerite, quartz, galena, anhydrite, rutile, barite, and grossular. [Crystals range in color from colorless through shades of green, blue, and purple. The crystals are found as modified cubes, octahedrons and rarely as dodecahedrons. Over the past few years several fine pockets of fluorite have been recovered. In 1976 and 1977 exceptional green fluorite crystals up to almost 8 cm were found. In 1978, a few remarkable green dodecahedrons up to about 3 cm in diameter associated with brillant dodecahedrons of galena were found. In the new mine on the Naica fault both blue and green fluorite octahedrons have been found, some reaching a size of 6 cm.]

COAHUILA

Municipio de Arteaga

San Vincente: Moderate distribution.

Municipio de Candela

Páncuco: Limited distribution. Noted from the Pánuco Mine.

Municipio de Melchor Múzquiz

Melchor Múzquiz: Widespread. Noted from the Sierra Grande and El Tule mines. Crystals to 9 cm. Associated with celestite and calcite.[Fluorite from here comes in two different colors, a dark blue/purple and a light blue with white interiors. The dark blue/purple crystals are from the Sierra Grande Mine and are found both as simple cubes and modified cubes. The white fluorite with the last stages of growth a bright blue is from the El Tule Mine and found as modified cubes. The fluorite from the El Tule Mine is found associated with large celestite crystals.]

Municipio de Ocampo

Santo Domingo: Moderate distribution within the Sierra de Aguachile.

DURANGO

Municipio de Cuencamé

Velardeña: Moderate distribution.

Municipio de Durango

Victoria de Durango: Limited distribution on Cerro de Los Remedeos. Associated with tridymite.

Municipio de Guanaceví

Guanaceví: Moderate distribution. Noted from the Socavón de San José Mine.

Municipio de Mapimí

Mapimí: Limited distribution. Noted from the Ojuela Mine. Crystals to 5 cm. Associated with barite, wulfenite, goethite, calcite, plattnerite, hemimorphite, murdockite,and aurichalcite. [It has been found as crystals of cubic, octahedron, and dodecahedron habits, with the later two being the rarer forms. The fluorite has been found in color ranging from colorless to shades of blue, purple, green, pink, and brown. The Guadalupe Mine has been listed as a producer, but this is part of the Ojuela Mine. At times some of the fluorite has a very bright red fluorescence.]

Municipio de Tamazula

Tamazula: Limited distribution.

GUANAJUATO

Municipio de Guanajuato

Guanajuato: Moderate distribution. Noted from the El Nopal Mine.

La Luz: Moderate distribution. Noted from the Santa Rita Mine/Bolañitos Mine. Associated with calcite and quartz.

Santa Rosa: Limited distribution within the Sierra de Santa Rosa.

Municipio de Santa Catarina

Santa Catarina: Limited distribution. Noted from the El Santín Mine.

GUERRERO

Municipio de Taxco

Taxco de Alcarón: Moderate distribution. Noted form the La

Azul, Xochicalco, and El Gavilán mines. Crystals to 3 cm. Associated with quartz, barite, calcite, and gearksutite. [The fluorite at the La Azul Mine is found associated with crystals of gearksutite, barite, quartz, and calcite. At the other mines only the quartz, barite, and calcite have been found associated with fluorite. The color of the fluorite varies from a brown to dark blue to purple.]

Municipio de Tehuilotepec

Tehuilotepec: Limited distribution.

HIDALGO

Municipio de Mineral del Monte

Mineral del Monte: Moderate distribution.

Municipio de Pachuca

Pachuca de Soto: Moderate distribution.

Municipio de Zimapán

Zimapán: Moderate distribution. Noted from the Santa clara, Chiquihuitemire, Lomo del Toro and Los Balcones mines. Crystals to 2 cm. Associated with galena and malachite.[Fluorite from a small prospect on the south side of Cerro de la Nopalera has produced small green fluorite crystals. The color was caused by malachite inclusions within the fluorite. The Santa Clara Mine has had light green fluorite found within its mine workings. Purple fluorite has been found at the Balcones Mine's San Rafael ore chimney.]

JALISCO

Municipio de Bolaños

Bolaños: Moderately distributed. Noted from the Santa Fe, San Cayetano, San José, and Vewtanilla mines. Associated with

sphalerite, galena, quartz, and calcite.

Municipio de Hostotipaquillo

Hostotipaquillo: Limited distribution.

MEXICO

Municipio de Temascltepec

Temascltepec de González: Moderately distributed.

Tequesquipan: Moderate distribution.

PUEBLA

Municipio de Tetela del Oro

Tetela del Oro: Limited distribution.

QUERETARO

Municipio de Caderayta

El Doctor: Moderate distribution.

SAN LUIS POTOSI

Municipio de Catorce

Catorce: Moderate distribution.

Municipio de Charcas

Charcas: Moderate distribution.

Municipio de Guadalcázar

Guadalcázar: Limited distribution within the Cerro San Cristóbal.

Municipio de Villa de Reyes

Reyes: Limited distribution within the Cerro de la Piedra del Molino.

Municipio de Zaragoza

Zaragoza: Widespread.Noted from the Consentida and Esperanza mines. Massive. [The Consentida

Mine is the largest open pit fluorite mine in the world. Fluorite is also mined underground at the Esperanza Mine. This location is slightly more than 50 km southeast of the capital city of San Luis Potosí.]

SONORA

Municipio de Agua Prieta

Agua Prieta: Limited distribution. South of Agua Prieta.

Municipio de Arizpe

Arizpe: Limited distribution. Noted from the Las Chipas Mine. Associated with polybasite.

ZACATECAS

Municipio de Chalchihuites

Chalchihuites: Limited distribution. Noted from the Santa María Dolores Mine.

Municipio de Mazapil

Mazapil: Limited distribution. Noted from the San José Nevada Mine.

Municipio de Zacatecas

Zacatecas: Moderate distribution. Noted from the La Cantera Mine.

FORSTERITE

Mg_2SiO_4. A member of the olivine mineral group found in igneous and metamorphic rocks. In its green gem form it is known as peridot.

BAJA CALIFORNIA SUR

Municipio de Ensenada

Isla Cedros: Limited distribution. Massive. (Peridot).

Isla Guadalupe: Limited distribution. Massive. (Peridot).

Associated with spinel and augite.

CHIHUAHUA

Municipio de Camargo

Santa Elena: Limited distribution. Noted from the Chavira Olivine-Peridote Mine. Massive. (Peridot). Associated with enstatite and spinel.

Municipio de Madera

Unknown location: Limited distribution. (Peridot).

FOWLERITE. See **RHODONITE** var. zincian

FRANKLINITE

$(Zn,Mn,Fe)(Fe,Mn)_2O_4$. A member of the spinel group, and found in the oxidized ore zone of zinc deposits.

CHIHUAHUA

Municipio de Aquiles Serdán

San Antonio el Grande: Limited distribution. Noted from the San Antonio Mine's 8th level. Associated with quartz and sphalerite.

FREIBERGITE

$(Ag,Cu,Fe)_{12}(Sb,As)_4S_{13}$. A rare sulfosalt found in hydrothermal veins.

CHIHUAHUA

Municipio de Aquiles Serdán

San Antonio el Grande: Limited distribution. Noted from the San Antonio Mine.

Municipio de Urique

Urique: Limited distribution.

ZACATECAS

Municipio de Concepción del Oro

Bonanza: Limited distribution. Noted from the Bonanza Mine. Crystals to 2 cm. Associated with quartz.

FREIESLEBENITE

$AgPbSbS_3$. A rare sulfosalt found in a few silver lead deposits.

CHIHUAHUA

Municipio de Uruáchic

Uruáchic: Limited distribution. Noted from the Bolas Mine.

DURANGO

Municipio de Guanaceví

Guanaceví: Limited distribution. Noted from the El Verde Mine.

Municipio de Tepeguanes

San Ignacio de la Sierra: Moderate distribution. Noted from the Huamuchil and El Salto mines.

GUERRERO

Municipio de Chilpancingo

Chilpancingo de los Bravos: Limited distribution. Noted from the Delfina Mine.

MEXICO

Municipio de Temascltepec

Temascltepec de González: Limited distribution. Noted from the Quebradillas Mine.

Municipio de Tlatlaya

Tlatlaya: Limited distribution. Most noted from the El Diablo Mine.

SAN LUIS POTOSI

Municipio de Catorce

Catorce: Moderate distribution.

SINALOA

Municipio de Cosalá

Guadalupe de Los Reyes: Limited distribution.

G

GALENA

PbS. A common primary mineral found in sedimentary, igneous, and metamorphic rocks. It is common in hydrothermal veins and can contain a high percentage of silver.

AGUASCALIENTES

Municipio de Asientos

Asientos: Widespread. Noted at the Santa Francisca, Santo Cristo, and Rosario mines. Associated with sphalerite, calcite, and pyrite.

Municipio de Tepezalá

Tepezalá: Widespread.

BAJA CALIFORNIA

Municipio de Ensenada

Calamajue: Moderately distributed. Noted from the Rey Salomón and San Antonio mines.

El Alamo: Moderately distributed. Noted at the Aurora and Escorpión mines.

Real del Castillo: Moderate distribution. Noted from the Ensueño Mine.

BAJA CALIFORNIA SUR

Municipio de San Antonio

Triunfo: Widespread. Most noted from the Gobernadora, Jesús María, Mina Rica, San Antonio, and Humbold mines. Massive. Associated with pyrite, sphalerite, and tetrahedrite.

CHIHUAHUA

Municipio de Aquiles Serdán

Francisco Portillo: Widespread. Noted from the El Potosí and Buena Tierra mines. Crystals to 2 cm. Associated with rhodochrosite, fluorite, sphalerite, pyrite, and calcite. [It has been found as fine crystal specimens from the 10th level of the El Potosí Mine associated with rhodochrosite and fluorite. Galena has been found altering to anglesite.]

San Antonio el Grande: Widespread. Noted from the San Antonio Mine. Crystals to 3 cm. highly distroted. Associated with fluorite, calcite, and sphalerite.

Municipio de Ascensión

Sabinal: Widespread. Noted from the Biznaga, Adventurera, Aztec, and San Blas mines. Massive.

Municipio de Batopilas

Batopilas: Moderately distributed. Noted from the Americana, San Miguel, and San Nestor mines.

Municipio de Cuidad Camargo

Sierra de las Encinillas: Widespread.

Municipio de Cusihuiriáchic

Cusihuiriáchic: Moderately distributed. Noted from the Soledad and La Reina mines. Massive. Associated with cerussite.

Guadalupe y Calvo: Widespread. Noted at the San Gil and Ahumada mines.

Municipio de Hidalgo del Parral

Hidalgo del Parral: Widespread. Noted from the El Tajo and Cinco de Mayo mines.

Municipio de Manuel Benavides

San Carlos: Widespread. Noted from the San Carlos Mine. Associated with sphalerite and calcite.

Municipio de Moris

Zahuayacan: Moderate distribution. Noted from the Santa Teresa Mine.

Municipio de Nuevo Casas Grandes

San Pedro Corralitos: Widespread. Noted from the San Pedro and Candelaria mines. Massive.

Municipio de San Francisco del Oro

San Francisco del Oro: Widespread. Noted at the San Diego and San Francisco mines. Massive.

Municipio de Santa Bárbara

Santa Bárbara: Widespread.

Municipio de Saucillo

Naica: Widespread. Noted from the Naica Mine. Brilliant crystals to 6 cm. [Crystals are simple and modified cubes and dodecahedrons. Associated with sphalerite, calcite, fluorite, and grossular. Much of the galena contains a very high percentage of silver.]

Municipio de Urique

Cerocahui: Moderately distributed. Noted from the Cerocahui Mine. Massive.

Piedras Verdes: Moderate distribution. Noted from the El Truyo Mine.

Urique: Widespread. Most noted from the Dulces Nombres, Colorado, Plomosa, Rosario, and Santo Niño mines. Massive.

COAHUILA

Municipio de Ramos Arizpe

Ramos Arizpe: Widespread. Noted from the La Placeras, Veta Rico, and Puerto Rico mines. Massive.

Municipio de Sierra Mojada

Sierra Mojada: Widespread. Most noted from the Santa María de Los Angles Mine.

Municipio de Matamoros

Matamoros de la Laguna: Moderately distributed. Noted from the Jimulco Mine.

DURANGO

Municipio de Cuencamé

Velardeña: Widespread.

Municipio de Guanaceví

Guanaceví: Widespread.

Municipio de Pánuco de Coronado

Pánuco de Coronado: Widespread. Most noted from the Avino Mine.

Massive. Associated with native silver. [This location was known formerly as San José de Avino.]

Municipio de Indé

Indé: Widespread. Noted from the Descubridora, Cabillo, Colorada, and La Cruz mines.

Municipio de Mapimí

Mapimí: Widespread. Noted from the Ojuela Mine. Massive. Associated with sphalerite, pyrite, marcasite, enargite, chalcopyrite, arsenopyrite, and tetrahedrite. [Much of the galena from here contains a high percentage of silver.]

Municipio de Mezquital

Mezquital: Widespread. Noted from the El Esfuerzo and Esplendida mines.

Municipio de Nombre de Dios

Nombre de Dios: Widespread. Most noted at the Dolores and El Carmen mines.

Municipio del Oro

Magistral: Widespread. Noted from the Porvenir Mine. Massive. Associated with chalcopyrite, chalcocite, and native silver.

Santa María del Oro: Widespread. Massive. Associated with native silver, cinnabar and native mercury.

Municipio de Poanas

Poanas: Widespread. Most noted from the Veta Grande and Rosario mines.

Municipio de San Dimas

Tayoltita: Widespread. Noted from the Candelaria Mine. Massive.

Municipio de San Pedro del Gallo

Peñoles: Widespread.

Municipio de Topia

Topia: Widely distributed. Massive.

GUANAJUATO

Municipio de Guanajuato

Guanajuato: Widely distributed. Massive.

La Luz: Widely distributed. Massive.

Municipio de San Luis de la Paz

Mineral de Pozos: Widespread. Massive.

Municipio de Xichú

Xichú: Widespread. Massive.

GUERRERO

Municipio de Chilpancingo

Carrizal: Moderately distributed. Noted from the Delfina Mine. Massive.

Municipio de Taxco

Taxco de Alcarón: Widespread. Massive. Noted from the Guerrero Mine. Associated with sphalerite, pyrite, and marcasite.

HIDALGO

Municipio de Cardonal

Cardonal: Widespread. Most noted from the La Punta Mine. Massive.

Municipio de Mineral del Chico

Mineral del Chico: Widespread. Massive.

Municipio de Mineral del Monte

Mineral del Monte: Widespread. Massive.

Municipio de Pachuca

Pachuca de Soto: Widespread.
Massive.

Municipio de Zimapán

Zimapán: Widespread. Noted from
the Lomo del Toro, Los Balcones,
and Miguel Hidalgo mines.
Crystals to 4 cm. Associated
with fluorite, jamesonite,
pyrite, and sphalerite. [Galena
has been found at the Lomo del
Toro Mine associated with
fluroite and at the Miquel
Hidalgo Mine with jamesonite. The
best specimen producing mine is
the Los Balcones Mine, where it
has been found as bright cubic
crystals up to 4 centimeter in
size associated with pyrite and
sphalerite.]

JALISCO

Municipio de Ameca

Ameca: Widespread. Noted from the
Cuauhtémoc Mine. Massive.

Municipio de Bolaños

Bolaños: Widespread. Massive.
Associated with sphalerite and
fluorite.

Municipio de Lagos de Moreno

Comanja: Moderate distribution.

Municipio de Tamazula de Gordiano

Villa de Tamazula: Moderate
distribution. Noted from the Las
Esperanzas Mine.

Municipio de Tequila

San Pedro Analco: Widespread.
Most noted from the Ajustadero
and Buenavista mines.

MEXICO

Municipio de Sultepec

Sultepec de Pedro Ascencio de
Alquisiras: Widespread. Most
noted at the Guaje, Salon, and
Sangre de Cristo mines. Massive.

Municipio de Temascaltepec

Temascaltepec de González:
Widespread.

Municipio de Zacualpan

Zacualpan: Widespread. Massive.

MICHOACAN

Municipio de Ario de Rosales

Ario de Rosales: Widespread.
Noted from the San Pazacuareta
and San Cristóbal mines.

Municipio de Angangueo

Mineral de Angangueo: Widespread.
Most noted from the El Carmen,
San Cristobal, and Trinidad
mines. Massive.

Municipio de Tlalpujahua

Tlalpujahua: Widespread.

MORELOS

Municipio de Tlaquiltenango

Huautla: Moderately distributed.
San Esteban Mine.

Municipio de Xochitepec

Cerros de Colotpec: Moderately
distributed.

Jamiltepec: Moderately distributed.

NAYARIT

Municipio de Santa María del Oro

Acuitlapilco: Moderate distribution.

NUEVO LEON

Municipio de Monterrey

Monterrey: Moderate distribution.
Noted at the Colorado Mine.
Associated with anglesite.

OAXACA

Municipio de Natividad

Natividad: Widespread. Noted from the Natividad Mine. Massive. Associated with native silver.

Municipio de San Jerónimo Taviche

San Jerónimo Taviche: Widespread. Most noted from the La Luz, La Cruz, and La Soledad mines.

Municipio de San Miguel Peras

San Miguel Peras: Widespread. Noted from the La Asunción Mine.

PUEBLA

Municipio de Chinantla

Cerro de Zompantitlán: Widespread.

Municipio de Tehuacán

Tehuacán: Moderate distribution. Most noted from the La Infanta Mine.

Municipio de Tetela de Ocampo

Tetela de Ocampo: Moderate distribution. Most noted from the Covadonga and Espejero mines.

QUERETARO

Municipio de Amoles

Ahuacatlan: Moderately distributed.

Municipio de Cadereyta

El Doctor: Widespread.

Municipio de Jalpan

Pinal de Amoles: Widespread. Most noted from the Escanela and Penasco mines.

Municipio de Tolimán

Río Blanco: Moderately distributed. Noted from the Santa Ana Mine.

SAN LUIS POTOSI

Municipio de Catorce

Catorce: Widespread.

Municipio de Cerro de San Pedro

Cerro de San Pedro: Widespread. Noted from the San Pedro and Socavon de la Victoria mines. Associated with sphalerite and pyrite.

Municipio de Charcas

Charcas: Widespread. Massive. Associated with pyrite and sphalerite.

Municipio de Guadalcázar

Guadalcázar: Moderately distributed.

Municipio de Villa la Paz

La Paz: Widespread.

Municipio de Villa de Reyes

Villa de Reyes: Moderate distribution. Noted from the Providencia Mine.

SINOLOA

Municipio de Cosalá

Cosalá: Widespread. Massive.

Municipio del Fuerte

Las Papas: Moderately distributed.

SONORA

Municipio de Alamos

Alamos: Moderately distributed. Noted from the Veta Grande Mine. Associated with native silver, pyrite, and sphalerite.

Municipio de Cananea

Cananea: Moderately distributed. Crystals to 2 cm. Associated with quartz, pyrite, and rhodochrosite.

Municipio de Carborca

Heróica Carborca: Widespread.
Most noted from the Amarilla and
Animas mines.

Municipio de Cucurpe

Cucurpe: Widespread within the
Cerro Prieto. Noted from the San
Francisco Mine. Massive. Associated with cerussite, wulfenite,
and mimetite.

VERACRUZ

Municipio de Las Minas

Zomelahuacan: Widespread. Most
noted from the San Miguel, San
Andres, and El Veladero mines.
Massive.

ZACATECAS

Municipio de Concepción
del Oro

Concepción del Oro: Widespread.
Massive.

Municipio de Fresnillo

Fresnillo de Gonzáles Echeverría:
Widespread within the Cerro de
Proano. Massive.

Plateros: Widespread. Most noted
from the Providencia Mine.
Massive.

Municipio de General
Francisco Hurguía

Nieves: Widespread. Most noted
from the San José, Santa Catrina,
and San Gregorio mines. Massive.

Municipio de Mazapil

Noche Buena: Widespread. Noted
from the Noche Buena Mine.
Massive.

Mazapil: Widespread. Noted from
Santa Rosa Mine. Crystals to 3
cm. Associated with sphalerite.

Municipio de Noria
de Angeles

Noria de Angeles: Widespread.
Massive.

Municipio de Sombrerete

Sombrerete: Widespread.

Municipio de Vetagrande

Vetagrande: Widespread. Massive.
Common.

Municipio de Zacatecas

Zacatecas: Widespread. Most noted
from the El Bote Mine. Massive.
Associated with sphalerite,
quartz, and calcite.

GALENOBISMUTITE

$PbBi_2S_4$. A rare mineral found in
lead bismuth deposits.

DURANGO

Municipio de Santa
María del Oro

Tepehuanes: Limited distribution.

GUANAJUATO

Municipio de Guanajuato

Santa Rosa: Limited distribution
within the Sierra de Santa Rosa.
Noted from the La Industrial
Mine.

GARNIERITE

A general term for a specific
group of minerals.

GEARKSUTITE

$CaAl(OH)F_4 \cdot H_2O$. An uncommon
mineral formed in sedimentary

rocks, pegmatites, hydrothermal veins, and hot springs.

GUERRERO

Municipio de Taxco

Taxco de Alcarón: Limited distribution. Noted from the La Azul Mine. Massive. Associated with fluorite.

GEHLENITE

$Ca_2Al(AlSi)O_7$. A common member of the melilite group found in metamorphic contact zones.

DURANGO

Municipio de Cuencamé

Velardeña: Limited distribution. Noted from the Velardeña Mine. Massive ("granular aggregates").

GERMANITE

$Cu_3(Ge,Fe)(S,As)_4$. A rare sulfosalt found in few base metal deposits.

GUANAJUATO

Municipio de Guanajuato

Guanajuato: Limited distribution. Rare.

SONORA

Municipio de Cananea

Cananea: Limited distribution.

GERSDORFFITE

NiAsS. An uncommon mineral found in hydrothermal veins.

CHIHUAHUA

Municipio de Maguarichic

Maguarichic: Limited distribution. Noted from the La Luz Mine.

SINOLOA

Municipio de Badriaguato

Alisitos: Moderately distributed. Noted from the Gloria Mine. Associated with gold, niccolite, erythrite, annabergite, pentlandite, pyrite, violarite, maucherite, and goethite.

GILLESPITE

$BaFeSi_4O_{10}$. A rare mineral found in few mineral deposits.

BAJA CALIFORNIA

Municipio de Tecate

Madrellana: Limited distribution. Associated with with sanbornite.

GISMONDINE

$CaAl_2Si_2O_8 \cdot 4H_2O$. A rare mineral of the Zeolite group, which forms in volcanic rocks.

ZACATECAS

Municipio de Concepción del Oro

Bonanza: Limited distribution. Massive. Associated with pyrite and tobermorite.

GLAUBERITE

$Na_2Ca(SO_4)_2$. An uncommon mineral found in sedimentary salt deposits and fumerolic activities.

CHIHUAHUA

Municipio de Jiménez

Laguna de Jaco: Limited distribution. Massive. Associated thenardite and halite.

GLAUCODOT

(Co,Fe)AsS. A rare mineral of the arsenopyrite mineral group found in cobalt deposits.

SONORA

Municipio de Alamos

Alamos: Moderately distributed. Associated cobaltite.

San Bernardo: Moderately distributed. Noted from the Sara Alica Mine. Associated with arsenopyrite and cobaltite.

GMELINITE

$(Na_2,Ca)Al_2Si_4O_{12} \cdot 6H_2O$. An uncommon member of the zeolite group found in ignerous rocks, especially basalt.

JALISCO

Municipio de Teocaltiche

Cerro de Narizón: Limited distribution.

GOETHITE

FeO(OH). A common mineral which forms under a wide variety of oxidizing conditions.

AGUASCALIENTES

Municipio de Asientos

Asientos: Widespread. Massive. Associated with all oxidized ore zone minerals.

Municipio de Tepezalá

Tepezalá: Widespread. Massive. Associated with all oxidized ore zone minerals.

BAJA CALIFORNIA

Municipio de Ensenada

Calmelli: Widespread. Massive.

BAJA CALIFORNIA SUR

Municipio de Mulegé

Santa Rosalía: Widespread. Massive.

Municipio de San Antonio

El Triunfo: Widespread. Masive.

San Antonio: Widespread. Massive.

CHIHUAHUA

Municipio de Ahumada

Constitución: Moderately distributed. Noted from the Mojina Mine. Massive

Municipio de Aquiles Serdán

Francisco Portillo: Widespread. Massive.

San Antonio el Grande: Widespread. Massive.

Municipio de Chihuahua

Labor de Terrazas: Widespread. Massive. Associated with azurite, malachite, and calcite.

Municipio de Coyame

Cuchillo Pardo: Widespread. Most noted from the Cobriza Mine. Massive.

Municipio de Nuevo Casas Grandes

San Pedro Corralitos: Widespread. Massive.

Municipio de Saucillo

Naica: Widespread. Massive.

COAHUILA

Municipio de Candela

Pánuco: Widespread. Noted from the Escondida and Pánuco mines. Massive. Associated with azurite and malachite.

Municipio de Castaños

Castaños: Moderately distributed. Noted from Mercedes Mine.

Municipio de Melchor Múzquiz

Melchor Múzquiz: Widespread. Most noted from the Cedral, Rosario, San Francisco, and San Gertrudis mines. Massive.

Municipio de Sierra Mojada

Sierra Mojada: Widespread. Massive. Associated with most of the minerals of the oxidized zone.

DURANGO

Municipio de Durango

Victoria de Durango: Moderately abundant. Most noted from the Cerro de Mercado Mine.

Municipio del Oro

Magestral: Widespread. Noted from the Povenar Mine. Massive.

Santa María del Oro: Widespread. Massive.

Municipio de Guanaceví

Guanaceví: Widespread. Massive.

Municipio de Indé

Indé: Widespread. Noted from the Buenavista Mine.

Municipio de Mapimí

Buruguilla: Widespread. Massive.

Mapimí: Widespread. Noted from the Ojuela Mine. Massive.

Municipio de San Dimas

Tayoltita: Widespread.

Municipio de Nombre de Dios

Nombre de Dios: Moderately distributed. Most noted from the Sacramento and Dolores y Santa Gertrudis mines.

GUANAJUATO

Municipio de San Luis de la Paz

Mineral de Pozos: Widespread. Most noted from the Santa Brígida, Juárez, Cinco Señores, and Trinidad mines.

Municipio de Xichú

Xichú: Widespread. Noted from the La Tribuna Mine.

GUERRERO

Municipio de Alcozauca de Guerrero

Ixcuinatoyac: Widespread.

HIDALGO

Municipio de Mineral del Monte

Mineral del Monte: Widespread. Most noted from the Jesús María, Dificultad and Carretera mines.

Municipio de Pachuca

Pachuca de Soto: Widespread. Noted from the Entrometida Mine.

Municipio de Zimapán

Zimapán: Widespread.

JALISCO

      Municipio de Bolaños

Bolaños: Widespread.

      Municipio de Magdalena

San Simon: Widespread. Noted from the El Aguila and Cinco mines.

MEXICO

      Municipio de Temascaltepec

Temascaltepec de González: Widespread. Noted from the El Rincon Mine.

MICHOACAN

      Municipio de Tlalpujahua

Tlalpujahua de Gonzales: Widespread.

      Municipio de Zitácuaro

Susupuato: Limited distibution.

PUEBLA

      Municipio de Tehuitzingo

Atopoltitlan: Moderate distribution on the Cerro del Guajolote.

SAN LUIS POTOSI

      Municipio de Catorce

Catorce: Moderate distribution.

      Municipio de Guadalcázar

Guadalcázar: Moderate distribution.

      Municipio de Villa de la Paz

La Paz: Widespread. Noted from the Santa María de la Paz Mine.

SINALOA

      Municipio de Badiraguato

Soyatita: Widespread. Noted from the San Antonio Mine.

SONORA

      Municipio de Alamos

Alamos: Moderately distributed.

San Antonio: Moderately distributed. Most noted from the Malaquita, Aduanas, and Chilla mines. Associated with most of the oxidized zone minerals.

      Municipio de Cananea

Cananea: Moerately distributed.

      Municipio de Cucurpe

Cerro Preita: Widespread. Associated with wulfinite, mimetite, and cerussite.

VERACRUZ

      Municipio de Las Minas

Zomelahuacan: Widespread.

ZACATECAS

      Municipio de Concepción del Oro

Aranzazú: Widespread.

Bonanza: Widespread.

      Municipio de Mazapil

Mazapil: Widespread.

GOLD

Au. A widely distributed native element, which is found typically in hydrothermal veins and in the oxidized zones of many metal deposits.

BAJA CALIFORNIA

Municipio de Ensenada

Bahia de Los Angelas: Limited distribution. Mina Las Flores.

Calamajue: Moderate distribution. (Veins and placers). Noted from the La Josefina, King Richard, Rey Salomón, and San Antonio mines. [Discovered in 1882 and produced over $250,000.]

Calmelli: Limited distribution.

El Alamo: Moderately distributed. (Placer/vein). Noted from the Aurora, Escorpión, and Biznaga mines. [Discovered in 1889 and within a few months over five thousand people were working and living near the Santa Clara placer fields.]

Isla de Cedros: Limited distribution. Noted from the Mascara de Hierro Mine. Associated with quartz, chalcocite, and hematite.

La Arrastras: Limited distribution. Noted from the Josefina Mine.

Las Palomas: Limited distribution. 24 km north of El Arco.

Real del Castillo: Moderate distribution. (Placer/lode). Noted from the América and Buena Vista mines. [Discovered in 1870 by Ambrosio Castillo as a placer deposit in the areas stream beds, but soon lode deposits were uncovered in the surrounding hills. From 1872 to 1882, it was the Territorial capital.]

San Borjas: Limited distribution. (Arroyo del Tule). San

Felipe: Moderate distribution. (Placer). Arroyo de Miramar. 16 km south of San Felipe.

Socorro: Moderate distribution. Most noted from the Garibaldi Mine.

BALA CALIFORNIA SUR

Municipio de Mulegé

San Andrés: (East of Punta de San Roque) Limited distribution.

Municipio de San Antonio

El Triunfo: Widely distributed. Associated with native silver. [Discovered in 1862 and by 1874 the thirty-six stamp mill was producing $50,000 monthly in gold and silver. The mines employed the local Yaquis Indians as miners. In 1918 all the mines were flooded by a hurricane. The French El Boleo Mining Company tried to rehibilitate the mines in 1926 but shortly pulled out of the district. The mines have since been abandoned.]

Municipio de los Cabos

Rancho San Felipe: Limited distribution. Noted from the San Juan and San Felipe mines. Associated with cervantite.

CHIHUAHUA

Municipio de Aldama

Aldama: Moderately distributed in the Placer fields of Santo Domingo and Placer de Guadalupe.

Puerto del Aire: (3 km east of Placer de Guadalupe). Moderately abundant. Most noted from the Puerto del Aire La Esperanza, and Virgin mines. Associated with uraninite.

Municipio de Allende

San Pedro de la Cienega: Moderately distributed. Noted from the Carniquena and San Pedro mines.

Municipio de Aquiles Serdán

Francisco Portillo: Limited distribution. Noted from the El Potosí Mine. (Santa Eulalia mining district.)

San Antonio el Grande: Limited distribution. Noted from the San

Antonio Mine's 14th and 15th level. Associated with sphalerite, chalcopyrite, galena, and pyrite. (Santa Eulalia mining district.)

Municipio de Batopilas

Batopilas: Limited distribution.

Cerro Colorado: Moderate distribution. Noted from the Nuesta Señora del Pilar Mine.

Municipio de Chihuahua

Labor de Terrazas: Moderate distribution. Noted from the La Prieta and Montecristo mines.

Municipio de Cusihuiriáchic

Cusihuiriáchic: Moderate distribution. Noted from the Princesa Mine. Associated with native silver.

Municipio de Guadalupe y Calvo

Guadalupe y Calvo: Moderate distribution.

Municipio de Guazapares

Las Hundidas: Moderate distribution.

San José: Moderate distribution. Noted from the Calabacillas and Monterde mines. Associated with hematite. [This location is famous for its specimens of gold and hemitite.]

Municipio de Ocampo

Pinos Altos: Moderate distribution.

Municipio de San Francisco del Oro

San Francisco del Oro: Widespread. Noted from the San Francisco del Oro Mine.

Municipio de Santa Bárbara

Santa Bárbara: Widespread.

Municipio de Saucillo

Naica: Moderately distributed.

Municipio de Urique

Piedras Verdes: Moderately distributed. Noted from the El Cobre, Lluvia de Oro, and Santa Maria mines.

Urique: Moderately distributed (Placer/lode).

DURANGO

Municipio de Pánuco de Coronado

Pánuco de Coronado: Moderately distributed. Noted from the Avino Mine. Associated with native silver and chalcopyrite. [This is the old Spanish gold/siver mining district of San Jose de Avino.]

Municipio Guanaceví

Guanaceví: Widely distributed.

Municipio de Indé

Villa Ocampo: Moderately distributed.

Municipio de Mapimí

Mapimí: Limited distribution. Noted from the Ojuela Mine. Associated with goethite, hematite, native silver, and various manganese oxides.

Municipio de San Pedro del Gallo

Peñoles: Limited distribution. Noted from the Descubridoro Mine.

Municipio del Oro

El Magistral: Limited distribution. Noted from the Porvenir Mine.

Santa María el Oro: Moderate distribution. Noted from the Santa Ana and Esperanza mines.

Municipio de San Dimas

Tayoltita: Widespread. Noted from the La Nuestra Señora de Candelaria, Bolaños, San Luis, Arana, and the Tayoltita mines. Associated with native silver, calcite, pyrite, and rhodenite. [Most of the original claims were discovered and denounced by a Spaniard by the name of Zambrano. His Candelaria Mine produced great wealth in gold and silver. Zambrano and his family opened other mines within the district. At the beginning of the Revolution in 1810, his nephew then in control of the Zambrano Estate collected what monies were available, quickly mined the rich ore left in the pillars of all the mines and fled México. The 25 years that Zambrano and his family operated the mines of San Dimas, The Kings taxes according to offcial records were $11 million dollars! This left Zambrano and his heirs $68.5 million to operate the mines and profits. Unfortunately the mining costs were extremely high, but still the profits were great. The mines of the area were extremely rich and produced many great bonanzas. The modern San Dimas Mining District is composed of three older and smaller districts; the San Dimas, Toyaltita, and Guarisemey. These districts produced over $150 million in gold and silver! They came from the complex system of 100 epithermal gold and silver veins. The gold is found at the present time in the mineral electrum.]

Municipio de Topia

San Antonio del Oro: Moderately distributed.

GUANAJUATO

Municipio de Guanajuato

Guanajuato: Moderately distributed within the Veta Madre. Noted from the San Juan de Rayas, El Cubo, and Peregrina mines. Associated with native silver.

La Luz: Moderately distributed. Noted from the Purísima, Santa Rita (Bolañitos) and Santa Cruz mines. [Gold at the Santa Cruz Mine was found as "rotten threads of rich gold ore."]

Municipio de San Luis de la Paz

Mineral de Pozos: Moderately distributed. Noted from the Santa Brígida, Trinidad, Cinco Señores, Angustias, and Escondida mines.

GUERRERO

Municipio de Taxco

Taxco de Alcarón: Moderate distribution. Noted from the La Cruz Mine.

Municipio unknown

Tepantitlán: Moderate distribution. Noted from the Dulces Hombres Mine.

HIDALGO

Municipio de Mineral del Chico

Mineral del Chico: Moderate distribution.

Municipio Mineral del Monte

Mineral del Monte: Widespread.

Municipio de Pachuca

Pachuca de Soto: Widespread.

Municipio de Zimapán

San José del Oro: Moderate distribution.

JALISCO

Municipio de Guachinango

El Barqueño: Moderately distributed.

Guachinango: Moderate distribution.

Municipio de Mascota

Mascota: Moderate distribution. Noted from the Jalpa, Jesús María, La Florida, and Providencia mines.

Municipio de Tequila

Anonas: Moderate distribution. Noted from the San Cayetano Mine.

MEXICO

Municipio del Oro

El Oro de Hidalgo: Widely distributed. Most noted from the Esperanza, El Oro, Dos Estrellas (all on the Veta San Rafeal), México and Nolan mines. Associated with quartz.

Municipio de Sultepec

Tlatlaya: Moderately distributed on the Cerro de los Ocotes. Most noted from the La Fama, Aguacate, Bella Manana, Santa Ana, and San Caralampio mines.

Municipio de Zacualpan

Zacualpan: Moderately distributed on the Cerro de Coronas. Noted from the Carboncillo Mine.

MICHOACAN

Municipio de Coalcomán

Placeres del Agua Hedionda: (Placer).

Municipio de Tlalpujahua

Tlalpujahua: Widespread. Veta Coronas. Most noted from the Concepción, San Hilario Borda, La Luz de la Cañada, Dolores, Providencia, Mina Dura, Santa Rita, Carmen, La Luz, Ocotes, Santos Mártires, Vírgenes, Charcas, El Muerto, San Antonio Comanja, Tetela, San Juan Nepomuceno, Socavón de la Casa, Coloradillas, Coronas, Manduermes, Socavón de los Mártires, and the Azteca mines. Associated with native silver. [This

location was formerly known as San Pedro and San Pablo and were combined into the large village and renamed Tlalpujahua. The Coronas vein was discovered in 1743 and produced millions in gold and silver. Over one eight year period, the veta Coronas produced in excess of $12 million dollars! The mines fell into ruins during the revolution, but were rebuilt and continued production for many years.]

NAYARIT

Municipio de Amatlán de Cañas

Barranca del Oro: Moderate distribution. Noted from the Rondanera, Soledad, La Chispa, and El Banco mines.

Municipio de Santa María del Oro

Acuitlapilco: Moderately distributed. Noted from the Esmeralda Mine.

OAXACA

Municipio de San Mateo Capulalpan

Capulapan de Méndez: Limited distribution. Associated with pyrite and stephanite.

Municipio de San Miguel Peras

San Miguel Peras: Moderate distribution. Noted from the Monserrate and Purísima mines. [This location has been referred to in early writings as Peras.]

Municipio de Santa María Yavesía

Santa María Yavesía: Limited distribution.

Municipio de Tetela de Ocampo

Tetela de Ocampo: Moderate distribution. Most noted from the Providencia and Covadonga mines.

QUERETARO

Municipio de Ezequiel Montes

Bernal: Moderate distribution.

Municipio de Tolimán

Río Blanco: Moderately distributed. Noted from the Cueva de Oro, Saucito, De la Grandeza, Trinidad, and Nueva California mines.

SAN LUIS POTOSI

Municipio de Cerro de San Pedro

Cerro de San Pedro: Moderately distributed. Most noted from the Gogorrón, El Ray, Victoria, and Baneno mines. Associated with hematite, calcite, and quartz. [This location when it was first discovered in the 1550s and produced gold nuggets from the surface up to 20 kg!]

Municipio de Guadalcázar

Guadalcázar: Moderately distributed. [This location, dating back to the 1620s, was first known as San Francisco de Guadalcázar, was worked first as a placer gold deposit. But as silver was discovered and mined, gold was recovered from the ores.]

SINALOA

Municipio de Badiraguato

Alisitos: Widely distributed. Noted from the gloria Mine. Associated with niccolite and millerite. [The gold is found in two types of vein assembledges. The first is gold in miccolite with lesser amounts of the jarosite, annabergite, gersdorffite, maucherite, and goethite. The other vein assembledge is gold in millerite with lesser amounts of erythrite, gersdorffite, jarosite, goethite, pentlandite, pyrite, and violarite.]

Municipio de Choix

Choix: Moderately distributed. Associated with native silver.

Huitis: Moderately distributed.

Municipio de Cosalá

Guadalupe de los Reyes: Widespread. Most noted from the Zapote, Estacas, Descubridora, Republicana, and Canadelaria mines. [This area was also worked as a placer gold district.]

Municipio de Culiacán

Tepuche: Moderately distributed. Associated with quartz.

Municipio del Fuerte

Sivirijoa: Moderately distributed. Associated with quartz.

Municipio del Rosario

Cacolotán: Moderately distributed. Noted from the Tajo Mine. Associated with quartz, hematite, acanthite, and sphalerite.

SONORA

Municipio de Arizpe

Arizpe: Moderately distributed. Noted from the Las Chipas and San Lorenzo mines.

Municipio de Cucurpe

Cucurpe: Moderately distributed throughout the Cerro Prieto. Noted from the San Francisco Mine. Associated with pyrite, goethite, quartz, mimetite, and wulfenite.

Municipio de Cumpas

Tepache: Moderately distributed. Most noted from the Coronado and La de Gracía mines.

Municipio de Hermosillo

La Prietas: Moderate distribution. Noted from the Las Prietas,

Las Amarillas, and Amargosa mines.

Municipio de Caborca

Heróica Carborca: Widespread. Most noted from the la Francesa, La Pinta, Vieja de Oro, Argentina, Cobriza Esmeralda, and Mina Grande mines.

Tajito: Moderately distributed. Noted from the Amarillas, San Buenaventura, Tajitos, Oro Blanco, Animas, and Abundancia mines.

Municipio de Magdalena

Magdalena: Moderate distribution. Noted from the Arca de Oro and Fourth of May mines.

Municipio de Pitiquito

La Cienega: Moderately distributed. Noted from the Deo Gratias Mine.

Municipio de Quiriego

Batacosa: Widespread. Most noted from the Creston, Colorada, Santa Cruz, Delfina, Colorada del Norte mines.

Municipio de Sahuaripa

Mulatos: Moderate distribution. Noted from the Santa Ana Mine. Associated with barite, quartz, and calcite. [This mine was known prior as the Mulatos Mine, but had its name changed with ownership.]

Municipio de Santa Ana

Santa Ana: Moderate distribution. (40 km southeast).

Municipio de Trincheras

La Mur: Limited distribution. Noted from the Las Animas Mine.

Trincheras: Moderately distributed. (Placer).

TAMAULIPAS

Municipio de Villagrán

San José: Moderately distributed. Most noted from the Santa Fe Mine.

VERACRUZ

Municipio de Las Minas

Zomelahuacan: Widespread.

ZACATECAS

Municipio de Fresnillo

Fresnillo de Gonzáles Echeverría: Limited distribution on Cerro de Proano.

Municipio de Mazapil

Mazapil: Moderately distributed. Most noted from the Santa Rosa Mine. Associated with native silver and acanthite.

Municipio de Mezquital del Oro

Mezquital del Oro: Moderately distributed. Noted from the Mezquital del Oro and San Diego Mine.

Municipio de Noria de Angeles

Noria de Angeles: Widespread. Most noted from the Matatuza and El Rufugio mines. [This mining district is now being developed into the largest open pit silver mine in the world.]

Municipio de Pinos

Pinos: Widespread. Most noted from the Candelaria, El Oro, María, San José, Mina Azul, and Infantita mines. Associated with native silver.

Municipio de Zacatecas

Zacatecas: Moderately distributed. Most noted from the El Bote, San Luis del Oro and La Estrella mines.

GOSLARITE

$ZnSO_4 \cdot 7H_2O$. An uncommon seconday mineral formed from the alteration of sphalerite.

CHIHUAHUA

Municipio de Aquiles Serdán

Francisco Portillo: Widespread. Massive. Uncommon. Santa Eulalia mining district.

San Antonio el Grande: Moderately distributed. Noted from the San Antonio Mine. Massive. Santa Eulalia mining district.

Municipio de Cusihuiriáchic

Cusihuiriáchic: Limited distribution. Noted from the Princeses Mine. Associated with native silver.

Municipio de Ocampo

Ocampo: Limited distribution.

DURANGO

Municipio de Guanaceví

Guanaceví: Limited distribution. Noted from the San Gill Mine. Massive.

Municipio de Mapimi

Mapimí: Limited distribution. Noted from the Ojuela Mine. Massive.

GRAPHITE

C. An uncommon mineral formed by metamophic or hydrothermal activities.

GUERRERO

Municipio de Copala

Jalapa: Moderate distribution.

OAXACA

Municipio de San Francisco Telixtalahuca

Las Sedas Teltlahuaca: Limited distribution.

SONORA

Municipio de La Colorada

San José de Pimas: Widespread. Most noted from the El Lápiz Mine. Massive.

Moradillas: Widespread. Noted from the Moradillas, San Francisco, La Fortuna, San Antonio, San Marcial de Arriba, San Marcial de Abajo, San José de los Nopales, La República, Santa Cecila, and Ogden mines. Massive. Associated with crystals of andalusite.

Municipio de Hermosillo

Villa de Coris: Moderately distributed. Noted from the El Lápiz Mine. Massive.

Municipio de Onavas

Santa María: Limited distribution. Noted from the La Barranca Mine.

GREENOCKITE

CdS. A rare mineral usually found coating sphalerite and in mafic igneous rocks.

CHIHUAHUA

Municipio de Aquiles Serdán

Limited distribution. Noted from the San Antonio Mine's 8th level. Associated with sphalerite and smithsonite. [The bright yellow smithsonite color that is found here is because of inclusions of greenockite.]

GUERRERO

Municipio de Taxco

Taxco de Alcarón: Limited distribution. Associated with sphalerite.

ZACATECAS

Municipio de Noria de Angeles

Noria de Angeles: Limited distribution.

GREIGITE

$FeFe_2S_4$. A rare mineral and member of the linnaeite group and found in sulfide zones of iron deposits.

ZACATECAS

Unknown location: Limited distribution.

GROSSULAR

$Ca_3Al_2(SiO_4)_3$. A common member of the garnet group formed by thermal and contact metamorphism of rocks rich in calcium and aluminum.

AGUASCALIENTES

Municipio de Tepezalá

Tepezalá: Moderate distribution. Crystals to 2 cm.

BAJA CALIFORNIA

Municipio de Ensenada

San Pedro Martier: Widespread. Cinnamon red to greenish crystals to 1 cm.

Municipio de Tecate

Jacume: Limited distribution. Cinnamon crystals to 1 cm.

Los Gavilanes: Moderate distribution. Crystals to 2 cm.

CHIHUAHUA

Municipio de Aquiles Serdán

Francisco Portillo: Limited distribution. Crystals to 2 cm.

San Antonio el Grande: Limited distribution. Most noted from the San Antonio Mine. Crystals to 2 cm. Associated with hedenbergite, quartz, and actinolite.

Municipio de Saucillo

Naica: Moderate distribution. Greenish to white crystals to 3 cm. Associated with calcite, anhydrite, and galena.

COAHUILA

Municipio de Candela

Sierra de la Candela: Moderate distribution.

Municipio de Sierra Mojada

Sierra de la Cruz: Moderate distribution within skarn zones. White, green, and pink crystals to 15 cm. Associated with vesuvianite. [This location often has been listed as Lake Jaco, Chihuahua but is actually a skarn and metamorphic contact zone here in the Sierra de la Cruz in Coahuila a few km east of lago jaco (Lake Jaco).]

DURANGO

Municipio de Mapimí

Mapimí: Limited distribution. Noted from the Ojuela Mine. Crystals to 3 cm.

Municipio de San Pedro del Gallo

Peñoles: Limited distribution. Noted from the Potosí Mine.

Municipio de Poanas

Poanas: Limited distribution on the Cerro del Sacrificio. Most noted from the La Candelaria Mine. Pale green to pink crystals to 4 cm.

MORELOS

Municipio de Ayala

Jalostoc: Moderately distributed throughout the Sierra Tlayecac on the Rancho San Juan. Rose, pink to white dodecahedral crystals to 10 cm and up to 11 kg. Associated with wollastonite and vesuvianite. [Occasionally the crystals are gemmy and yield fine cut stones. This material has been slabbed and polished. In older literature this mineral was known as Landerite, Xalostocite, or rosolite. This location has had a spelling change from Xalostoc.]

PUEBLA

Municipio de Zapotitlán Salinas

Salinas Grandes: Limited distribution.

SAN LUIS POTOSI

Municipio de Guadalcázar

Guadalcázar: Moderate distribution. Noted from the San Rafael Mine.

Municipio de Villa de Reyes

Villa de Reyes: Limited distribution. Noted from the Providencia Mine.

SONORA

Municipio de Alamos

Alamos: Limited distribution.

TAMAULIPAS

Municipio de San Carlos

San Carlos: Limited distribution. Most noted at the El León Mine.

Municipio de Villagrán

San José: Moderately distributed on the Cerro de la Piedra Iman.

VERACRUZ

Municipio de Las Minas

Las Minas: Limited distribution. Most noted from the La Rinconada Mine.

Municipio de Profesor Rafael Ramirez

Piedras Parada: Moderately distributed. Gemmy green crystals to 2 cm. Associated with quartz amethyst. [This mineral is rather common here, but it was not until the 1980s that it became known to collectors. This municipio was formally known as Las Vigas and was recently changed.]

ZACATECAS

Municipio de Concepción del Oro

Concepción del Oro: Moderate distribution. Noted from the La Perla Mine.

GROUTITE

$MnO(OH)$. A rare mineral found in manganese deposits.

DURANGO

Municipio de Mapimí

Mapimí: Limited distribution. Noted from the Ojuela Mine.

Massive. Groutite has been found
here as an alteration product of
cryptomelane.

GUADALACAZARITE. See
METACINNABAR

GUANAJUATITE

Bi_2Se_3. A rare mineral found in
few metal deposits.

AGUASCALIENTES

Municipio de Tepezalá

El Cobre: Limited distribution.
Noted from the Santa Bárbara
Mine.

Tepezalá: Limited distribution.

GUANAJUATO

Municipio de Guanajuato

Santa Rosa: Limited distribution
within the Sierra de Santa Rosa.
(Rancho Calvillo) Santa Catarina
and La Industria mines. Crystals
to 2 cm. Associated with paraguanajuatite, bismuth, and
pyrite. [Guanajuatite was discovered at the Santa Catarina
Mine on the Rancho Calvillo and
named after the state that it was
found in. This is the type
location for the mineral.]

GYPSUM

$CaSO_4 2H_2O$. A common mineral
forms in sedimentary, volcanic,
and hydrothermal veins.

BAJA CALIFORNIA

Municipio de Mexicali

San Felipe: Moderately distribu-
ted 40 km south. Noted from the
La Delicias and San Carlos mines.
Massive. Associated with sulfur.

A 12 cm long gypsum "ram's horn" from
Francisco Portillo, Chihuahua. Romero
Mineralogical Museum, Tehuacan Puebla.

[Found here as "ram's horns" to
6 cm on the sulfur.]

BAJA CALIFORNIA SUR

Municipio de Mulegé

Isla San Marcos: Widespread.
Massive. Gypsum has been mined
here for the smelter at Santa
Rosalía.

Santa Rosalía: Widespread.
(Selenite). Crystals to 2 m.
Most noted from the Amelia and
Curuglu mines. Associated with
boleite, cumengite, and psudo-
boleite.

CHIHUAHUA

Municipio de Ahumada

Ahumada: Widespread within the
Ahumada playa. (Selenite)
Crystals to 40 cm. Clusters of
crystals have been found over 2 m
in size. [This location has
produced tons of of these crystal
clusters and will continue as
long as the playa receive enough
water enriched with sulfate to
allow the crystals to develop
within the playa sands. Ahumada
was formally known as Villa
Ahumada.]

Los Lamentos: Moderately distributed. (Selenite). Noted from the
Erupción/Ahumada Mine.

Municipio de Aldama

Estacion de Calera: Widespread.
Crystals to 60 cm. Common. [In

the mid 1970s, a small cavern was uncovered near here full of water, when pumped dry it was found to be lined with selenite crystals. The crystals were tanish in color and reached a length of just over 60 cm long and 5 cm in diameter. The cave was named for its crystals, the "Caverna de Velas" or Cave of the Candles. Because the water could not be completely pumped out, the cave was never fully explored. In 1984, the local water table lowered and collecting began again.]

San Sóstenes: Moderate distribution.

Municipio de Aquiles Serdán

Francisco Portillo: Widespread. Crystals to 12 m. Most noted at the El Potosí, Buena Tierra, and Las Animas mines. Associated with calcite, fluorite, rhodochrosite, sklodowskite, creedite, pyrite, and novacekite. [The gypsum here is found in massive form, fine "rams horns," and fine crystals. The snow white "rams horns" have been found here have been up to over 30 cm. Crystals of gypsum (selenite) found here have a color range from water clear to a snow white and cream. In 1978, when in developing the Santo Domingo Ramp, a cavern was discovered. The cave was named after the patron saint of the district, Caverna de Santo Dominigo, and is one of Mexico's largest selenite caves. The cavern is composed of three rooms. The main gallery, is approximately 100 m long and 10 m wide and at its highest point 15 m high! Crystals in the cave are of two types, short, blocky prisms and long, well terminated arm size crystals. The crystals in the cave range from 1 to 2 m in length and have a diameter of about 12 cm. About 50 to 60 crystals have the same diameter, but have lengths from 2 to 3 m. Six crystals were measured and averaged 3.7 m in length. A longer crystal could not be reached for direct measurement, but was at least 3.9

m long. All the long crystals were water clear. The short prismatic crystals were water clear when small but became snow white as they became larger in size. The largest of these stubby crystals was 65 cm long and 50 cm in diameter. All the selenite crystals had developed from the floor of the cave, mostly in large sprays of crystals. The walls and ceilings were lined with snow white stalatic gypsum and large "rams horns" of gypsum. Associated with the gypsum in the ceiling are very tiny fluorite crystals. These bright blue, 2 to 3 mm crystals, exhibited both cubic and octahedral crystal habit.

The credit for protecting the Caverna de Santo Domingo has to go to the superintendent of mines for IMMSA, Amando Ibara. He has done all that is possible to save this unique mineralogical occurrence. But with all the heavy mining vehicle traffic on the ramp a few meters away, and the blasting of ore a few hundred meters below, the larger crystals are beginning to break, slowly and sadly the cave is beginning to go into the last stages of any cave development, death and final collapse.

Gypsum from the other mines of the west camp have different associations: at the El Potosí Mine, it is found with creedite, rhodochrosite, and calcite; at Buena Tierra, with calcite and fluorite; at the Las Animas Mine, with sklodowkite, novacekite, and calcite.]

San Antonio el Grande: Widespread. (Selenite). Most noted at the San Antonio Mine.

Municipio de Camargo

La Negra: Moderate distribution. Noted from the La Negra Mine. Associated with hematite and magnetite.

La Perla: Moderate distribution. Noted from the La Perla Mine. Associated with hematite and magnetite.

Municipio de Galeana

Galeana: Moderate distribution. (Selenite). Water clear crystals to 1 m. [Many of the crystals had bubbles of gas that had been trapped inside when the crystal was developing. Some of the gas bubbles would move almost half the length of the crystal they were in. Many of the crystals, especially the smaller ones exhibited the common fish tail twinning on their terminations.]

Municipio de Meoqui

Cerro del Mármol: Moderately distributed. (Alabaster).

Sierra del Carrizo: Moderately distributed. (Alabaster).

Municipio de Saucillo

Naica: Widespread. (Selenite). Noted from the Gibralter and Naica mines. Crystals to 2 m. Associated with anhydrite, fluorite, galena, pyrite, an sphalerite. [A most remarkable selenite discovery was made in the 1940s when on the 120 m mine level of the Gibralter Mine, two caves lined with selenite were uncovered. The smaller of the of the two, the Xochitl cave, contained crystals up to 40 cm in length many had movable bubbles and phantoms. The larger cave, the Cave of Swords (over 100 m long and 10 m high) was lined with crystals of selenite. The crystals reached a length of 2.5 m and were about 15 to 20 cm in diameter. The larger crystals were not totally water clear, but had a somewhat cloudy interior. The terminations often exhibited typical fish tail twinning. The smaller crystals exhibited rounded terminations because of water variations within the cave during formation. When the cave was first seen by the miners, they thought they had broken into an old mine working and the crystals were mine timbers. Because the hot, 52°C, water contains much dissolved sulfate within the mine, the sumps within the mine usually are lined with crystals of selenite. This is such a problem, that periodically the sumps, pumps, and pipes have to be stripped of their crystal lining. The miners have fun hanging nuts, bolts, and even geologist picks into the sumps and letting selenite crystals develop on them.]

COAHUILA

Municipio de Ocampo

Ocampo: Moderately distributed. (Selenite). Crystals to 70 cm. [Crystals of selenite have been found near here in small caves as frosted fish tail selenite. Many kilos of crystal clusters were mined from this locaity, but at the present time the caves appear to have played out. The crystals that were heavily twinned all had frosted surfaces and reached a maxmium length of 70 cm and a diameter of 20 cm. The longer crystals had a much smaller diameter and coversely the larger diameter crystals were shortened. The clusters of crystals were mined up to 1.5 m in length and 2 m in width.]

COLIMA

Municipio de Coquimatlán

Hacienda de Magdalena: Widespread.

DURANGO

Municipio de Cuencamé

Pedricena: Widespread on the Rancho de Fernandez.

Municipio de Durango

Victoria de Durango: Moderately distributed. Noted from the Cerro de Mercado Mine. Associated with fluorapatite.

Municipio de Mapimí

Cerro Jaboncillo: Moderately distributed. Noted from the La Tenebrosa Mine.

Sierra de Banderas: Moderately distributed. Noted from the San José de Banderas Mine.

Mapimí: Moderately distributed. (Selenite). Noted from the Ojuela Mine. Crystals to 1 m. Associated with goethite, koettigite, and copper. [At the end of 1983, a large water course was discovered in the upper areas of the Ojuela Mine that was filled with fine crystals of selenite. The crystals ranged from a few cm in length to large fish tailed twinned crystals of 1 m. The crystal surfaces were slightly etched giving them a frosted appearance.]

Municipio de San Pedro del Gallo

Peñoles: Moderate distribution. Noted from the La Bufa Mine.

GUANAJUATO

Municipio de Guanajuato

Guanajuato: Moderately distributed on the Veta Madre.

GUERRERO

Municipio de Huitzuco

Huitzuco de los Figueroa: Moderate distribution. (Selenite). Noted from the La Cruz Mine. Crystals. Associated with crystals of sulfur and livingstonite enclosed within the selenite.

Municipio de Taxco

Taxco de Alcarón: Moderate distribution. (Selenite). Crystals. Associated with native silver. [In the early years of mining, crystals of selenite were found containing wires of native silver.]

HIDALGO

Municipio de Mineral del Chico

Mineral del Chico: Moderate distribution.

Municipio de Pachuca

Pachuca de Soto: Moderate distribution. Noted from the Encino Mine. Crystals containing wires of native silver.

Municipio de Zimapán

Bonanza: Moderate distribution. Noted from the San Judas Mine.

JALISCO

Municipio de Tecolotlán

Ameca: Moderately distributed on the Cerro del Vallado.

Rancho Tamazula: Moderate distribution.

MEXICO

Municipio de Texcoco

Lago de Texcoco: Widespread. (Selenite). Crystals to 18cm.

MICHOACAN

Municipio de Tlalpujahua

Tlalpujahua: Moderately distributed.

MORELOS

Municipio de Jonacatepec

Jonacatepec: Moderately distributed.

QUERRETARO

Municipio de Tolimán

Sayarea: Moderately distributed.

SAN LUIS POTOSI

Municipio de Catorce

Catorce: Widespread. (Selenite).
Noted from the San Pedro,
Dolores, and Anexas mines.

Municipio de Guadalcázar

Guadalcázar: Widespread.
(Selenite). Noted from the Jesús,
San Juan, and La Trinidad
mines.

Municipio Villa de la Paz

La Paz: Moderate distribution.
(Selenite). Noted from the Sante
Fe Mine.

SINALOA

Municipio de Cosalá

Guadalupe de los Reyes: Moderately distributed.

TAMAULIPAS

Municipio de Ciudad
Victoria

Rancho del Progreso: Moderately
distributed.

ZACATECAS

Municipio de Mazapil

San Pedro Ocampo: Moderate
distribution. Crystals.

H

HAKITE

$(Cu,Hg,Ag)_{12}Sb_4(Se,S)_{13}$. A rare mineral and member of the tetrahedrite group, which is found in selenium deposits.

SONORA

Municipio de Moctezuma

Moctezuma: Limited distribution. Noted from the Moctezuma Mine.

HALITE

NaCl. A common mineral formed in sedimentary environments.

BAJA CALIFORNIA SUR

Municipio de Comondú

Isla del Carmen: Moderately distributed near Punta Candelcos. Crystals to 3 cm. [This is a sunken volcanic crater, which has been mined commercially for halite.]

Municipio de La Paz

Isla de San José: Moderately distributed.

Municipio de Mulegé

Guerrero Negro: Widespread. Crystals to 10 cm. [Halite is still mined commercially in evaporation ponds.]

Santa Rosalía: Moderately distributed. Crystals to 1 cm.

CHIHUAHUA

Municipio de Aquiles Serdán

Francisco Portillo: Limited distribution.

San Antonio el Grande: Limited distribution. Noted at the San Antonio Mine.

Municipio de Jiménez

Laguna de Jaco: Moderate distribution. Crystals to 2 cm. Associated with thenardite and glauberite. [Halite has been mined at this playa lake.]

DISTRITO FEDERAL

Atzacoalco: Limited distribution. Moderate distribution. Massive.

MEXICO

Municipio de Texcoco

Lago de Texcoco: Limited distribution. Associated with trona.

SAN LUIS POTOSI

Municipio de Salinas

Peñon Blanco: Moderate distribution. Noted from the El Refugio Mine.

YUCATAN

Municipio de Hunucmá

Sisal: Moderate distribution. Crystals to 2 cm.

HIDALGO

Municipio de Mineral de Chico

Mineral de Chico: Moderate distribution. Massive.

QUERETARO

Municipio de Tolimán

Soyatal: Moderate distribution. Massive.

ZACATECAS

Municipio de Fresnillo

Plateros: Limited distribution.

HALLOYSITE

$Al_2Si_2O_5(OH)_4$. an uncommon member of the kaolinite group formed by actions of sulfate rich waters on kaolinite and in hydrothermal alterations of other minerals.

BAJA CALIFORNIA SUR

Municipio de Mulegé

Santa Rosalía: Widespread. Massive.

CHIHUAHUA

Municipio de Saucillo

Naica: Widespread. Massive.

DISTRITO FEDERAL

Azcapotzalco: Moderate distribution. Massive.

Villa de Guadalupe Hidalgo: Moderate distribution. Massive.

HALOTRICHITE

$FeAl_2(SO_4)_4 \cdot 22H_2O$. An uncommon secondary mineral formed by the weathering of rocks containing pyrite and aluminum.

BAJA CALIFORNIA SUR

Municipio de La Paz

La Paz: Limited distribution.

DURANGO

Municipio de Mapimí

Mapimí: Limited distribution. Noted from the Ojuela Mine. Massive.

GUANAJUATO

Municipio de Guanajuato

Guanajuato: Limited distribution on the Veta Madre.

MEXICO

Municipio de Temascaltepec

Temascaltepec de González: Moderate distribution. Noted from the El Rosario Mine.

HARMOTOME

$(Ba,K)_{1-2}(Si,Al)_8O_{16} \cdot 6H_2O$. An uncommom member of the zeolite mineral group and found in igneous rocks.

CHIHUAHUA

Municipio de Batopilas

Batopilas: Limited distribution. Associated with native silver and heulandite.

MICHOACAN

Municipio de Angangueo

Mineral de Angangueo: Moderately distributed.

SONORA

Municipio de Alamos

Alamos: Limited distribution. (Small manganese prospect northwest).

San Bernardo: Limited distribution. Noted from the Sara Alica Mine. Crystals to 2 cm. Associated with stilbite and calcite.

HAUSMANNITE

Mn,Mn_2O_4. An uncommon mineral found in high temperature veins or in contact metamorphic rocks.

DURANGO

Municipio de Mapimí

Mapimí: Limited distribution. Noted from the San Pedro and Ojuela mines. Massive. Associated with pyrolusite and cryptomelane.

HIDALGO

Municipio de Zimapán

Zimapán: Limited distribution.

HEDENBERGITE

$CaFeSi_2O_6$. An uncommon mineral of the pyroxene group which forms in metamorphosed limestones.

AGUASCALIENTES

Municipio de Asientos

Asientos: Moderately distributed.

Municipio de Tepezalá

Tepezalá: Moderately distributed.

BAJA CALIFORNIA

Municipio de Tecate

Rosa de Castilla: Moderately distributed. Noted from the El Fenomeno Mine.

CHIHUAHUA

Municipio de Aquiles Serdán

Francisco Portillo: Moderate distributon. Noted from the El Potosí Mine's silicate ore body. Associated with ilvaite and fayalite.

San Antonio el Grande: Limited distribution. Noted from the San Antonio Mine's skarn zone. Associated with grossular, quartz, and actinolite.

Municipio de Saucillo

Naica: Limited distribution.

SONORA

Municipio de Cananea

Cananea: Limited distribution.

ZACATECAS

Municipio de Concepción
del Oro

Aranzazú: Limited distribution.
Noted form the San Antonio Mine.

Municipio de Fresnillo

Fresnillo de González Echeverría:
Limited distribution.

HEDYPHANE

$(Ca,Pb)_5(AsO_4)_3Cl$. An uncommon
member of the apatite mineral
group found in lead deposits.

CHIHUAHUA

Municipio de Manuel
Benavides

San Carlos: Limited distribution.
Noted from the San Carlos Mine.

DURANGO

Municipio de Mapimí

Mapimí: Limited distribution.
Noted from the Ojuela Mine's 14th
level(San Juan Poniente stope)
White crystals to 4 mm. Associated with mimetite, bindheimite,
and goethite.

HELVITE

$Mn_4Be_3(SiO_4)_3S$. A rare member of
the helvite mineral group formed
in contact zones of metasomatic
rocks, in granites, and pegmatites.

CHIHUAHUA

Municipio de Aquiles Serdán

Francisco Portillo: Moderately
distributed. Noted from the El
Potosí Mine's 5th level (silicate
ore body). Crystals to 4 mm.
Associated with manganite and
rhodochrosite.

HEMATITE

α-Fe_2O_3. An uncommon mineral
found in metamorphic, igneous,and
sedimentary rocks also hydrothermal veins and in grossans of
many base metal deposits. Some
hematite varieties (specularite
and martite) have also been found
within Mexico.

AGUASCALIENTES

Municipio de Asientos

Asientos: Moderate distribution.

Municipio de Tepezalá

Tepezalá: Moderately distributed
on the Veta Nopal.

BAJA CALIFORNIA

Municipio de Ensenada

El Socorro: Moderately distributed.

Isla de Cedros: Limited distribution. Noted from the Mascara de
Hierro Mine. Crystals to 1
cm. Associated with quartz.

Las Palomas: Moderately distributed.

San Fernando: Moderately distributed.

Real del Castillo: Widespread.

BAJA CALIFORNIA SUR

Municipio de San Antonio

El Triunfo: Widespread.

CHIAPAS

Municipio de Acacoyagua

Acacoyagua: Moderately distributed. Noted from the El Porvenir
Mine.

Municipio de Pichucalco

Pichucalco: Moderately distributed. Noted from the Sante Fe
Mine.

CHIHUAHUA

Municipio de Allende

Talamantes de Arriba: Widespread. Associated with magnetite, rhodonite, and cryptomelane.

Municipio de Aquiles Serdán

Francisco Portillo: Widespread.

San Antonio el Grande: Widespread. Noted from the San Antonio Mine.

Municipio de Camargo

La Negra: Widespread. Noted from the La Negra Mine. Crystals to 2 cm. Associated with magnetite and gypsum. (Specularite) is noted from here.

La Perla: Widespread. Noted from the La Perla Mine. Crystals to 2 cm magnetite and gypsum. (Specularite) is noted from here.

Municipio de Cusihuiriáchic

Cusihuiriáchic: Moderately distributed. (Specularite) has been found here.

Municipio de Guazapares

San José: Moderately distributed. Noted from the Calabacillas and Montrede mines. Massive. (Specularite). Associated with gold.

Municipio de Hidalgo del Parral

Hidalgo del Parral: Widespread. (Specularite).

Municipio de San Francisco del Oro

San Francisco del Oro: Widespread. (Specularite).

Municipio de Santa Bárbara

Santa Bárbara: Widespread. (Specularite).

Municipio de Saucillo

Naica: Moderately abundant throughout the Sierra de Naica. (Specularite).

COAHUILA

Municipio de Candela

Pánuco: Moderate distribution. Noted from the Pánuco Mine.

Municipio de Castaños

Castaños: Moderately distributed. Noted from the Mercedes Mine.

Municipio de Melchor Múzquiz

La Encantada: Widespread. Most noted from the La Encantada Mine.

Municipio de Monclova

Monclova: Widespread.

Municipio de Sierra Mojada

Sierra Mojada: Moderate distribution. Noted from the La Fortuna Mine.

DURANGO

Municipio de Coneto de Comonfort

Amerila: Moderate distribution. Noted at the Twentyninth of June Mine. Massive. Associated with cassiterite.

Municipio de Cuencamé

Cuencamé de Ceniceros: Widespread. Most noted from the San Antonio, Azul Grande, Azul Chica, San Juan, San José, and Guardarraya mines.

Municipio de Durango

Cacaria: Widespread within the Sierra de Cacaria. Noted from the La Martao Mine. (Northwest of

Victoria de Durango) Massive.
Associated with cassiterite.

Cerro de los Remedios: Moderate
distribution. Noted from the
Remedios and La Vanguardia
mines. [This location is now
within the city of Victoria de
Durango.]

Victoria de Durango: Widespread
throughout Cerro de Mercado.
Noted from the Cerro de Mercado
Mine. (Martite and specularite).
Crystals to 5 cm. Associated with
magnetite and fluorapatite.
[Martite has been found replacing
magnetite crystals. Cerro de
Mercado was discovered in 1552 by
Gines Vásquez de Mercado. He had
hoped to find gold and silver but
instead discovered a 700 foot
hill of iron. The Cerro, named
for its discoverer, was not
mined until 1828 when the Cerro
de Mercado Mine began. The ore
body of Cerro de Mercado is
typical of a displacement deposit
and one of México's major iron
deposits.]

Municipio de Gomez Palacio

Dinamita: Moderate distribution.
Noted from the La Lucha Mine.

Municipio de Guanaceví

Guanaceví: Widespread.

Municipio de Mapimí

Buruguilla: Moderately distributed. Noted from the Descubridora
Mine. Associated with native
copper, cuprite, tenorite,
delafossite, and goethite.

Mapimí: Limited distribution.
Noted from the Ojuela Mine.

Municipio del Oro

El Oro: Moderate distribution.
Most noted from the Candela Mine.

Municipio de Poanas

Poanas: Moderately distributed.
Noted from the Santa Gertrudis
Mine.

Municipio de Topia

Topia: Moderate distribution.
Noted from the Provedora Mine.

GUANAJUATO

Municipio de Allende

Cerro la Mina: Limited distribution. Associated with cassiterite.

Municipio de Guanajuato

La Luz: Limited distribution.
Most noted at the Santa Rita
Mine. Bolañitos Mine.

Municipio de San Luis de la Paz

Mineral de Pozos: Moderate
distribution. Most noted from the
Santa Brígida and Escondida
mines.

Municipio de San Miguel Allende

Cerro de la Mina: Limited
distribution. Associated with
cassiterite.

Municipio de Santa Catarina

Santa Catarina: Moderately
distributed throughout the Mesa
de la Fajas. (Specularite).
(Eight km south southeast of
Santa Catarina). Most noted from
the San Santín Mine.

GUERRERO

Municipio de Ahuacuotzingo

Cerro de Tecpanchihui: Limited
distribution.

Municipio de Chilapa de Alvarez

Tacualoya: Moderate distribution.

Municipio de La Union

Zapotillo: Widespread. (Pluton
Iron deposit). Massive. Associated with magnetite.

Municipio de Taxco

Taxco de Alcarón: Limited distribution. Massive.

HIDALGO

Municipio de Cardonal

Cardonal: Moderate distribution. Most noted from the La Pinta and Dulces Nombres mines.

Municipio de Mineral del Monte

Mineral del Monte: Widespread. Most noted from the Dificultad, Santa Brígida, Santa Inés, Vizcaina, Jesús María, and Maravillas Mines.

Municipio de Pachuca

Pachuca de Soto: Widespread. Noted from the El Bordo, San Perdro, Rosario, Entrometida, San Nicanor, and San Antonio mines.

Municipio de Zimapán

Zimapán: Widespread. Noted from the Los Balcones, San Miguel, Santa Inés, Guadalupe, and Santo Tomás mines.

JALISCO

Municipio de Tonila

Pihuamo: Moderately distributed.

MEXICO

Municipio del Oro

El Oro: Moderate distribution.

Municipio de Temascaltepec

Temascaltepec de González: Moderate distribution. Noted from the Capitana Mine.

Municipio de Zacualpan

Zacualpan: Moderate distribution. Noted from the Santa Inés Mine.

MICHOACAN

Municipio de La Huacana

Inguaran: Moderate distribution.

Municipio de Lázaro Cárdenas

El Mango: Widespread. Most noted from the Las Truchas Mine. Massive. Associated with magnetite.

NAYARIT

Municipio Santa María del Oro

Rancho Mojarritas: Limited distribution.

NUEVO LEON

Municipio de Lampazos de Naranjo

Rancho Golondrinas: Limited distribution.

OAXACA

Municipio de San Miguel Peras

Santa María Zaniza: Limited distribution.

PUEBLA

Municipio de Jolalpan

Coyuca: Limited distribution on Cerro de Guayabo.

Municipio de Tepeyahualco

Tepeyahualco: Limited distribution. Noted from the La Hucha Mine.

QUERETARO

Municipio de Cadereyta

El Doctor: Widespread. Most noted from the San Juan Nepomuceno Mine.

SAN LUIS POTOSI

Municipio de Catorce

Catorce: Limited distribution. Noted from the La Concepción Mine.

Municipio de Cerro de San Pedro

Cerro de San Pedro: Widespread.

Municipio de Charcas

Charcas: Moderately distributed.

Municipio de Guadalcázar

Guadalcázar: Limited distribution. Noted from the San Rafael Mine.

SONORA

Municipio de Arizpe

Sinoquipe: Widespread. Most noted from the Omega, Ceres, Bonanza, Mercedes, Fierro y Plata, and Dos Republicas mines.

Municipio de Nacozari de García

Nacozari de García: Moderate distribution. (Specularite). Most noted from the Pilares de Nacozari and Las Pilares mines.

Municipio de Navojo

Aduana: Limited distribution. Noted from the Quintera Mine.

TAMAULIPAS

Municipio de Villagrán

San José: Limited distribution.

VERACRUZ

Municipio de Tatatila

Tatatila: Moderate distribution. Most noted from the Providencia and San Francisco mines.

ZACATECAS

Municipio de Concepción del Oro

Aranzazú: Moderate distribution. Noted from the San Antonio Mine. Crystals to 2 cm. Associated with quartz and garnet.

Municipio de Mezquital del Oro

Muzquital del Oro: Limited distribution. Noted from the Mezquital Mine.

Municipio de Zacatecas

Zacatecas: Moderate distribution.

HEMIMORPHITE

$Zn_4Si_2O_7(OH)_2 \cdot H_2O$. An uncommon secondary mineral found in the oxidized zones of zinc deposits.

CHIHUAHUA

Municipio de Aquiles Serdán

Francisco Portillo: Limited distribution. Noted from the El Potosí Mine. Clear to white crystals to 12 cm. Associated with calcite, hematite, and goethite. [A crystal lined pocket was discovered in the 1970s that contained crystals up to 13 cm in

length and 4 cm in width. The
crystals were white, but had
reddish tips due to hematite
inclusions.]

San Antonio el Grande: Limited
distribution. Noted from the San
Antonio Mine. [Above the 8th
level white crystals to 4
cm. Associated with aurichalcite,
rosasite, calcite, hematite, and
goethite. In 1983 a large vug was
uncovered that contained white
crystals to 4 cm in length and
goethite.]

Municipio de Coyame

Coyame: Limited distribution.
Noted from the Faivre Mine.

Municipio de Hidalgo
del Parral

Hidalago del Parral: Moderately
distributed.

Municipio de Saucillo

Naica: Moderately distributed
throughout the Sierra de Naica.
Crystals to 5 cm. Associated with
calcite and goethite.

COAHUILA

Municipio de Melchor
Múzquiz

La Encantada: Moderately distributed. Most noted from the La
Encantada Mine. (La Priete Mine).
Crystals to 2 cm. Associated with
goethite.

Municipio de Saltillo

Gomez Farias: Limited distribution on the Cerro de Toro. Noted
at the Alpha Mine.

DURANGO

Municipio de Guadalupe
Victoria

Guadalupe Victoria: Limited
distribution. Noted from the

Santo Niño Mine. Light blue
botryoidal masses. [During the
late 1970s a large area of
hemimorphite was found at
the Santo Niño Mine. It was
sky blue in color and found in
large botryoidal masses.]

Municipio de Mapimí

Mapimí: Widespread. Most noted
from the Ojuela Mine. Colorless
to white crystals to 4 cm in
length. Associated with adamite,
calcite, goethite, aurichalcite,
and rosasite. [This is Mexico's
leading producer of hemimorphite. It is found almost at the
surface to just above the water
table at the 12th to 13th
level. Rosettes of crystals have
been found up to 8 cm in size.]

Municipio del Oro

Magistral: Limited distribution.
Noted from the Esmeralda Mine.

HIDALGO

Municipio de Mineral
del Chico

Mineral del Chico: Limited
distribution.

NUEVO LEON

Municipio de Lampazos
de Naranjo

La Fraternal: Limited distribu-
tion. (Fifteen km east of Santa
Catarina) Noted from the Montaña
Mine.

ZACATECAS

Municipio de Concepción
del Oro

Concepción del Oro: Limited
distribution. Associated with
wulfenite.

HERCYNITE

$FeAl_2O_4$. An uncommon member of the spinel mineral group found in metamorphosed arigillaceous sediments.

CHIHUAHUA

Municipio de Cuauhtémoc

Cuauhtémoc: Limited distribution.

HERDERITE

$CaBe(PO_4)F$. A rare mineral found in igneous rocks.

CHIHUAHUA

Municipio de Santa Bárbara

Santa Bárbara: Limited distribution.

HESSITE

Ag_2Te. A rare tellurium mineral found in hydrothermal veins.

JALISCO

Municipio de San Sebastián

San Sebastián: Limited distribution. Noted from the Quintería and El Refugio mines. Associated with gold.

NAYARIT

Municipio de Santiago Ixcuintla

Rancho Acaponetilla: Limited distribution. Associated with gold.

SONORA

Municipio de Moctezuma

Moctezuma: Limited distribution. Noted from the Bambollita Mine.

HETEROGENITE

$CoO(OH)$. A rare mineral found in cobalt deposits.

SONORA

Municipio de Alamos

San Bernardo: Limited distribution. Noted from the Sara Alica Mine. Associated with erythrite and cobaltian mansfieldite.

HEULANDITE

$(Ca,Na_2)_4[Al_8Si_{28}O_{72}]\cdot24H_2O$. An uncommon mineral of the zeolite group formed in igneous rocks.

BAJA CALIFORNIA SUR

Municipio de Mulegé

Santa Rosalía: Limited distribution.

CHIHUAHUA

Municipio de Batopilas

Batopilas: Limited distribution.

GUANAJUATO

Municipio de Guanajuato

Guanajuato: Moderate distribution. Noted from the Napol Mine. Associated with chabazite.

Santa Rosa: Moderate distribution throughout the Sierra Santa Rosa.

[This location is on the Rancho Calvillo.]

HIDALGO

Municipio de Mineral del Chico

Mineral del Chico: Limited distribution.

Municipio de Pachuca

Pachuca de Soto: Limited distribution.

Municipio de Zimapán

Zimapán: Moderate distribution.

MICHOACAN

Municipio de Tlalpujahua

Tlalpujahua: Moderately distributed.

OAXACA

Municipio de la Pe

La Planchita: Limited distribution. Most noted from the La Planchita Mine. Associated with chabazite.

HEXAHYDRITE

$MgSO_4 \cdot 6H_2O$. An uncommon secondary mineral found in caves and mines.

DURANGO

Municipio de Mapimí

Mapimí: Moderate distribution. Noted from the Ojuela and San Juan mines. Massive (Efflorescences). Associated with epsomite, goslarite, and bianchite.

HIDALGOITE

$PbAl_3(SO_4)(AsO_4)(OH)_6$. A rare mineral and member of the bendantite group, which is found in lead deposits.

HIDALGO

Municipio de Zimapán

Zimapán: Limited distribution. Noted from the San Pascual Mine. [This mineral was discovered at this mine and is the type location.]

HILLEBRANDITE

$Ca_2SiO_3(OH)_2$. A rare mineral found in contact metamorphic environments.

DURANGO

Municipio de Cuencame

Valardena: Limited distribution. Noted from the Terneras Mine. [This is the type location for this mineral.]

HISINGERITE

$Fe_2Si_2O_5(OH)_4 \cdot 2H_2O$. A secondary mineral formed as an alteration product.

CHIHUAHUA

Municipio de Aquiles Serdán

Francisco Portillo: Moderate distribution. Noted from the El Potosí Mine's silicate ore body. Associated with rhodochrosite.

San Antonio el Grande: Limited distribution. Noted from the San Antonio Mine.

Municipio de Hidalgo
del Parral

Hidalgo del Parral: Limited
distribution.

HOLLANDITE

$Ba(Mn,Mn)_8O_{16}$. An uncommon
mineral and member of the
cryptamelane group, which is
found in manganese deposits.

CHIHUAHUA

Municipio de Allende

Talamantes: Moderate distribution.

HORNBLENDE. See FERRO-
HORNBLENDE

HUEBNERITE

$MnWO_4$. An uncommom member of the
wolfamite group and is found in
tungsten deposits.

CHIHUAHUA

Municipio de Aquiles Serdán

Francisco Portillo: Limited
distribution. Most noted from the
El Potosí Mine's silicate
ore body. Crystals to 8 cm.
Associated with rhodochrosite,
galena, and fluorite.

DURANGO

Municipio de Rodeo

Rodeo: Limited distribution.
Noted from the La Flaca Mine.

SONORA

Municipio de Altar

Altar: Limited distribution.

HYDROBIOTITE

$(Mg,Fe)_{3-4}(Al,Fe)Si_4O_{10}$
$(OH)_{1-3}\cdot4H_2O$. A rare and
questionable mineral of the
mica group.

HIDALGO

Municipio de Zimapán

Zimapán: Limited distribution.
Noted from the Lomo del Toro
Mine.

HYDROHETAEROLITE

$Zn_2Mn_4O_8-H_2O$. A rare mineral
found in zinc and manganese
deposits.

DURANGO

Municipio de Mapimí

Mapimí: Limited distribution.
Noted from the San Pedro and
Ojuela mines. Elongated crystals
to 8 mm. Associated with cryptomelane and manganite.

HYDRONIUM JAROSITE

$(H_3O)Fe_3(SO_4)_2(OH)_6$. A uncommon
mineral and member of the alunite
group, which form by the breakdown of sulfide ores by mine
waters, previously known as
carphosiderite.

GUERRERO

Municipio de Leonardo Bravo

Xochipala: Limited distribution.
[The name of this municipio was
recently changed from Chichi-
hualco.]

HYDROXYAPOPHYLLITE

$KCa_4Si_8O_{20}(OH,F)\cdot8H_2O$. An
uncommon member of the apophyll-

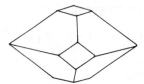

A 1 cm hydroxyapophyllite crystal from the El Potosi Mine, Francisco Portillo, Chihuahua. (Modified by Shawna Panczner from Sinkankas, 1964)

ite mineral group and occurs as a secondary mineral in basalts in amygdules and in contact metamorphic limestones bordering intrusives.

CHIHUAHUA

Municipio de Aquiles Serdán

Francisco Portillo: Limited distribution. Most noted from the El Potosí and Buena Tierra mines. Crystals to 2 cm. Associated with pyrrhotite, pyrite, arsenopyrite, galena, sphalerite, and calcite. [In 1983, a small pocket was uncovered on the 10th level of the El Potosí Mine that produced 2 cm in length. The crystals exhibited both the prismatic and pyramidal habit.]

GUANAJUATO

Municipio de Guanajuato

Guanajuato: Moderately distributed within the Veta Madre. Most noted from the Valenciana Mine. Pink crystals to 4 cm. Associated with amethyst. [Plates of crystals of this mineral have been found up to 10 cm in diameter.]

La Luz: Moderate distribution. Most noted from the La Luz and El Refugio mines. Rose to white crystals to 6 cm. Associated with amethyst.

HIDALGO

Municipio de Pachuca

Pachuca de Soto: Limited distribution. Noted from the El Rosario Mine. White crystals to 6 cm.

HYDROZINCITE

$Zn_5(CO_3)_2(OH)_6$. An uncommon secondary mineral fomed by the alterations of sphalerite and found in the oxidized ore zone.

CHIHUAHUA

Municipio de Ahumada

Los Lamentos: Limited distribution. Noted from the Erupción/Ahumada Mine. Associated with hematite.

Municipio de Aldama

Las Plomosas: Limited distribution. Noted from the La Plomosas and El Lago mines. [This old mining area is on the Rancho Viejo, 8 km southwest of the estación Picachos.]

Municipio de Aquiles Serdán

Francisco Portillo: Moderate distribution. White tuffs of crystals to 4 mm. Associated with hemimorphite, hematite, and goethite.

San Antonio el Grande: Moderate distribution. Noted from the San Antonio Mine. Associated with hemimorphite, calcite, and goethite.

Municipio de Manuel Benavides

San Carlos: Limited distribution. Noted from the San Carlos Mine.

Municipio de Saucillo

Naica: Moderate distribution. Crystals to 4 mm. Associated with goethite.

DURANGO

Municipio de Mapimí

Mapimí: Moderate distribution. Noted from the Ojuela Mine. White to colorless crystals to 6 mm. Associated with aurichalcite, goethite, hemimorphite, and plattnerite. [Under the UV light, this mineral from here has a pale blue fluorescence.]

HYPERSTHENE

$(Mg,Fe)_2Si_2O_6$. An uncommon mineral and member of the pyroxene group that is found in ultrabasic rocks.

HIDALGO

Municipio de Zimapán

Zimapán: Limited distribution. Noted from the Lomo del Toro Mine.

I

ICE

H_2O. A common mineral formed only at low temperatures in the form of snow, hail, and frost from our atmospheric moisture.

MEXICO

Municipio de Atlautla

Volcán Popocatepetl: Moderately distributed. Massive.

Municipio de Amecameca

Volcán Iztacihuatl: Moderately distributed. Massive.

Municipio de Temascaltepec

Volcán Nevado de Toluca (Xinantecatl) Moderately distributed. Massive.

VERACRUZ

Municipio de la Perla

Volcan Orizaba: Moderately distributed. Massive.

IDOCRASE. See VESUVIANITE

ILLITE is a group name, not a specific mineral.

ILMENITE

$FeTiO_3$. A common mineral in igneous and metamorphic rocks, also found in sediments.

CHIHUAHUA

Municipio de Aquiles Serdán

Francisco Portillo: Limited distribution.

San Antonio el Grande: Limited distribution. Noted from the San Antonio Mine.

DURANGO

Municipio de Mapimí

Mapimí: Limited distribution. Noted from the Ojuela Mine.

OAXACA

Municipio de San Francisco Talixtlahuaca

San Francisco Talixtlahuaca: Moderate distribution. Most noted from the Santa Ana Mine. Associated with orthoclase, albite, allanite, ferrocolumbite, and fergusonite.

SAN LUIS POTOSI

Municipio de Guadalcázar

Guadalcázar: Limited distribution.

ILSEMANNITE

$Mo_3O_8 \cdot nH_2O$. A rare secondary mineral formed by the oxidation of molybdenum minerals.

DURANGO

Municipio de Gómez Palacio

Sierra de Berejillo: Limited distribution. Noted from the Merced Mine.

ILVAITE

$CaFe_2Fe(SiO_4)_2(OH)$. An uncommon mineral found in contact metamorphic deposits.

CHIHUAUA

Municipio de Aquiles Serdán

Francisco Portillo: Limited distribution. Noted from the El Potosí Mine's silicate ore body. Associated with fayalite and rhodochrosite.

Municipio de Saucillo

Naica: Moderate distribution.

SINALOA

Municipio de Badiraguato

Alisitos: Limited distribution. Noted from the Gloria Mine.

INESITE

$Ca_2Mn_7Si_{10}O_{28}(OH)_2 \cdot 5H_2O$. An uncommon mineral found in few metal deposits.

DURANGO

Municipio de Tamazula

Rancho Ventana: Limited distribution. Noted from the San Cayetano Mine.

IODARGYRITE

AgI. A rare secondary mineral found in the oxidized ore zones of silver deposits.

CHIHUAHUA

Municipio de Aquiles Serdán

Francisco Portillo: Limited distribution. [This mineral was found in the early years of mining here in the Santa Eulalia mining district.]

San Antonio el Grande: Limited distribution. Noted from the San Antonio Mine.

Municipio de Cuidad Carmargo

Sierra Encinillas: Widespread. Most noted from the Barril, Guadalupe, La Fe, Mallen, Santo Niño, Veta Grande, Carmen, and Providencia mines.

Municipio de Cusihuiriáchic

Cusihuiriáchic: Widespread throughout district. Crystals.

Municipio de Urique

Urique: Limited distribution. Noted from the Rosario Mine.

DURANGO

Municipio de San Padro de Gallo

Peñoles: Limited distribution. Noted from the San Rafael Mine.

GUANAJUATO

Municipio de San Luis de la Paz

Mineral de Pozos: Moderate distribution. Noted from the Garibaldi and El Carmen mines.

MEXICO

Municipio de Temascaltepec

Temascaltepec de Gonzáles: Moderate distribution.

PUEBLA

Municipio de Tetela del Oro

Tetela del Oro: Limited distribution.

QUERETARO

Municipio de Cadereyta

Cadereyta de Montes: Moderate distribution. Most noted from the Santa Inés mine.

SAN LUIS POTOSI

Municipio de Catorce

Catorce: Moderate distribution. Most noted from the La Purísima Concepción Mine. Associated with chlorargyrite, hematite, and cerussite. Crystals.

Municipio de Cerro de San Pedro

Cerro de San Pedro: Moderate distribution.

SONORA

Municipio de Guaymas

San Marcial: Limited distribution. Noted from the Grande Mine.

ZACATECAS

Municipio de Mazapil

Albarradon: Moderate distribution. Noted from the Albarradon Mine. Crystals. [This mineral was discovred at the Albarradon Mine, which is the type location for this mineral.]

Municipio de Zacatecas

Zacatecas: Moderate distribution. Most noted from the Quebradilla Mine.

IODYRITE. See **IODARGYRITE**

IOLITE. See **CORDIERITE**

IRIDIUM

Found in meteorite at several locations within México.

IRON

Found in meteorites at various locations within México.

Mimetite, approx. 50 cm × 35 cm, from El Potosí Mine, Francisco Portillo, Chihuahua, in a private collection.

Mimetite, approx. 12 cm × 5 cm, from Congreso-León Mine near San Pedro Corralitos, Chihuahua, in the William Larson collection, Fallbrook, California.

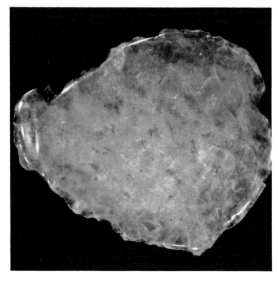

Opal (precious), approx. 3.4 cm × 2.4 cm and weighing 34 cts., from near Magdalena, Jalisco, in the collection of William Larson, Fallbrook, California. *(Photograph copyright © 1984 by Harold and Erica Van Pelt, reproduced with permission.)*

Opal (precious), approx. 6 cm × 3 cm, in rhyolite from near Magdalena, Jalisco, in the collection of William Larson, Fallbrook, California. *(Photograph copyright ©1984 by Harold and Erica Van Pelt, reproduced with permission.)*

Pyrargyrite, with crystals up to approx. 5 cm long, associated with calcite from Guanajuato, Guanajuato, in the William Larson collection, Fallbrook, California.

Pyromorphite on goethite, approx. 12 cm × 5 cm, from San Luis Mine near San José, Chihuahua, in the collections of the Museo Mineralógico de Romero, Tehuacán, Puebla.

Pyrrhotite with galena and sphalerite, approx. 20 cm × 9 cm overall with pyrrhotite crystals up to 9 cm in length, from El Potosí Mine, Francisco Portillo, Chihuahua, in the collections of the Museo Mineralógico de Romero, Tehuacán, Puebla.

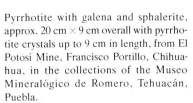

Quartz (amethyst), approx. 40 cm × 30 cm from near Piedras Parado, Veracruz, in the J. P. Cand collection, Switzerland. The Amethyst crystals are up to approx. 5 cm long.

Quartz (amethyst), with crystals up to approx. 10 cm long, from near Amatitlán, Guerrero, in the Carl Faddis collection, Seattle, Washington. *(Photograph by Carl Faddis of an oil painting by Susan Robinson, Ottawa, Canada.)*

Rhodochrosite on calcite from El Potosí Mine, Francisco Portillo, Chihuahua, in the John Barlow collection, Appleton, Wisconsin. The rhodochrosite crystals are approx. 5 cm long. *(Photograph by Wendell E. Wilson.)*

Silver associated with acanthite, approx. 15 cm × 5 cm, from Batopilas, Chihuahua, in the collections of the Museo Mineralógico de Romero, Tehuacán, Puebla.

Smithsonite, approx. 15 cm × 10 cm, from San Antonio Mine, San Antonio el Grande, Chihuahua, in the collections of the Museo Mineralógico de Romero, Tehuacán, Puebla.

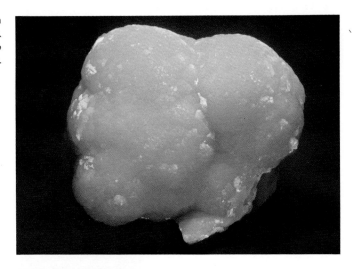

Wulfenite, with crystals up to approx. 5 cm long, associated with mimetite and goethite from near San Pedro Corralitos, Chihuahua, in the Benny Fenn collection, Colonia Juárez, Chihuahua.

Wulfenite associated with mimetite from San Francisco Mine near Cucurpe, Sonora, in the William Larson collection, Fallbrook, California. The crystal is approx. 3 cm × 4 cm.

J

JADE. See **JADEITE** or **ACTINOLITE**

JADEITE

$Na(Al,Fe)Si_2O_6$. An uncommon member of the pyroxene group that forms under high pressure and low temperature metamorphism.

GUERRERO

    Municipio de Arcelia

Almaloya: Limited distribution.

    Municipio de Texco

Texco de Alcarón: Limited distribution.

MEXICO

    Municipio de Tejupilco

Tejupilco de Hidalgo: Limited distribution.

JALPAITE

Ag_3CuS_2. A rare mineral found in few metal deposits.

JALISCO

    Municipio de Hostoti-
    paquillo

Hostotipaquillo: Limited distribution. Noted from the Mololoa Mine. Associated with acanthite and covellite.

QUERETARO

    Municipio de Jalpan

Jalpan: Limited distribution.

ZACATECAS

    Municipio de Jalpa

Jalpa: Limited distribution. Noted from the La Leonora Mine. [This is the type location for this mineral named for the location.]

JAMESONITE

$Pb_4FeSb_6S_{14}$. An uncommon mineral found in medium to low temperature veins of lead/antimony deposits.

BAJA CALIFORNIA SUR

Municipio de San Antonio

Triunfo: Widespread. Most noted from the Gobernadora, Espinosena, Humboldt, Nacimiento, Soledad, San José, Comstock, Reforma, San Antonio, and Tiro ninty-six mines. Crystals to 2 cm. Associated with stibnite, pyrite, and berthierite.

CHIHUAHUA

Municipio de Aquiles Serdán

San Antonio el Grande: Limited distribution. Noted from the San Antonio Mine.

Municipio de Hidalgo del Parral

Hidalgo del Parral: Limited distribution. Noted from the La Pietá Mine. Associated with calcite and galena.

DURANGO

Municipio de Cuencamé

Velardeña: Limited distribution.

Municipio de Mapimí

Mapimí: Limited distribution. Associated with boulangerite and bindheimite.

Municipio de Nombre de Dios

Nombre de Dios: Limited distribution. Noted at the Las Vacas Mine.

Municipio de Santiago Papasquiaro

San Andrés de la Sierra: Limited distribution. Noted from the Santa Rita Mine.

GUERRERO

Municipio de Taxco

Taxco de Alcarón: Moderate distribution.

HIDALGO

Municipio de Zimapán

Zimapán: Widespread. Most noted from the Miguel Hidalgo, El Afloramiento, Chiquihuites, Tecomates, San Vicente, and La Sirena mines. Associated with sphalerite, pyrite, galena, chalcopyrite, and pyrrhotite.

MEXICO

Municipio de Tlatlaya

Los Ocotes: Limited distribution. Noted from the Sacavón de los Platanos Mine.

Municipio de Zacualpan

Zacualpan: Moderate distribution. Noted from the Resguardo mine.

MICHOACAN

Municipio de Huetamo

Huetamo: Limited distribution. Los Dolores Mine.

NAYARIT

Municipio de Compostela

Compostela: Moderate distribution. Most noted from the Negociación de Huitzizila Mine.

QUERETARO

Municipio de Cadereyta

El Doctor: Moderate distribution.

SAN LUIS POTOSI

Municipio de Catorce

Catorce: Limited distribution. Crystals to 2 cm.

SONORA

Municipio de Navojoa

Navojoa: Moderate distribution. Noted from the La Gloria Mine.

Municipio de Sahuaripa

Mulatos: Limited distribution. Noted from the San Francisco Mine.

Municipio de Onavas

La Barranca: Moderate distribution. Noted from the Tajos, Belem, Santa Fe, and San José de las Arenillas mines.

VARACRUZ

Municipio de Las Minas

Zomelahuacan: Limited distribution. Noted from the Zomelahuacan Mine.

ZACATECAS

Municipio de General Francisco Murguía

Nieves: Widespread. Most noted from the Santa Rita Mine. Crystals to 10 cm. Associated with pyrite, barite, valentinite, and stibnite. [The mines are southwest of Nieves and at the present only one is in operation, the Santa Rita Mine. The primary ore mineral here is jamesonite, which is found as intergrown crystal masses in the ore veins. Crystals have been found from 2 to 4 mm in diameter and up to 10 cm in length. The jamesonite from here has often been mislabled as being from Sombrerete, which is some 90 km southwest of the Santa Rita Mine.]

Municipio de Mazapil

Noche Buena: Moderate distribution. Most noted from the Noche Buena Mine. Crystals to 4 cm. Associated with calcite and boulangerite. [Much of what has been labled as jamesonite from the Noche Buena Mine is Boulang-

errite. The jamesonite from here is found as mats of hair like crystals. The jamesonite crystals are small, 2 mm in diameter and 2 cm in length.]

Municipio de Sombrerete:

San Martín: Moderately distributed. Noted from the San Martín Mine. Crystals to 6 cm. Associated with stibnite and calcite.

San Pantaleón de la Noria: Moderate distribution. Noted from the La Noria Mine. Crystals to 10 cm. [This mine, which has been closed and under water for over 50 years, will reopen in the mid 80s. The crystals were found in sprays of match size crystals to 10 cm in length and were perhaps the best ever found in México.]

JAROSITE

$KFe_3(SO_4)_2(OH)_6$. A common mineral of the alunite mineral group, found in the oxidized ore zone of base metal deposits.

DURANGO

Municipio de Mapimí

Mapimí: Widespread. Golden brown crystals to 4 mm. Associated with bindheimite, goethite and mimetite.

HIDALGO

Municipio de Mineral del Monte

Mineral del Monte: Moderate distribution.

Municipio de Pachuca

Pachuca de Soto: Moderte distribution.

SONORA

Municipio de Moctezuma

Moctezuma: Limited distribution.
Noted from Moctezuma Mine.
Associated with emmonsite.

JENNITE

$Ca_9H_2Si_6O_{18}(OH)_8 \cdot 6H_2O$. An
uncommon mineral found in
metamorphic environments.

MICHOACAN

Municipio de Zitácuaro

Susupuato: Moderately distributed
throughout Cerro Mazahua, 7 km
southwest.

JOHANNSENITE

$CaMnSi_2O_6$. An uncommon mineral
of the pyroxene group found in
metamorphic rocks.

HIDALGO

Municipio de Mineral
del Monte

Mineral del Monte: Moderate
distribution. Noted from the
Espiritu Santo Mine.

PUEBLA

Municipio de Tetla
de Ocampo

Tetla de Ocampo: Limited distribution. Associated with
xonotlite.

JORDANITE

$Pb_{14}(As,Sb)_6S_{23}$. A rare mineral
found in lead/antimony deposits.

ZACATECAS

Municipio de Mazapil

Mazapil: Limited distribution.
Noted from the Noche Buena Mine.

K

KALINITE

$KAl(SO_4)_2 \cdot 11H_2O$. An uncommon seconday mineral.

HIDALGO

Municipio de Singuilucan

Cuyamaloya: Limited distribution.

KALIOPHILITE

$KAlSiO_4$. An uncommon mineral found in few mineral deposits.

HIDALGO

Municipio de Huasca

San Miguel de Regla: Moderately distributed within the Barranca de Regla.

KAOLINITE

$Al_2Si_2O_5(OH)_4$. A common clay mineral formed by the hydro-thermal and alteration by weathering of other minerals.

CAMPECHE

Municipio de Champotón

San José Carpiza: Widespread.

CHIHUAHUA

Municipio de Camargo

Santa Rosalía: Moderately distributed.

Municipio de Urique

Urique: Widespread.

COAHUILA

Municipio de Candela

Pánuco: Moderately distributed. Noted from the Pánuco Mine.

DURANGO

Municipio de Mapimí

Mapimí: Moderately distributed. Noted from the Ojuela Mine.

Municipio de Rodeo

Yerbabuena: Moderate distribution.

DISTRITO FEDERAL

San Angel: Limited distribution

Villa de Guadalupe Hidalgo: Limited distribution.

HIDALGO

Municipio de Mineral
del Chico

Mineral del Chico: Moderate
distribution. Noted from the
Capula Mine.

Municipio de Mineral
del Monte

Mineral del Monte: Moderate
distribution. Most noted from the
Jesús María and Dificultad mines.

OAXACA

Municipio de San Francisco
Telixtlahuaca

Telixtlahuaca: Moderate distribution. Noted from the Santa Cruz
Mine. Associated with orthoclase,
albite, muscovite, and biotite.

QUERETARO

Municipio de Cadereyta

Cadereyta de Montes: Moderate
distribution. Noted from the
Santa Inés Mine.

SONORA

Municipio de Sahuaripa

Sahuaripa: Limited distribution. Associated with dickite.

KASOLITE

$Pb(UO_2)SiO_4H_2O$. A rare mineral
found as a oxidation product of
primary uranium minerals.

CHIHUAHUA

Municipio de Aquiles Serdán

Francisco Portillo: Limited
distribution.

KERMACITE

Found in meteorites at many
locations within México.

KERMESITE

Sb_2S_2O. An uncommon mineral found
in antimony deposits.

CHIHUAHUA

Municipio de Uruáchic

Arechuybo: Limited distribution.

GUERRERO

Municipio de Huitzuco

Huitzuco de los Figueroa:
Moderate distribution. Noted from
the La Cruz, Trinidad, Tumbaga,
Victoria and the La Sorpresa
mines.

OAXACA

Municipio de San Juan
Mixtepec

Tejocotes: Moderate distribution.
Noted from the Yucunani Mine.

QUERETARO

Municipio de Cadereyta

El Doctor: Moderate distribution.

SAN LUIS POTOSI

Municipio de Guadalcázar

Guadalcázar: Limited distribution. Noted from the La Trinidad
Mine.

KILLALAITE

$2Ca_3Si_2O_7 \cdot H_2O$. An uncommon
mineral found in metamorphic
environments.

MICHOACAN

Municipio de Zitácuaro

Susupuato: Limited distribution on the Cerro Mazahua.

KLOCKMANNITE

CuSe. A rare mineral found in few metal deposits.

SONORA

Municipio de Moctezuma

Moctezuma: Limited distribution. Noted from the Moctezuma Mine. Associated with chalcomenite, paratellurite, bambollaite, and selenium.

KOBELLITE

$Pb_5(Bi,Sb)_8S_{17}$. An uncommon mineral found in lead deposits.

JALISCO

Municipio de Mascota

Rancho de San Rafael: Limited distribution.

Municipio de Tapalpa

Sierra de Tapalpa: Limited distribution.

KOETTIGITE

$Zn_3(AsO_4)_2 \cdot 8H_2O$. A rare mineral and member of the vivianite group, which is found in the oxidized ore zones of zinc deposits.

DURANGO

Municipio de Mapimí

Mapimí: Moderately distributed. Noted from the Ojueal Mine. Light

blue needlelike crystals to 6 cm. Associated with symplesite, gypsum, parasymplesite, adamite, legrandite, goethite, and metakoettigite. [In 1974, a pocket was discovered with a few of the crystal sprays up to 6 cm in length. Because this mineral is found in the area of the mine near where the legrandite is found, the Mexican miners called it "legrandita azul'' or blue legrandite. It is found near the bottom of the oxidized ore zone, between sulfide and oxide ores. Because the water table is found here, this area is usually under water and only during the dry seasons of the year is collecting possible.]

KRAUSITE

$KFe(SO_4)_2 \cdot H_2O$. An uncommon secondary mineral found in few metal deposits.

DURANGO

Municipio de Cuencamé

Velardeña: Limited distribution. Noted from the Santa María Mine. Associated with voltarite. [This mineral was first found at the Santa María Mine and is the type location.]

KURANAKHITE

$PbMnTeO_6$. A rare mineral found in few tellurium-rich metal deposits.

SONORA

Municipio de Moctezuma

Moctezuma: Limited distribution. Most noted from Moctezuma and Bambolla mines. Associated with emmosite, schmitterite, cuzticite, and eztlite.

KUTNOHORITE

Ca(Mn,Mg,Fe)(CO$_3$)$_2$. An uncommon member of the dolomite mineral group.

CHIHUAHUA

Municipio de Aquiles Serdán

Francisco Portillo: Limited distribution. Most noted from the El Potosí Mine's 7th level. Crystals. Associated with rhodochrosite, galena and fluorite.

HIDALGO

Municipio de Molango

Otongo: Limited distribution. Noted from the Nonoalco and Tetzintla mines. Associated with rhodochrosite.

L

LABRADORITE

$(Ca,Na)Al(Al,Si)Si_2O_8$. A common
mineral of the feldspar mineral
group that is found in volcanic
and other igneous rocks, also
found in metamorphic and
sedimentary rocks.

SONORA

Municipio de Puerto Peñasco

Santo Domingo: Limited distribution within the Pinacate volcanic
field. Massive. (Crystal grains to
6 cm).

LAUMONTITE

$CaAl_2Si_4O_{12} \cdot 4H_2O$. An uncommon
member of the zeolite mineral
group that is found in igneous
rocks, skarns, and metamorphosed
sediments.

GUANAJUATO

Municipio de Guanajuato

Guanajuato: Moderately distributed on the Veta Madre. Noted
from the Valenciana Mine.
Associated with stilbite and
quartz.

HIDALGO

Municipio de Mineral
del Chico

Rio Hondo: Moderately distributed.

SONORA

Municipio de Alamos

Alamos: Moderately distributed,
20 km east. Crystals to 6 cm.
Associated with stilbite and
leonhardite. [The laumontite
has been found here as long
match like crystals, 4 to 6 cm
long. When exposed to the
atmosphere, the clear crystals
will slowly turn milky and alter
to leonhardite.]

LAZURITE

$(Na,Ca)_{7-8}(Al,Si)_{12}O_{24}(S,SO_4)$. A
rare mineral and member of the
sodalite group, which is found in
granite pegmatites, quartzites,
and quartz veins.

BAJA CALIFORNIA SUR

Municipio de Mulegé

Santa Rosalía: Limited distribution.

HIDALGO

Municipio de Mineral del Monte

Mineral del Monte: Limited distribution.

LEAD

Pb. A rare mineral (element) found in few metal deposits.

VERACRUZ

Municipio de Jalapa

Jalapa-Enríquez: Limited distribution. Massive.

Municipio de Las Minas

Zomelahuacan: Limited distribution. Noted from the San Guillermo Mine. Massive.

LEADHILLITE

$Pb_4(SO_4)(CO_3)_2(OH)_2$. A rare mineral of secondary origin found in oxidized lead deposits.

SONORA

Municipio de Alamos

Alamos: Limited distribution. Crystals to 3 mm. Associated with wulfenite and alamosite.

LEGRANDITE

$Zn_2(AsO_4)(OH)\cdot H_2O$. A rare mineral found in zinc deposits rich in arsenic.

DURANGO

Municipio de Mapimí

Mapimí: Limited distribution. Noted from the Ojuela Mine's 12th and 17th levels. Crystals to 28 cm. Associated with goethite, smithsonite, paradamite, koettigite, gypsum, scorodite and arsenopyrite. [This mineral was discovered in the lower portion of the oxidized ore zone, on the 12th and 13th levels, of the Ojuela Mine in the early 1960s. It is probable that the miners who had worked in this area of the mine in the early 1910s and 1920s had seen traces of this yellow mineral, but did not bring it to the attention of the mine geologist. When the Peñoles stopped major operations in the mid 1940s, the pumping of water from the mine also ceased which allowed the water to fill the lower portions of the mine. It depended on the local rainy season was as to the levels of the mine that would be flooded. This usually fluctuated from between the 12th to 18th levels. of the mine. This unfortunately is the zone where the legrandite is found. The first pocket or vug uncovered contained very small crystals of lagrandite associated with crystals of adamite, but several months later, in 1962, a second and larger pocket was discovered producing crystals of legrandite up to 6 cm in length and with a diameter of 7 to 8 mm. In 1965, the next major pocket was mined out, this time the bright yellow crystals reached lengths of 8 to 10 cm and in beautiful crystal sprays on matrix with crystals up to 4 or 5 cm long! In 1968, The next legrandite pocket produced beautiful sprays of crystals with crystals 5 cm or more in length. It was not until 1979 did the mineral world have to change its record books. In the summer of that year, several of the licensed gambosenos, specimens collectors, had been waiting for the water in the mine to lower in the 17th level, so they could follow the traces of lagrandite they had worked on and had been forced to leave because of the raising water many months before. As the water slowly lowered several of them began to explore the walls and ceilings in the knee deep water. Soon the exceptionally dry summer caused the water table to lower and work could begin again. Late one afternoon after the gambosenos had left for the day, following the streaks of yellow in the mine floor, one of the older gambosenos returned, as he said, "... I could feel it in my bones..," after digging deeper into the wet floor, he opened a cavity several meters long and what he saw even took his breath away. There inside the pocket were crystals of legrandite as big as his fingers. He worked through the night without stopping, which proved to be unfortunate for. in his haste, many of the larger crystals were removed from the matrix. His tired eyes saw that the pocket was pinching off and after carefully packing his remarkable find, he reached the mine entrance just as the sun rose above the desert floor. From this pocket came the famous Aztec Sun legrandite with its two 20 cm double complex crystal cluster and many smaller finger size crystals and sprays. About four weeks later another pocket was discovered that produced even larger crystals. The largest was a complex single crystal club 28 cm long!]

NUEVO LEON

Municipio de Lampazos de Naranjo

Lampazos de Naranjo: Limited distribution. Noted from the Flor de Peña Mine. Crystals to 4 cm. Associated with sphalerite, pyrite, goethite, mimetite and calcite. [Legrandite was discovered here in the 1920s from the Flor de Peña Mine. Crystal has been found up to 4 cm in length. This is the type location for this mineral.]

LEONHARDITE. See LAUMONTITE

LEPIDOCROSITE

α-FeO(OH). An uncommon mineral formed by the weathering of other iron containing minerals.

DURANGO

Municipio de Mapimí

Mapimí: Limited distribution. Noted from the San Rafael Mine. Crystals rosettes and platelets to 1 mm.

LEPIDOLITE

$K(Li,Al)_3(Si,Al)_4O_{10}(F,OH)_2$. An uncommon mineral and member of the mica group that is found in lithium rich pegmatites.

BAJA CALIFORNIA

Municipio de Ensenada

La Huerta: Moderate distribution. La Verde Mine. Associated with elbaite, quartz and various feldspars.

Rancho Viejo: Moderate distribution. Noted from the El Socorro and Las Delicias mines. Associated with elbaite, quartz, and various feldspars.

LEUCOXENE. See ILLMENITE

LIBETHENITE

$Cu_2(PO_4)(OH)$. A rare secondary mineral found in the oxidized zones of copper deposits.

DURANGO

Municipio de Mapimí

Mapimí: Limited distribution. Noted from the Ojuela Mine's 5th

level, Santo Domingo stope.
Crystals to 4 mm. Associated with
hemimorphite.

LIMONITE. See GOETHITE

LINARITE

$PbCu(SO_4)(OH)_2$. An uncommon
secondary mineral found in the
oxidized zones of metal deposits.

JALISCO

Municipio de Bolaños

Bolaños: Limited distribution.
Noted from the Santa Fe Mine's
2nd level. Associated with quartz
and calcite.

Municipio de Mascota

Bramador: Limited distribution.
Noted from the El Refugio Mine.

LITHARGE

PbO. An uncommon mineral forming
in the alkaine and oxidizing
environments of lead deposits.

AGUASCALIENTES

Municipio de Asientos

Asientos: Moderate distribution.
Noted from the El Orito Mine.
Massive.

BAJA CALIFORNIA

Municipio de Ensenada

El Alamo: Moderate distribution.
Noted from the Escorpión Mine.
Massive.

BAJA CALIFORNIA SUR

Municipio de San Antonio

El Triunfo: Widespread. Massive.

CHIHUAHUA

Municipio de Batopilas

Batopilas: Moderately distributed. Massive.

Municipio de Cuishuiráchic

Cuishuiriáchic: Limited distribution. Noted from the La Reina
Mine. Massive.

Municipio de Saucillo

Naica: Moderate distribution.

DURANGO

Municipio de Guanaceví

Guanaceví: Moderately distributed. Most noted from the Caldas,
Rosario, and Capuzaya mines.

Municipio de Mapimí

Mapimí: Moderately distributed.
Noted from the Ojuela Mine.
Associated with wulfenite.
[This mineral has been found both
in and on wulfenite.]

GUANAJUATO

Municipio de San Luis
de la Paz

Mineral de Pozos: Moderate
distribution. Noted from the
Santa Brígida and La Trinidad
mines. Massive.

GUERRERO

Municipio de Iguala

Platanillo: Limited distribution.

Municipio de General
Heliodoro Castillo

Campo Morado: Moderate distribution. Most noted from the La
Reforma and El Naranjo mines.

HIDALGO

Municipio de Pachuca

Pachuca de Soto: Moderate
distribution.

Municipio de Zimapán

Zimapán: Moderately distributed. Noted from the Guadalupe Mine.

JALISCO

Municipio de Bolaños

Bolaños: Moderately distributed. Noted from the Santa Fe Mine. Massive.

NUEVO LEON

Municipio de Cerralvo

Villa Cerralvo: Limited distribution.

OAXACA

Municipio de Ocotlán de Morelos

Ocotlán: Limited distribution. Noted from the La Soledad Mine. [This location is known locally as just Ocotlán.]

SAN LUIS POTOSI

Municipio de Catorce

Catorce: Moderate distribution. Most noted from the San Pedro and La Purísmia Concepción mines. Massive.

Municipio de Villa de la Paz

La Paz: Moderately distributed. Most noted from the La Paz and Santa Ana mines. Massive.

SONORA

Municipio de Cumpas

Cumpas: Limited distribution. Noted from the Tobacachi Mine.

Municipio de Sahuaripa

Mulatos: Moderately distributed. Most noted from the San Francis-co, Campero and San Francisco mines.

VERACRUZ

Municipio de Las Minas

Zomelahuacan: Limited distribution.

ZACATECAS

Municipio de Concepción del Oro

Aranzazú: Moderately distributed. Most noted from the Albarradón, San Eligio, Los Galemes, San Pedro, and the La Nieva mines.

Municipio de Zacatecas

Zacatecas: Limited distribution. Noted from the San Fernando Mine. Massive.

LIVINGSTONITE

$HgSb_4S_8$. A rare mineral found in mercury deposits rich in antimony.

12 cm long livingstonite crystals in gypsum associated with sulfur from the La Cruz Mine, Huitzuco de la Figueroa, Guerrero, in the Pinch Mineralogical Museum, Rochester, New York.

GUERRERO

Municipio de Huitzuco

Huitzuco de Los Figueroa: Moderately distributed. Most noted from the La Cruz Mine's Espíritu Santo Rebaja stope, Rosita Rebaja stope and the El Carmen level, San Agustín, Agua Salada, and La Unión mines. Crystals to 12 cm. Associated with gypsum, sulfur, cinnabar, calcite, stibnite and metacinnabar. [The best specimens of livingstonite have been found in the upper areas of the Espíritu Santo Rebaja stope, within 20 to 50 meters of the surface. Livingstonite is found as needlelike crystals up to 12 cm in length. These long needle crystals are often found enclosed in selenite. Crystals are also found with crystallized stibnite, but the livingstonite crystals are smaller in size.]

QUERETARO

Municipio de Tolimán

Soyatal: Moderate distribution.

SAN LUIS POTOSI

Municipio de Guadalcázar

Guadalcázar: Limited distribution.

LOELLINGITE

$FeAs_2$. An uncommon mineral found in mesothermal veins in mineral deposits.

HIDALGO

Municipio de Zimapán

Zimapán: Limited distribution. Noted from the Santa Rita Mine.

MICHOACAN

Municipio de Angangueo

Angangueo: Moderately distributed. Most noted from the Carmen and San Cristóbal mines.

QUERETARO

Municipio de Cadereyta

El Doctor: Limited distribution. Noted from the El Doctor Mine.

SONORA

Municipio de Alamos

San Fernanado: Limited distribution. Noted from the Sara Alicia Mine.

LOTHARMEYERITE

$CaZnMn(AsO_4)_2(OH)\cdot2H_2O$. A rare secondary mineral found in the oxidized zone of arsenic rich ore bodies.

DURANGO

Municipio de Mapimí

Mapimí: Limited distribution. San Judas stope between the 13th and 14th levels of the Ojuela Mine. Dark reddish orange crystals to 6 mm. Associated with adamite, goethite and cryptomelane. [This mineral was discovered here in the later part of 1981 during the period of time when the pockets of purple adamite were being uncovered.]

LUICIANITE. See VARISITE

LUDLAMITE

$(Fe,MgMn)_3(PO_4)_2\cdot4H_2O$. A rare mineral found in phosphate rich deposits.

CHIHUAHUA

Municipio de Aquiles Serdán

San Antonio el Grande: Limited distribution. Most noted from the San Antonio Mine. Crystals to 8 cm. Associated with vivianite, pyrite, goethite, dolomite, and various manganese oxides. [In October 1982, in the newly developed ramp project at the San Antonio Mine, stringers of ludlamite were consistently being uncovered. They produced little to speak of until November 10, when another page of the mineral record book had to be rewritten. The sinking of the ramp had progressed down to the 13th level and had intersected the level alongside one of the stopes. The first shift had just set off a round of explosives at the end of their shift and left for the day, leaving the second shift the job to inspect and muck the area out. The first miner into the area was to inspect for loose rocks and any unfired explosives. As the light from his mining light illuminated the wall, he saw a cavity about the size of a basketball lined with green crystals! In the bottom of the pocket, broken loose by the blast were four finger size ludlamite crystals! Not having any tools with him, he put the four crystals in his shirt pocket and quickly hurried off to the tool box a few hundred meters away. Upon his return, the miner was surprised to find the pocket cleaned of its gemmy green lining! His fellow workers had arrived and began not to muck out the area, but rather the pocket! Over the next two days several smaller pockets were encountered and mined out. Over 50 specimens with crystals over 1 cm in length were found. Only about a dozen specimens were found with crystals over 4 cm in length! Four specimens had record setting crystals which were up to 8 cm long! Many of The crystals from this discovery exhibited a complex parallel crystal develop-ment. Some of the crystals had one side of them lightly coated with a manganese oxide and pyrite. A week or so after this the mine was completely flooded and after several months of pumping, the mine water was lowered. Nothing was found upon return to the area.]

M

MACKAYITE

$FeTe_2O_5(OH)$. A rare secondary mineral found in the oxidized zone of tellurium deposits.

SONORA

Municipio de Moctezuma

Moctezuma: Limited distribution. Noted from the Moctezuma and San Miguel mines. Associated with barite and tellurium.

MAGNESITE

$MgCO_3$. An uncommon mineral and member of the calcite mineral group found in metamorphic and sedimentary rocks, also in hydrothermal veins.

BAJA CALIFORNIA

Municipio de Encenada

Isla de Cedros: Limited distribution. Massive.

BAJA CALIFORNIA SUR

Municipio de Comondú

Bahía Magdalena: Limited distribution.

Isla Santa Margarita: Limited distribution.

Municipio de Mulegé

Santa Rosalía: Moderate distribution. Massive.

CHIHUAHUA

Municipio de Aquiles Serdán

Francisco Portillo: Limited distribution.

MICHOACAN

Municipio de Angangueo

Mineral de Angangueo: Moderate distribution. Most noted from the Dolores, Carrillos and San Antonio mines.

ZACATECAS

Municipio de Concepción del Oro

Aranzazú: Limited distribution.

MAGNETITE

Fe,Fe_2O_4. A common mineral of the spinal group, which is found in most rock environments.

AGUASCALIENTES

Municipio de Asientos

Asientos: Moderate distribution.

BAJA CALIFORNIA

Municipio de Ensenada

Isla de Cedros: Widespread. Noted from the Máscara de Hierro Mine. Associated with gold and chalcocite.

San Fernando: Widespread. Most noted from the San Fernando and Chalcocita mines. Associated with native copper, cuprite, bornite, and malachite.

Real del Castillo: Moderately distributed. Associated with gold.

BAJA CALIFORNIA SUR

Municipio de Comondú

Isla Margarita: Moderately distributed.

CHIPAS

Municipio de Cintalapa

La Razon: Widespread.

CHIHUAHUA

Municipio de Allende

Talamantes: Moderately distributed. Associated with hematite and pyrolusite.

Municipio de Aquiles Serdán

Francisco Portillo: Moderate distribution.

San Antonio el Grande: Moderate distribution. Noted from the San Antonio mine's skarn zone. Crystals to 2 cm. Associated with quartz and hedenbergite.

Municipio de Camargo

La Negra: Widespread. Noted from the La Negra Mine (now the Hercules Mine). Crystals to 4 cm. Associated with hematite. [This mineral has been found as replacements of large, 2 to 4 cm in diameter cyrstals of fluoapatite.]

La Perla: Widespread. Noted from the La Perla Mine. Crystals to 4 cm. Associated with hematite.

Municipio de Saucillo

Naica: Moderate distribution.

COLIMA

Municipio de Manzanillo

Peña Colorada: Widespread. Noted at the Peña Colorada Mine.

DURANGO

Municipio de Durango

Victoria de Durango: Widespread. Noted at the Cerro de Mercado Mine. Associated with hematite and fluorapatite. [First seen by the Spanish in 1552 by an expedition sent to find what had been reported earlier as a hill of solid silver and gold! Gines Vasquez de Mercado lead the Spanish conquistador's and after arriving at the hill he found not silver or gold, but iron! The Spanish not needing an iron mine this far removed from Mexico City did not develop the discovery and it was not until 1828 that mining began on a limited basis. The technologies at that time could not properly smelt the ore and it was not until 1888 did large scale mining begin.]

Municipio de Gómez Palacio

Dinamita: Moderate distribution. Noted at the La Lucha Mine. Associated with hematite.

Municipio de Mapimí

Mapimí: Moderately distributed. Noted at the Ojuela Mine. Associated with pyrrhotite, hematite, vesuvianite, grossular and titanite.

GUERRERO

Municipio de La Union

La Palama: Widespread. Common.

Zapotillo: Widespread. Associated with hematite.

Municipio de Taxco

Buenavista: Moderately distributed. Most noted from the Natividad and Anexas mines.

HIDALGO

Municipio de Mineral del Monte

Mineral del Monte: Moderately distributed.

Municipio de Pachuca

Pachuca de Soto: Moderately distributed. Noted from the Rosario Mine.

JALISCO

Municipio de Tonila

Pihuamo: Widespread on Cerro Piedra Iman.

MICHOACAN

Municipio de Lázaro Cárdenas

El Mango: Widespread. Associated with hematite. [This is one of Mexico's developing iron areas.]

Las Truchas: Widespread. Noted from the Las Truchas Iron Mine. Associated with hematite.

MORELOS

Municipio de Villa de Ayala

Jalostoc: Moderately distributed. [This location was formerly known as Xalostoc.]

NAYARIT

Municipio de Ruiz

El Zopilote: Moderately distributed. Noted from the Nueva Vizcaya Mine.

NUEVO LEON

Municipio de Lampazos de Naranjo

Sierra del Carrizal: Moderate distribution. Noted from the La Fraternal Mine.

OAXACA

Municipio de Zimatlán de Alvarez

Santa Cruz Mixtepec: Moderate distribution.

PUEBLA

Municipio de Tlanalapan

Cerro de Galván: Moderately distributed. Associated with goethite and minerals of the chlorite group. Common.

QUERETARO

Municipio de Cadereyta

Cedereyta de Montes: Moderate distribution. Noted from the El Doctor Mine.

SAN LUIS POTOSI

Municipio de Charcas

Charcas: Widespread.

Municipio de Villa de Ramos

Ramos: Moderate distribution.

SINOLOA

Municipio de Cosalá

Cosalá: Moderate distribution.

TAMAULIPAS

Municipio de San Carlos

San José: Moderate distribution on Cerro del Diente.

VERACRUZ

Municipio de Las Minas

Zomelahuacan: Moderate distribution.

ZACATECAS

Municipio de Concepción del Oro

Aranzazú: Widespread. Noted from the Cerro Prieto Mine.

MALACHITE

$Cu_2(CO_3)(OH)_2$. An uncommon mineral found in the oxidized zone of copper deposits.

AGUASCALIENTES

Municipio de Asientos

Asientos: Widespread. Most noted from the Santa Francisca, No Pensada, and El Cristo mines. Associated with azurite and goethite. Crystals to 8 mm.

Municipio de Tepezalá

El Cobre: Widespread. Noted from the San Bartolo, Santa Clara, and Copper Queen mines. Associated with goethite and azurite. Crystals to 8 mm.

Tepezalá: Widespread. Associated with azurite and goethite.

BAJA CALIFORNIA

Municipio de Ensenada

Calamajoe: Moderate distribution. Noted from the San Antonio Mine.

El Socorro: Moderately distributed. Noted from the Gabribaldi Mine.

Las Palomas: Moderate distribution. Noted from the Alejandra Mine.

Real del Castillo: Widespread. Most noted from the San Felipe and Eloisa mines.

San Fernando: Widespread. Most noted from the San Fernando del Cobre, Copper Queen and Adela mines.

BAJA CALIFORNIA SUR

Municipio de Mulegé

Santa Rosalía: Moderate distribution. Noted from the Santa Rita Mine. Associated with azurite and goethite.

Municipio de San Antonio

El Triunfo: Widespread. Associated with goethite and azurite.

CHIAPAS

Municipio de Pichucalco

Pichucalco: Moderately distributed. Noted from the Santa Fe Mine.

CHIHUAHUA

Municipio de Aquiles Serdán

Francisco Portillo: Limited distribution. Noted from the Buana Tierra Mine. Associated with calcite and goethite.

San Antonio el Grande: Limited distribution. Noted from the San Antonio Mine. Associated with azurite and goethite.

Municipio de Chihuahua

Labor de Terrazas: Widespread. Most noted from the Río Tinto Mine. Associated with azurite, cuprite, native copper, goethite and chalcopyrite.

Municipio de Camargo

Sierra de las Encinillas: Widespread.

Municipio de Coyame

Cuchillo Parado: Widespread at the Cerro de la Empumosa. Crystal replacements of Azurite to 1 cm. Associated with azurite and goethite.

Las Vigas: Widespread at the Sierra de la Boquilla. Associated with azurite, cuprite and goethite.

Municipio de Cusihuiriáchic

Cusihuiriáchic: Moderately distributed. Most noted from the Princesa and La Reina mines. Associated with azurite and goethite.

Municipio de Saucillo

Naica: Moderate distribution.

COAHUILA

Municipio de Múzquiz

La Encontada: Moderately distributed. Noted at the La Encontada Mine. Associated with azurite and goethite.

Municipio de Sierra Mojada

Sierra Mojada: Moderate distribution. Most noted from the El Porvenir and Fortuna mines.

COLIMA

Municipio de Coquimatlán

Hacienda Magdalena: Limited distribution.

DURANGO

Municipio de Cuencamé

Velardeña: Moderate distribution. Most noted at the Velardeña and Santa Juana mines.

Municipio de Indé

Indé: Limited distribution. Noted from the Terrible Mine.

Municipio de Mapimí

Buruguilla: Moderate distribution. Noted from the La Descubridora Mine.

Mapimí: Moderately distributed. Most noted from the Ojuela Mine's Cumbres stope. Tuffs of crystals to 8 mm. Associated with azurite, cuprite, native copper, and goethite.

Municipio del Oro

El Oro: Moderate distribution. Most noted from the Promontorio, La Cruz Negra, Santo Niño, Reina Victoria, and El Gran Lucero mines.

Municipio de Pánuco de Coronado

Pánuco de Coronado: Moderate distribution. Noted from the Avino Mine. Associated with azurite and goethite. [This location was known as the old Spanish mining area of San Jose de Avino.]

Municipio de San Luis de Cordero

San Luis de Cordero: Moderate distribution. Most noted from the Rosario, Boca del Cobre, and San Luis mines.

Municipio de San Dimas

Tayoltita: Moderate distribution.

Municipio de San Pedro
del Gallo

Peñoles: Moderate distribution.
Most noted from the Jesús María
and Corazón de Jesús mines.
Associated with azurite and
goethite.

GUANAJUATO

Municipio de San Luis
de la Paz

Mineral de Pozos: Moderate
distribution. Noted from the
Santa Brígida, Progreso,
Mascota and Trinidad mines.

Municipio de Xichú

Xichú: Moderately distributed.
Noted from the Tribuna Mine.

GUERRERO

Municipio de Huitzuco

Tlaxmalac: Limited distribution.

Municipio de Zumpango
del Río

Mezcala: Moderate distribution.
Associated with azurite.

HIDALGO

Municipio de Pachuca

Pachuca de Soto: Moderate
distribution. Noted from the
Moctezuma, Rosario, La Constancia, and El Candado mines.

Municipio de Zimapán

Zimapán: Moderate distribution.
Noted from the Lomo del Toro and
Guadalupe mines.

JALISCO

Municipio de Bolaños

Bolaños: Moderately distributed.
Most noted from the Santa Fe,

Barranco, Iguana, Verde and
Descubridora mines.

Municipio de Chiquilistlán

Chiquilistlán: Moderate distribution.

Municipio de Mescota

Mescota: Moderately distributed.
Noted from the Purísima Mine.

MEXICO

Municipio de Tlatlaya

Rancho Los Ocotes: Moderate
distribution. Noted from the La
Fama Mine.

Municipio de Zacualpan

Zacualpan: Widespread.

MICHOACAN

Municipio de La Huacana

Inguaran: Moderately distributed.

Oropeo: Moderately distributed.
Most noted from the San Cristó-
bal, China, San Enrique, and
Puerto de Mayapito mines.

NAYARIT

Municipio de San Pedro
Langunillas

San Pedro Langunillas: Limited
distribution. Noted from the
Guadalupe Mine.

PUEBLA

Municipio de Chirantla

Atopoltitlán: Moderately distributed on the Cerro del Guajolote.
Associated with azurite.

QUERETARO

Municipio de Cadereyta

El Doctor: Limited distribution.
Noted from the Maravillas Mine.

SAN LUIS POTOSI

Municipio de Charcas

Charcas: Limited distribution. Noted from the Guadalupe Mine.

Municipio de Villa de la Luz

La Luz: Moderately distributed. Most noted from the Santa María de la Paz, Concepción, Hidalgo and Dolores y Anexas mines.

Municipio de Villa de Ramos

Ramos: Moderate distribution. Noted from the Cocinera·Mine. Associated with azurite and goethite.

SONORA

Municicpio de Alamos

Alamos: Widespread, 23 km east. Noted from the Empaco, La Aztecita, La Azteca and Malaquita. Crystal tuffs to 8 mm. Associated with azurite, smithsonite, and goethite.

Municipio de Bacerac

Bacerac: Moderate distribution. Noted from the Herreria Mine.

Municipio de Cananea

Cananea: Widespread.

Municipio de Yécora

Yécora: Limited distribution. Crystal replacements of azurite to 3 cm. Associated with azurite and goethite.

TAMAULIPAS

Municipio de Miquihuana

Miquihuana: Moderate distribution. Noted from the San Francisco and Santa Cruz mines.

Municipio de Villagrán

San José: Moderate distribution.

ZACATECAS

Municipio de Chalchihuites

Chalchihuites: Moderate distribution. Noted from the Jesús María Mine.

Municipio de Concepción del Oro

Aranzazú: Widespread. Associated with azurite and goethite.

Concepción del Oro: Widespread. Most noted from the Cata Arroyo and El Tajo mines. Associated with azurite and goethite.

El Cobre: Moderate distribution. Noted from the El Cobre Mine. Crystal tuffs to 8 mm and crystal replacements of Azurite to 12 cm. Associated with azurite and goethite. [This location has produced crystals of malachite after azurite to 12 cm in length!]

Municipio de Mazapil

Mazapil: Widespread. Most noted from the San Carlos, Santa Rosa and Nieva mines. Associated with azurite and goethite.

Municipio de Pinos

Pinos: Moderate distribution. Associated with azurite.

Municipio de Sombrerete

La Noria de San Pantaleón: Limited distribution. Noted from the Sacramento Mine.

MANGANITE

MnO(OH). An uncommon mineral found in hydrothermal ore veins.

CHIHUAHUA

Municipio de Allende

Talamantes: Widespread. Associated with pyrolusite and rhodochrosite.

Municipio de Aquiles Serdán

Francisco Portillo: Limited
distribution. Noted from the
Buena Tierra and El Potosí mine's
silicate ore body. Crystals to
6 mm. Associated with fayalite,
helvite, rhodochrosite, and
ilvaite.

San Antonio el Grande: Limited
distribution. Noted from the San
Antonio Mine.

DURANGO

Municipio de Mapimí

Mapimí: Moderately distributed. Most noted from the Ojuela
and San Pedro mines. Crystals to
8 mm. [This mineral is the last
manganese mineral to form here
within the Mapimi mining district.]

HIDALGO

Municipio de Mineral
del Monte

Mineral del Monte: Moderate
distribution. Most noted from the
Ahuichote, Jesús María, Aviadero,
Dificultad, Santa Brígida and
Carretera mines.

JALISCO

Municipio de Lagos
de Moreno

Comanja de Corona: Moderately
distributed.

PUEBLA

Municipio de Piaxtla

Poaxtla: Limited distribution on
the Cerro de Tlapantepetl, 2 km.

SONORA

Municipio de Cucurpe

Cucurpe: Moderately distributed.
Noted at the Santa Apolonia Mine.

ZACATECAS

Municipio de Mazapil

Mazapil: Limited distribution.

MANSFIELDITE

$AlAsO_4 \cdot 2H_2O$. An uncommon mineral
found in arsenic-rich deposits.

SONORA

Municipio de Alamos

San Bernardo: Limited distribution. Noted from the Sara Alica
Mine. Associated with erythrite.

MAPIMITE

$Zn_2Fe_3(AsO_4)_3(OH)_4 \cdot 10H_2O$.d A rare
mineral found in zinc deposits
rich in arsenic.

DURANGO

Municipio de Mapimí

Mapimí: Limited distribution.
Noted from the Ojuela Mine.
Crystals to 4 mm. Associated with
goethite. [This was discovered
here and named after the location, which is the type
location for this mineral.]

MARCASITE

FeS_2. A common mineral and a
member of the marcasite group
that forms as a replacement
mineral in sedimentary rocks,
from acid solutions at low
temperatures, and as a supergene
mineral in lead and zinc deposits.

AGUASCALIENTES

Municipio de Asientos

Asientos: Moderately distributed.

BAJA CALIFORNIA SUR

Municipio de San Antonio

El Triunfo: Widespread.

San Antonio: Widespread.

CHIHUAHUA

Municipio de Aquiles Serdán

Francisco Portillo: Widespread. Crystals to 3 cm. Associated with pyrrhotite, pyrite, galena and sphalerite.

San Antonio el Grande: Moderately distributed. Noted from the San Antonio Mine. Crystals to 2 cm. Associated with pyrite, galena, sphalerite, calcite, and quartz.

DURANGO

Municipio de Mapimí

Mapimí: Widespread. Noted from the Ojuela and Monterrey mines. At the Ojuela and Monterrey mines, it is found as replacement of pyrrhotite.

GUANAJUATO

Municipio de Dolores Hidalgo

Ciudad de Dolores Hidalgo, Cuna de la Independencia Nacional: Limited distribution. Crystals to 1 cm.

Municipio de Guanajuato

Guanajuato: Moderate distribution. Noted from the Cata, Santa Gertrudis, Secho, San Eligio and Sirena mines.

La Luz: Limited distribution. Noted from the San Cayetano Mine.

GUERRERO

Municipio de Coyuca de Catalán

Puerto del oro: Widespread. Most noted from the Pinzón, San José, and Trinidad mines.

Municipio de Huitzuco

Huitzuco de la Figueroa: Limited distribution.

Municipio de Taxco

Taxco de Alcarón: Widespread. Most noted at the El Pedregal, San Antonio, Jesús, and Guerrero mines. Crystals to 1 cm. Associated with quartz and calcite. [The Guerrero Mine has produced large smoky quartz crystals coated with small marcasite crystals.]

HIDALGO

Municipio de Pachuca

Pachuca de Soto: Widespread.

JALISCO

Municipio de La Yesca

La Yesca: Moderately distributed.

MEXICO

Municipio de Zacualpan

Zucualpan: Widespread.

MICHOACAN

Municipio de Angangueo

Mineral de Angangueo: widespread. Most noted from the El Carmen, Catingón, San Cristóbal, and Carrillos mines.

Municipio de Susupuato de Guerrero

Cerro Teamaro: Moderate distribution. Noted from the Los Reyes Mine.

NAYARIT

Municipio de Amatlán de Cañas

Barranca del Oro: Moderately distributed. Most noted from the Rondanera and Zertuchena mines.

OAXACA

Municipio de Natividad

Natividad: Moderate distribution. Noted from the Natividad Mine.

PUEBLA

Municipio Tetela del Ocampo

Tetela del Oro: Limited distribution.

SONORA

Municipio de Alamos

Alamos: Moderately distributed. Noted from the Tesache Mine.

Municipio de Onavas

La Barranca: Widespread. Most noted from the Belem and José de las Arenillas mines.

ZACATECAS

Municipio de Fresnillo

Fresnillo de González Echeverría: Widespread. Crystals to 10 mm in botryoidal masses. Associated with quartz and pyrite. [Since the early 1980s, this location has produced many fine specimens of marcasite and marcasite coating quartz (amethyst and rock crystal) crystals. Some of the botryoidal masses have been found up to 60 cm in diameter.]

Municipio de Mazapil

Mazapil: Widespread.

Municipio de Noria de Angels

Noria de Angeles: Moderately distributed.

Municipio de Sombrerete

Sombrerete: Moderately distributed. Noted from the La Blanca Mine.

Municipio de Vetagrande

Vetagrande: Widespread.

Municipio de Zacatecas

Zacatecas: Moderately distributed.

MARGARITASITE.

$(Cs,H_3O,K)(UO_2)_2(VO_4)_2 \cdot nH_2O$. A rare mineral found in uranium deposits.

CHIHUAHUA

Municipio de Chihuahua

Sierra Pena Blanca: Moderate distribution. Noted from the Margaritas Mine. Associated with carnotite, weeksite, and uraninite. [This mineral was discovered here and named after the mine. It is the type location for this mineral.]

MASSICOT

PbO. An uncommon mineral found in the oxidized zone of lead deposits.

AGUASCALIENTES

Municipio de Asientos

Asientos: Moderate distribution. Noted from the El Orito Mine.

BAJA CALIFORNIA

Municipio de Ensenada

Alamo: Limited distribution. Noted from the Escorpión Mine.

BAJA CALIFORNIA SUR

Municipio de San Antonio

El Triunfo: Widespread.

CHIHUAHUA

Municipio de Aquiles Serdán

San Antonio el Grande: Moderately distributed. Noted at the San Antonio Mine.

Municipio de Cusihuiriáchic

Cusihuiriáchic: Moderate distribution. Noted from the La Reina Mine.

Municipio de Santa Bárbara

Santa Bárbara: Moderate distribution. Noted from the Santa Barbara Mine.

Municipio de Saucillo

Naica: Moderately distributed.

DURANGO

Municipio de Guanaceví

Guanaceví: Moderate distribution. Most noted from the Caldas, Rosario, Capuzaya and Sierra de San Lorenzo mines.

Municipio de Mapimí

Mapimí: Limited distribution. Noted from the Ojuela Mine. Associated with wulfenite. Massive.

GUANAJUATO

Municipio de San Luis de la Paz

Mineral de Pozos: Moderate distribution. Noted from the Santa Brígida and La Trinidad mines.

GUERRERO

Municipio de General Heliodoro Castillo

Campo Morado: Moderate distribution. Noted from the La Reforma and El Naranjo mines. Mines.

HIDALGO

Municipio de Pachuca

Pachuca de Soto: Moderately distributed.

Municipio de Zimapán

Zimapán: Moderate distribution. Noted from the Guadalupe and Lomo del Toro mines.

JALISCO

Municipio de Bolaños

Bolaños: Moderate distribution. Most noted from the Santa Fe and La Blanguita mines.

NUEVO LEON

Municipio de Cerralvo

Villa Cerralvo: Limited distribution.

OAXACA

Municipio de San Jerónimo Taviche

San Jerónimo Taviche: Moderately distributed. Note from the La Soledad and Negociación de Cinco Señores mines.

PUEBLA

Municipio de Tlanalapan

Matamoros: Moderate distribution. Noted from the Esperanza Mine.

QUERETARO

Municipio de San Juan del Río

Ahuacatlan: Moderately distributed. Most noted from the El Carmen, Penasco and Santiago mines.

SAN LUIS POTOSI

Municipio de Catorce

Catorce: Moderate distribution. Noted from the San Pedro and Concepción mines.

Municipio de Villa de la Paz

La Paz: Moderately distributed. Most noted from the La Paz and Santa Ana mines.

SINALOA

Municipio de Cosalá

Guadalupe de los Reyes: Limited distribution.

SONORA

Municipio de Cumpas

Cumpas: Moderate distribution. Noted from the Tobacachi Mine.

Municipio de Sahuaripa

Mulatos: Moderately distributed. Most noted from the San Francisco, Campero and El Potrero mines.

Municipio de Soyopa

San Antonio de la Huerta: Limited distributed. Noted at the Rosarena Mine.

VERACRUZ

Municipio de Las Minas

Zomelahuacan: Moderately distributed.

ZACATECAS

Municipio de Mazapil

Mazapil: Moderately distributed. Most noted from the Santa Rosa, Albarradon, San Eligio, Los Galemes, San Pedro and La Nieva mines.

Municipio de Zacatecas

Zacatecas: Limited distribution. Noted from the San Fernando Mine.

MATILDITE

$AgBiS_2$. A rare mineral found in silver deposits.

CHIHUAHUA

Municipio de Saucillo

Naica: Limited distribution. Noted from the Naica Mine.

SAN LUIS POTOSI

Municipio de Guadalcázar

Guadalcázar: Limited distribution.

ZACATECAS

Municipio de Fresnillo

Fresnillo de González Echeverría: Moderately distributed.

MAUCHERITE

$Ni_{11}As_8$. A rare mineral found in nickel deposits.

SINALOA

Municipio de Badiraguato

Alisitos: Limited distribution. Noted from the Gloria Mine. Associated with gersdorffite and violarite.

MEIONITE

$3CaAl_{12}Si_2O_8 \cdot CaCO_3$. An uncommon mineral and member of the scapolite group, which is found in metamorphic rocks.

HIDALGO

Municipio de Pachuca

Cerro de las Navajas: Limited distribution.

NUEVO LEON

Municipio de Lampazos de Naranjo

Sierra del Carrizal: Moderate distribution.

OAXACA

Municipio de La Pe

La Planchita: Limited distribution, west of Ayoquezco, north of La Planchita. La Planchita Mine. White crystals to 8 cm. Associated with augite, zircon, and calcite.

MELACONITE. See **TENORITE**

MELANTERITE

$FeSO_4 \cdot 7H_2O$. An uncommon mineral found as a secondary mineral formed from the alteration of pyritic ores.

CHIHUAHUA

Municipio de Aquiles Serdán

Francisco Portillo: Moderate distribution.

San Antonio el Grande: Moderate distribution. Noted at the San Antonio Mine.

DURANGO

Municipio de Mapimí

Mapimí: Moderate distribution. Noted fromt the Ojuela Mine.

GUANAJUATO

Municipio de Guanajuato

Villapando: Limited distribution. Noted from the La Loda Mine.

HIDALGO

Municipio de Tepeji del Río

Tepeji del Río: Moderately distributed.

Municipio de Zimapán

Zimapán: Limited distribution. Noted from the Lomo del Toro Mine. Greenish efflorescences.

MICHOACAN

Municipio de Angangueo

Mineral de Angangueo: Moderate distribution. Most noted from the San Cristóbal and San Cayetano mines.

SONORA

Municipio de Onavas

La Barranca: Moderately distributed.

ZACATECAS

Municipio de Fresnillo

Fresnillo de González Echeverría: Moderately distributed throughout the Cerro de Proaño.

MENDOZAVILITE

$NaAl(SO_4) \cdot 11H_2O$. A rare mineral found in the oxidized ore zone of mineral deposits.

SONORA

Municipio de Cumpas

Cumobabi: Limited distribution,
27 km southwest of Cumpas. San
Judas Mine. Yellow masses.
Associated with muscovite and
paramendozavilite. [This location
is within the La Verde mining
district and is the type locality
for this mineral.]

MENEGHINITE

$Pb_{13}CuSb_7S_{24}$. A rare mineral
found in lead/copper deposits.

HIDALGO

Municipio de Zimapán

Zimapán: Limited distribution.
Noted from the Concordia and Lomo
del Toro mines.

MERCURY

Hg. An uncommon mineral (element) of secondary origins
forming from the alteration of
other mercury minerals.

CHIHUAHUA

Municipio de Saucillo

Naica: Limited distribution.

DURANGO

Municipio de Canatlán

Palomas: Limited distribution.
Noted from the Guadalupe Mine.

Municipio del Oro

El Oro: Moderate distribution.
Associated with cinnabar.

Magistral: Moderately distributed. Noted from the Porvenir
Mine. Associated with cinnabar.

GUANAJUATO

Municipio de Guanajuato

Guanajuato: Limited distribution.
On the Veta Madre. [In the early
years of mining it was found in
most mines of the Veta Madre.]

Municipio de San Luis de la Paz

Cerro Blanco: Moderate distribution.

Cerro Gordo: Moderately distributed. Associated with cinnabar.

GUERRERO

Municipio de Huitzuco

Huitzuco de Los Figueroa:
Widespread. Most noted from the
La Cruz, San Agustin and Agua
Salada mines. Associated with
cinnabar.

Municipio de Taxco

Huahuaxtla: Widespread. Most
noted from the Cinco Cervezas,
San Luis, Aurora and Esperanza
mines. Associated with cinnabar,
egglestonite, montroydite, and
terlinguite.

HIDALGO

Municipio de Zimapán

Zimapán: Moderately distributed.
Associated with cinnabar.

JALISCO

Municipio de Mascota

El Moral: Moderately distributed.
Associated with cinnabar.

Municipio de Talpa de Allende

Bramador: Moderate distribution.
Associated with cinnabar.

QUERETARO

Municipio de Cadereyta

El Doctor: Moderate distribution. Noted from the Las Cabras Mine. Associated with cinnabar.

SAN LUIS POTOSI

Municipio de Cerro de San Pedro

Cerro de San Pedro: Moderately distributed. Associated with cinnabar.

Municipio de Charcas

Charcas: Limited distribution. Associated with cinnabar.

Municipio de Guadalcázar

Guadalcázar: Moderately distributed. Most noted from the Las Animas, Los Barros, El Escarabajo, El Refugio, San Augustín, San José, La Trinidad, and Santa Lucía mines. Associated with cinnabar. [This area has been a major producer of mercury since the 1600s.]

Municipio de Mexquitic

Bocas: Moderate distribution. Associated with cinnabar.

Municipio de Moctezuma

San Antonio Rul: Limited distribution. Associated with cinnabar.

ZACATECAS

Municipio de Fresnillo

Plateros: Moderate distribution. Associated with cinnabar.

Municipio de Mazapil

Mazapil: Moderate distribution. Most noted from the Cinabrio Mine. Associated with cinnabar.

San Pedro Ocampo: Moderate distribution. Associated with cinnabar.

MERWINITE

$Ca_3Mg(SiO_4)_2$. An uncommon mineral found in few mineral deposits.

DURANGO

Municipio de Cuencamé

Velardeña: Limited distribution.

MICHOACAN

Municipio de Zitácuaro

Susupuato: Moderately distributed wthin the Cerro Mazahua.

MESOLITE

$Na_2Ca_2Al_6Si_9O_{30} \cdot 8H_2O$. An uncommom mineral and a member of the zeolite mineral group and found in igneous rocks and in hydothermal veins.

GUANAJUATO

Municipio de Guanajuato

Cerro del Capulin: Limited distribution.

Sierra de Santa Rosa: Moderately distributed.

JALISCO

Municipio de Guadalajara

Barranca de San Cristobal: Moderate distribution.

Municipio de Tapalpa

Tapalpa: Moderately distributed throughout the Sierra de Tapalpa.

MORELOS

Municipio de Ayala

Jalostoc: Limited distribution. Noted from the San Luis and San

Felipe mines. [In the order records, this location was spelled Xalostoc.]

SINALOA

Municipio de Culiacán

Río Culiacán: Moderate distribution.

META-AUTUNITE

$Ca(UO_2)_2(PO_4)_2 \cdot 2-6H_2O$. An uncommon secondary unranium mineral.

DURANGO

Municipio de Tamazula

El Mezquite: Limited distribution. Associated with autunite.

METACINNABAR

HgS. A common secondary mineral and member of the sphalerite group, which is found in the upper zone of mercury deposits.

CHIHUAHUA

Municipio de Camargo

Sierra de Encinillas: Limited distribution. [This mineral from the location was originally called Guadalacazarite, which was found to be metacinnabar.]

GUERRERO

Municipio de Taxco

Buenavista de Cuellar: Moderate distribution.

Huahuaxtla: Moderate distribution. Most noted from the San Luis, Aurora and Esperanza mines. Associated with mercury.

Municipio de Teloloapan

Teloloapan: Moderate distribution. Noted from the Concepción Mine.

SAN LUIS POTOSI

Municipio de Guadalcázar

Guadalcázar: Moderately distributed. (Selenian Metacinnabar). Most noted from the La Providencia and Trinidad mines. [The Trinidad Mine was the discovery site of the mineral guadalcazarite, which was found to be metacinnabar.]

ZACATECAS

Municipio de Fresnillo

San Onofre: Moderately distributed. Associated with mercury. [This location, near Plateros, produced a new mineral that was identified and named Onofreite. But later studies proved it to be a selenium metacinnabar.]

METAKOETTIGITE

$(Zn,Fe)_3(AsO_4)_2 \cdot 8H_2O$. A rare mineral found in few metal deposits rich in arsenic.

DURANGO

Municipio de Mapimí

Mapimí: Limited distribution. Noted from the Ojuela Mine. Crystals.
Associated with koettigite and goethite. [This is the type location for this mineral.]

METATORBERNITE

$Cu(UO_2)_2(PO_4)_2 \cdot 8H_2O$. An uncommon secondary mineral found in copper/uranium deposits.

SONORA

Municipio de Soyopa

San Antonio de la Huerta: Limited distribution. Most noted from the El Triunfo Mine. Associated with torbernite.

METATYUYAMUNITE

$Ca(UO_2)_2(VO_4)_2 \cdot 3-5H_2O$. An uncommon secondary mineral found in the oxidized zones of uranium/vanadium deposits.

CHIHUAHUA

Municipio de Aldama

Sierra del Curvo: Limited distribution. Noted from the El Calbario Mine.

METAVOLTINE. Specimens exist but location data is not available.

MIARGYRITE

$AgSbS_2$. A rare mineral found in low temperature hydrothermal veins.

AGUASCALIENTES

Municipio de Asientos

Asientos: Moderately distributed. Most noted from the Santo Cristo, Santa Francisca, and Descubridora mines.

CHIHUAHUA

Municipio de Cuishuiráchic

Cusihuiriáchic: Moderate distribution.

DURANGO

Municipio de San Pedro del Gallo

Peñoles: Limited distribution.

GUANAJUATO

Municipio de Guanajuato

Guanajuato: Moderately distributed within the Veta Madre.

GUERRERO

Municipio de Taxco

Acamixtla: Moderate distribution.

HIDALGO

Municipio del Mineral del Monte

Mineral del Monte: Moderately distributed.

Municipio de Pachuca

Pachuca de Soto: Moderate distribution.

MEXICO

Municipio de Sultepec

El Cristo: Limited distribution.

Sultepec de Pedro Ascencio de Alquisiras: Moderately distributed.

OAXACA

Municipio de San Pedro Teviche

San Pedro Teviche: Limited distribution.

Municipio de Zimatlán de Alvarez

Las Peras: Moderately distributed within the Cañada de Las Peras.

SAN LUIS POTOSI

Municipio de Catorce

Catorce: Moderate distribution.

SONORA

Municipio de Alamos

Alamos: Moderately distributed. Most noted from the La Dura, Doctor Brown, Mezquite, Prieta, Animas, and Cuchillo mines.

Municipio de Sahuaripa

Trinidad: Moderately distributed. Noted from the Bufa, Dios Padre, Del Arco, and Espiritu mines.

Municipio de Onavas

La Barranca: Moderate distribution. Noted from the Los Tajos Mine.

ZACATECAS

Municipio de Sombrerete

Sombrerete: Moderate distribution. Noted from the Veta Negra mine. Associated with proustite.

Municipio de Vetagrande

Vetagrande: Moderate distribution.

Municipio de Zacatecas

Zacatecas: Moderate distribution.

MICROCLINE

$KAlSi_3O_8$. A rather common mineral of the feldspar mineral group formed in igneous rocks, in pegmatites, and in hydrothermal solutions on the country wall rocks.

DURANGO

Municipio de Mapimí

Mapimí: Limited distribution. Noted at the Ojuela Mine.

SONORA

Municipio de Santa Cruz

Santa Cruz: Widespread throughout the Cerro de Chivto. Noted from the Guadalupaña Mine. White to tan crystals to 8 cm. Associated with dravite, scheelite, and fluorapatite.

MILARITE

$K_2Ca_4Al_2Be_4Si_{24}O_{60} \cdot H_2O$. A rare mineral and member of the Osumitite group and found in few mineral depsoits.

GUANAJUATO

Guanajuato: Moderate distribution. Noted from the Valenciana Mine. Yellowish green crystals to 3 cm. Associated with adularia and quartz.

MILLERITE

NiS. An uncommon mineral found in low temperature veins.

JALISCO

Municipio de Pihuamo

Tonila: Limited distribution.

MICHOACAN

Municipio de La Huacana

Inguaran: Moderate distribution.

SINALOA

Municipio de Badiraguato

Alisito: Moderate distribution. Noted from the Gloria Mine. Associated with gold, erythrite, gersdorffite, jarosite, goethite, pyrite, pentlandite, and violarite.

MIMETITE

$Pb_5(AsO_4)_3Cl$. An uncommon secondary mineral and member of the apatite group and is found in oxidized zones of lead deposits.

CHIHUAHUA

Municipio de Aquiles Serdán

Francisco Portillo: Moderately distributed. Most noted from the El Potosí, Santo Domingo, Inglaterra, and Velardeña mines. Crystals to 1 cm. Associated with calcite, goethite and creedite. [In the late 1970s, a large cavity lined with mimetite was dicovered in the Inglaterra Mine between the 8th and 9th level. The botryoidal mimetite ranged in color from shades of gray, tan, brown, and orange. Since the Inglaterra Mine is part of the El Potosí Mine system, the mimetite was found in the Inglaterra Mine section of the El Potosí Mine.]

San Antonio el Grande: Moderate distribution. Noted from the San Antonio Mine. Bright yellow botryoidal masses. Associated with wulfenite and pyromorphite.

Municipio de Manuel Benavides

San Carlos: Moderate distribution. Noted from the San Carlos Mine. Botryoidal masses. Associated with vanadinite and calcite.

Municipio de Nuevo Casas Grandes

San Pedro Corralitos: Moderate distribution on the Cerro del Capulin. Most noted from the Congresso-León and San Pedro mines. Bright yellow to orange botryoidal masses. Associated with goethite. [The San Pedro Mine is now interconected with the larger Congresso-León Mine and is now known as the Congresso-León Mine. In 1970, a series of pockets lined with bight yellow botryoidal mimetite were discovered on the 19th level of the old San Pedro Mine section of the Congresso-León Mine. A very pale yellow mimetite has been found at a small unnamed prospect a few killometers north of the Congresso-León Mine. It was composed of small crystal aggragrates associated with wulfenite on goethite.]

Municipio de Saucillo

Naica: Moderate distribution.

COAHUILA

Municipio de Melchor Múzquiz

La Encantada: Moderate distribution. Noted from the La Encantada Mine. Associated with wulfenite.

DURANGO

Municipio de Cuencamé

Velardeña: Moderate distribution. Yellow botryoidal masses. Associated with goethite. [In the late 1960s a large cavity was discovered and, unfortunately, was listed as being from Mapimi.]

Municipio de Durango

Sierra de Cacaria: Moderate distribution. Noted from the El Diablo Mine. Associated with cerussite and anglesite. Mimetite was found often as replacements of cerussite from here.

Municipio de Mapimí

Mapimí: Moderate distribution. Noted from the Ojuela and San Juan mines. Pale yellow to greenish yellow crystals to 6mm. Associated with wulfenite, goethite, cerussite, plumbojarosite, bindheimite and anglesite.

GUERRERO

Municipio de Arcelia

Compo Morado: Limited distribution. Noted from the El Naranjo Mine.

Municipio de Teloloapan

Teloloapan: Moderately distributed. Noted from the 8th level of
the El Naranjo Mine.

HIDALGO

Municipio de Zimapán

Zimapán: Moderately distributed.
Noted from the Guadalupe Mine.
Greenish crystals to 1 cm.
Associated with pyromorphite.

JALISCO

Municipio de Bolaños

Bolaños: Moderate distribution.
Noted from the La Iguana Mines's
1st level and the 2nd level of
the Santa Fe Mine. Associated
with hematite.

SAN LUIS POTOSI

Municipio de Cerro
de San Pedro

Cerro de San Pedro: Moderately
distributed. Red-orange botryoidal masses. Associated with
goethite. [In 1970, several
pockets of bright red-orange
botryodial mimetite was discovered here.]

Municipio de Guadalcázar

Guadalcázar: Moderately distributed throuhout the Sierra de San
Cristóbal. Most noted from the
Promontorio Mine. Associated with
wulfenite and phosgenite.

SONORA

Municipio de Cucurpe

Cucurpe: Moderate distribution
within Cerro Prieto. Noted from
the San Francisco Mine 2nd to
10th level. Yellow to red-orange
botryoidal masses. Associated
with wulfenite, quartz, and
gold. [It has been found here as
replacements of wulfenite
crystals.]

Municipio de Rayón

Rayón: Moderate distribution.
Pale yellow botryodial masses.
Associated with wulfenite.

ZACATECAS

Municipio de Concepción
del Oro

Bonanza: Moderate distribution.
Noted from the El Refugio Mine.

Municipio de Fresnillo

Plateros: Moderately distributed.

Municipio de Mazapil

Mazapil: Moderate distribution.
Noted from the Santa Rosa Mine.
Yellow botrytoidal masses.
Associated with wulfenite and
pyromorphite.

Municipio de Ojo Caliente

La Blanca: Moderate distribution
within the Sierra de Santiago.
Noted from the Bilbao Mine.
Yellow to orange botryodial
masses.

Municipio de Zacatecas

Zacatecas: Limited distribution.
Scarce.

MINIUM

Pb_2PbO_4. An uncommon secondary
mineral formed in extreme
oxidizing environments in lead
deposits.

BAJA CALIFORNIA SUR

Municipio de San Antonio

El Triufo: Moderate distribution.

CHIHUAHUA

Municipio de Aquiles Serdán

Francisco Portillo: Limited
distribution. Noted from the 4th

level of the El Potosí Mine. Associated with calcite.

San Antonio el Grande: Limited distribution. Noted from the San Antonio Mine.

Municipio de Manuel Benavides

Manuel Benavides: Moderate distribution. Noted from the Las Dos Marias Mine. Associated with anglesite, cerussite and hemimorphite.

COAHUILA

Municipio de Melchor

Cedral: Limited distribution.

DURANGO

Municipio de Mapimí

Mapimí: Limited distribution. Noted from the Ojuela Mine. Associated with wulfenite, adamite, mimetite, goethite, galena, and barite.

HIDALGO

Municipio de Zimapán

Zimapán: Limited distribution. Noted from the Lomo del Toro Mine.

JALISCO

Municipio de Bolaños

Bolaños: Limited distribution. Noted from the Santa fe Mine.

SONORA

Municipio de Álamos

Río Chico: Moderate distribution.

MIRABILITE

$Na_2SO_4 \cdot 10H_2O$. An uncommon secondary mineral found in arid environments.

CHIHUAHUA

Municipio de Aquiles Serdán

Francisco Portillo: Limited distribution.

MIXITE

$BiCu_6(AsO_4)_3(OH)_6 \cdot 3H_2O$. A rare secondary mineral found in few metal deposits.

ZACATECAS

Municipio de Concepción del Oro

Bonanza: Limited distribution. Noted from the Bonanza Mine. Pale whitish-blue to green botryoidal masses.

MIZZONITE

$mCa_4(Al_6Si_6O_24)CO_3nNa_4(Al_3Si_9-O_24)Cl$. A member of the scapolite mineral group and is an intermediate mineral between marialite and meionite. Found in metamorphic rocks, contact zones, altered basic igneous rocks, and volcanic bombs.

HIDALGO

Municipio de Zimapán

Zimapán: Limited distribution.

MOCTEZUMITE

$Pb(UO_2)(TeO_3)_2$. A rare mineral found in few metal deposits.

SONORA

Municipio de Moctezuma

Moctezuma: Limited distribution. Noted from the Moctezuma Mine. Associated with burckhardtite and zemannite. [This mineral was named after the location, which is the type locality for this mineral.]

MOLYBDENITE

MoS_2. An uncommom mineral found in igneous and metamorphic rocks, especially along contact zones.

CHIAPAS

Municipio de Pichucalco

Pichucalco: Limited distribution. Noted from the Santa Fe Mine.

CHIHUAHUA

Municipio de Guadalupe

Guadalupe: Limited distribution. Associated with quartz.

Municipio de Saucillo

Naica: Limited distribution. Noted from the Naica Mine.

COAHUILA

Municipio de Candela

Pánuco: Moderate distribution. Noted from the Pánuco Copper Mine.

DURANGO

Municipio de Mapimí

Mapimí: Limited distribution. Noted from the Ojuela Mine. Crystal plates and rosettes to 4 mm. Associated with pyrite, sphalerite, and galena.

GUANAJUATO

Municipio de Guanajuato

Guanajuato: Moderate distribution. Most noted from the El Nopal and Santa Inés mines.

Santa Rosa: Moderate distribution throughout the Sierra de Santa Rosa. Noted from the La Industria Mine (on the Rancho Calvillo). Crystals to 2 cm. Associated with dolomite, calcite, and chalcopyrite.

HIDALGO

Municipio de Zimapán

Zimapán: Moderate distribution. Noted from the Lomo del Toro Mine. Associated with molybdite.

JALISCO

Municipio de Mascota

San Sebastián: Moderately distributed within the arroyo del Calera. Associated with molybdite.

MEXICO

Municipio de Temascaltepec

Temascaltepec de González: Moderate distribution.

MICHOACAN

Municipio de Huetamo

Baztan: Limited distribution.

NUEVO LEON

Municipio de Lampozos de Naranjo

El Carrizal: Limited distribution within the Sierra del Carrizal. Noted from the El Porvenir Mine.

PUEBLA

Municipio de Tetela de Ocampo

Tetela del Oro: Moderate distribution.

SINALOA

Municipio de San Ignacio

Sindicatura de san Juan: Moderate distribution throughout the arroyo de Teccuitapa.

SONORA

Municipio de Cananea

Cananea: Moderate distribution.

Municipio de Cumpas

Cumpas: Moderate distribution.

Jecori: Limited distribution. Noted from the San Julián Mine.

Municipio de Nacozari de García

Nacozari de García: Moderate distribution. Noted from the Canutillo Mine.

Municipio de Onavas

Tonichi: Moderate distribution.

Municipio de Sahuaripa

El Encinal: Moderately distributed Noted from the Lydia Mine.

Sahuaripa: Moderate distribution. Noted from the La Trinidad Mine.

ZACATECAS

Municipio de Zacatecas

Zacatecas: Limited distribution. Noted from the Saltillito Mine.

MOLYBDITE

MoO_3. An uncommom mineral found in molybdenum deposits.

CHIHUAHUA

Municipio de Guadalupe y Calvo

Guadalupe y Calvo: Moderate distribution. Noted from the Hidalgo Mine.

HIDALGO

Municipio de Zimapán

Zimapán: Limited distribution. Noted from the Lomo del Toro

Mine. Associated with molybdenite.

JALISCO

Municipio de Mascota

San Sebastián: Limited distribution within the arroyo del Calera. Associated with molybdenite.

PUEBLA

Municipio de Tlanalapantla

Matamoros: Moderate distribution.

SONORA

Municipio de Nacozari de García

Nacozari de García: Limited distribution. Noted from the Canutillo Mine.

MONAZITE

$(Ce,La,Nb,Th)PO_4$. A rather uncommon mineral and member of the monazite group and is found in igneous, metamorphic and sedimentary rocks.

DURANGO

Municipio de Mapimí

Mapimí: Limited distribution. Noted from the Ojuela Mine.

SAN LUIS POTOSI

Municipio de Guadalcázar

Guadalcázar: Moderately distributed.

MONTICELLITE

$CaMgSiO_4$. An uncommon mineral of the olivine group formed in metamorphic environments and in ultramafic rocks.

HIDALGO

Municipio de Zimapán

Zimapán: Limited distribution. Noted from the Lomo del Toro Mine.

MONTMORILLONITE

$Na_{0.7}(Al_{3.3}Mg_{0.7})(Si_8O_{20})(OH)_4 \cdot n-H_2O$. A common mineral of the smectite group, which forms from the alteration of volcanic tuffs and ash and also from hydro-thermal activities.

CHIHUAHUA

Municipio de Saucillo

Naica: Widespread. Massive.

VERACRUZ

Municipio de Tatatila

Tatatila: Moderately distrib-uted. Massive.

MONTROYDITE

HgO. A rare mineral found in the oxidized zones of a mercury deposits.

GUERRERO

Municipio de Taxco

Huahuaxtla: Moderately distrib-uted. Noted from the San Luis Mine. Crystals to almost 4 cm. Associated with eglestonite, terlinguaite, calomel, cinnabar and metacinnabar. [Mercury ore was discovered here in 1924 and by November 1925 mining had began. During the early days of development at the San Luis mine in 1930 and 1931, pockets were discovered above the 30 meter level containing crystals of montroyed up to almost 4 cm long! Mine records indicate that the miners sent sacks full of

these crystals to the resorts to recover the mercury!]

QUERETARO

Municipio de Cadereyta

El Doctor: Limited distribution. Noted from the San Onofre Mine. Associated with mosesite.

MORDENITE

$(Ca,Na_2K_2)Al_2Si_{10}O_{24} \cdot 7H_2O$. An uncommom mineral of the zeolite group and are found in volcanic rocks.

GUANAJUATO

Municipio de Guanajuato

Guanajuato: Limited distribution.

MOSESITE

$Hg_2N(SO_4,MoO_4,Cl) \cdot H_2O$. A rare mineral found in mercury de-posits.

GUERRERO

Municipio de Taxco

Huahuaxtla: Limited distribution. Noted from the San Luis Mine.

QUERETARO

Municipio de Cadereyta

El Doctor: Limited distribution. Noted from the San Orofre Mine. Associated with montroydite.

MOTTRAMITE

$PbCu(VO_4)(OH)$. An uncommon secondary mineral and member of the olivenite mineral group that is found in the oxidized ore zones of metal deposits.

SONORA

Municipio de Imuris

Imuris: Limited distribution.

MROSEITE

$CaTe(CO_3)O_2$. A rare mineral found in few metal deposits.

SONORA

Municipio de Moctezuma

Moctezuma: Limited distribution. Noted from the Moctezuma and San Miguel mines. Associated with zemanite, spiroffite, tellurium, and denningite.

MURDOCHITE

$PbCu_6(O,Cl,Br)_8$. A rare secondary mineral found in the oxidized ore zones of copper/lead deposits.

DURANGO

Municipio de Mapimí

Mapimí: Limited distribution. Noted from the Ojuela Mine. Crystals to 3 mm. Associated with plattnerite, hemimorphite and goethite.

MUSCOVITE

$KAl_2(Si_3Al)O_{10}(OH,F)_2$. A common rock forming mineral of the mica group and is found in all rock types.

BAJA CALIFORNIA

Municipio de Ensenada

El Alamo: Widespread.

CHIHUAHUA

Municipio de Aquiles Serdán

Francisco Portillo: Limited distribution.

HIDALGO

Municipio de Zimapán

Zimapán: Limited distribution. Noted from the Lomo del Toro Mine.

JALISCO

Municipio de Bolaños

Bolaños: Limited distribution.

OAXACA

Municipio de San Francisco Telixtlahuaca

San Francisco Telixtlahuaca: Moderately distributed. (Sericite). Noted from the Santa Ana Mine. Associated with orthoclase, albite, and biotite.

SAN LUIS POTOSI

Municipio de Guadalcázar

Realego: Moderate distribution.

SONORA

Municipio de Caborca

Heróica Caborca: Moderate distribution. Noted from the El Negro and La Nahuila mines.

VERACRUZ

Municipio de Profesor Rafael Ramirez

Profesor Rafael Ramirez: Widespread within the Barranca de Tatatila.

N

NAGYAGITE

$Pb_5Au(Te,Sb)_4S_{5-8}$. A rare mineral found in metal deposits.

CHIHUAHUA

Municipio de Moris

Sahuayacan: Limited distribution. Santo Niño and Sahuayacan mines. Associated with galena, pyrite, and petzite.

NANTOKITE

$CuCl$. An uncommon secondary mineral found in arid copper deposits.

DURANGO

Municipio de Mapimí

Mapimí: Limited distribution. Noted from the Ojuela Mine. Associated with rosasite, aurichalcite, plattnerite, cerussite, and malachite.

TAMAULIPAS

Municipio de Villagrán

San José: Limited distribution. Noted from the La Piedra Iman Mine.

NATANITE

$FeSn(OH)_6$. A rare mineral found in few metal deposits.

CHIHUAHUA

Municipio de Aquiles Serdán

Francisco Portillo: Limited distribution. Tenth level of the El Potosí Mine's silicate ore body. Honey-yellow crystals from 0.5 to 1.0 mm. Associated with rhodochrosite.

NATROJAROSITE

$NaFe_3(SO_4)_2(OH)_6$. An uncommon secondary mineral and member of the alunite group that is found in metal deposits.

CHIHUAHUA

Municipio de Aquiles Serdán

Francisco Portillo: Limited distribution.

NATROLITE

$Na_2Al_2Si_3O_{10} \cdot 2H_2O$. An uncommon mineral and member of the zeolite mineral group that is found in igneous rocks and as an alteration product in metamorphic rocks.

CHIHUAHUA

Municipio de Aquiles Serdán

Francisco Portillo: Limited distribution.

HIDALGO

Municipio de Huasca de Ocampo

San Miguel Regla: Moderate distribution throughout the Barranca de Regla.

JALISCO

Municipio de San Cristóbal de la Barranca

San Gasper: Moderate distribution.

MORELOS

Municipio de Ayala

Jalostoc: Limited distribution. Noted from the San Felipe Mine.

SAN LUIS POTOSI

Municipio de Charcas

Charcas: Limited distribution. Crystals to 3 cm. Associated with calcite.

ZACATECAS

Municipio de Mazapil

Noche Buena: Limited distribution. Noted from the Noche Buena Mine. Associated with hydroxy-apophyllite and pyrite.

NATRON

$Na_2CO_3 \cdot 10H_2O$. An uncommon mineral found in playa lake deposits.

MEXICO

Municipio de Texcoco

Lago de Texcoco: Moderately distributed.

NAUMANNITE

Ag_2Se. A rare mineral found in few silver deposits.

GUANAJUATO

Municipio de Guanajuato

Guanajuato: Moderate distribution. Most noted from the El Capulín, San Bernabé, and San Juan de Rayas mines.

La Luz: Limited distribution. Noted from the Santa Rita Mine, previously known as the Bolañitos Mine.

GUERRERO

Municipio de Taxco

Taxco de Alcarón: Limited distribution.

NEPHELINE

$(Na,K)AlSiO_4$. An uncommon mineral and member of the feldspathoid mineral group which forms in alkaline igneous rocks.

HIDALGO

Municipio de Zimapán

Zimapán: Limited distribution. Noted from the Lomo del Toro Mine.

NICCOLITE. See NICKELINE

NICKEL

Found at several locations in meteorites within Mexico.

NICKELINE

NiAs. An uncommon mineral found in nickel deposits. Formally known as niccolite.

SINALOA

Municipio de Badiraguato

Alisitos: Moderate distribution. Noted from the Gloria Mine. Massive. Associated with gold, anabergite, gersdoffite, jarosite, and maucherite.

NIFONTOVITE

$Ca_3B_6O_6(OH)_{12} \cdot 2H_2O$. A rare mineral found in few mineral deposits.

SAN LUIS POTOSI

Municipio de Charcas

Charcas: Limited distribution. Crystals to 6 cm.

NITER

KNO_3. An uncommon mineral of the aragenite mineral group forming in arid regions. Also known as saltpeter.

CHIHUAHUA

Municipio de Camargo

Laguna de Jaco: Moderate distribution. Massive.

DURANGO

Municipio de Peñón Blanco

Peñón Blanco: Moderate distribution. Massive.

NONTRONITE

$Na_{0.33}Fe_2(Si,Al)_4O_{10}(OH)_2 \cdot nH_2O$. An uncommon mineral of the smectite group found in veins as an alteration of volcanic glass.

CHIHUAHUA

Municipio de Aquiles Serdán

Francisco Portillo: Limited distribution. Noted from the Inglaterra Mine.

Municipio de Saucillo

Naica: Limited distribution.

OAXACA

Municipio de San Juan Mixtepec

Tejocotes: Limited distribution. Noted from the Yucunani Mine.

SAN LUIS POTOSI

Municipio de Villa de la Paz

La Paz: Limited distribution. Noted from the Dolores Mine.

ZACATECAS

Municipio de Concepción del Oro

Aranzazú: Limited distribution. Noted from the San Antonio Mine. Associated with garnet.

NOVACEKITE

$Mg(UO_2)_2(AsO_4)_2 \cdot 12H_2O$. A rare mineral and member of the autunite group that is found in few metal deposits.

CHIHUAHUA

Municipio de Aldama

Puerte de Aire: Limited distribution. Most noted from the Virgen Mine. [This location is also known as Placer de Guadalupe.]

Municipio de Aquiles Serdán

Francisco Portillo: Limited distribution. Noted from the Las

Animas Mine. Tuffs of crystals to 8 mm. Associated with gypsum and calcite.

NSTUTITE

$Mn_xMn_{1-x}O_{2-2x}(OH)_{2x}$. An uncommon mineral found in manganese deposits.

HIDALGO

Municipio de Molango

Molango: Widespread. Most noted from the Nonoalco Mine.

O

OGDENSBURGITE

$Ca_3ZnFe_6(AsO_4)_5(OH)_{11}\cdot5H_2O$. A rare secondary mineral found in the oxidized zone of arsenic rich ore bodies.

DURANGO

Municipio de Mapimí

Mapimí: Limited distribution. Between the 13th and 14th levels of the San Judas stope of the Ojuela Mine. Crystals to 2 mm. Associated with goethite and adamite. [This mineral was found on only a few specimens in the San Judas stope of the Ojuela Mine. It was found in the later part of 1981 during the period of time when several pockets of large purple adamite crystals were being uncovered. It was found near the area where the lotharmeyerite was discovered.]

OJUELAITE

$ZnFe_2(AsO_4)_2(OH)_2.4H_2O$. A rare secondary mineral found in arsenic-rich metal deposits.

DURANGO

Municipio de Mapimí

Mapimí: Limited distribution. Noted from the Ojuela Mine.

Crystals to 4 mm. Associated with scorodite and goethite. [This mineral was discovered here and named for the mine. This is the type location.]

OKENITE

$CaSi_2O_4(OH)_2\cdot H_2O$. An uncommon mineral and member of the zeolite mineral group that is found in volcanic rocks and hydrothermal veins.

ZACATECAS

Municipio de Mazapil

Noche Buena: Limited distribution. Noted from the Noche Buena Mine. Associated with galena.

OLIGOCLASE

$NaCaAlSi_2O_8$. A common rock forming mineral and member of the feldspar group that can be found in all rock types.

SAN LUIS POTOSI

Municipio de Guadalcázar

Realejo: Widespread.

OLIVENITE

$Cu_2AsO_4(OH)$. An uncommon secondary mineral found in the oxidized zones of base metal deposits.

DURANGO

Municipio de Mapimí

Mapimí: Limited distribution. Noted from the Ojuela Mine. Crystals to 2 mm. Associated with goethite. [There is still some question as to correctness of the identification of this mineral from here.]

HIDALGO

Municipio de Zimapán

Zimapán: Limited distribution. Most noted from the Miguel Hidalgo and Dolores mines. Crystals to 2 mm. Associated with scorodite.

OLIVINE is a group name, not a specific mineral.

ONOFRITE. See **METACINNABAR** var. selenian

OPAL

$SiO_2 \cdot nH_2O$. A common mineral found in volcanic rocks of metastable nature. This is a questionable mineral and is often referred to as a mineraloid.

CHIHUAHUA

Municipio de Bocoyna

Creel: Moderately distributed. Massive. Cherry opal-red and yellow colors.

Sisoguichic: Moderate distribution. Massive. (Common opal). Salmon colored.

Municipio de Hidalgo del Parral

Cañada del Negro: Moderate distribution. Massive. (Common opal).

Municipio de Ojinaga

Hacienda Santa Isabel: Limited distribution. Massive. (Precious opal).

COAHUILA

Municipio de Sierra Mojada

Sierra Mojada: Limited distribution. Noted from the Fortuna Mine. Massive. (Wood and Hyalite opal).

DURANGO

Municipio de Mezquital

Mezquital: Moderate distribution. Noted from the San Francisco del Oro Mine. Massive. (Wood opal).

DISTRICTO FEDERAL

Azcapotzalco: Limited distribution on the Hacienda de Aspeita. Massive. (Wood opal).

GUANAJUATO

Municipio de Cuerámaro

Cuerámaro: Limited distribution. Noted from the Buenos Aires Mine. Massive. (Precious and cherry opal).

Municipio de Guanajuato

Cerro del Rosa de Castilla: Moderate distribution. Massive. (hyalite).

El Cubo: Limited distribution. Massive. (Precious opal).

Sierra de Santa Rosa: Limited distribution on the Rancho de Emmedio. Massive. (Precious and cherry opal). Moderate distribution on the Rancho de

Calvillo. Massive. (Common opal).

Sierra de Guanajuato: Moderate distribution. Massive. (Common opal).

Ovejeras: Limited distribution. Massive. (Tripolite).

Vilaseca: Moderate distribution. Massive (Common opal).

Municipio de Guanimaro

Cerro Campuzano: Limited distribution. Massive. (Precious opal).

Municipio de Pénjamo

Pénjamo: Limited distribution. Most noted at the Buenos Aires Mine. Massive. (Precious and cherry).

Municipio de San Luis de la Paz

Mineral de Pozos: Limited distribution on the Cerro del Aguila. Massive. (Hylite).

Municipio de Santa Catarina

Santa Catarina: Limited distribution. Noted from the Santin Mine. Massive. Associated with cassiterite. Massive.

GUERRERO

Municipio de Cocula

Coacoyula: Limited distribution. Massive. (Precious opal-milky).

Municipio de Huitzuco

Huitzuco de los Figueroa: Limited distribution. Massive. (Precious opal-black with extremely good fire).

Municipio de San Miguel Totolapan

San Nicolás del Oro: Limited distribution. Massive. (Precious opal-pale yellow and red with moderate fire and common opal).

HIDALGO

Municipio de Pachuca

Cerro de San Cristóbal: Moderate distribution. Massive. (Common opal).

Municipio de Huasca de Ocampo

San Miguel Regla: Moderately distributed throughout the Barranca de Regla. Massive. (Common opal).

Tepezalá: Limited distribution within the arranca de Tepezalá on the Cerro de las Fajas. Massive. (Precious opal).

Municipio de Tulancingo

San Miguel: Limited distribution within the Barranca Agua Dulce. Massive. (Precious opal).

Municipio de Zacualtipán

Zacualtipán: Moderate distribution. Massive. (Common opal and hydrophane).

Municipio de Zimapán

Zimapán: Moderate distribution. Noted from the Lomo del Toro Mine. Massive. (Common opal) Limited distribution. Massive. (Precious opal and hydrophane).

JALISCO

Municipio de Magdalena

La Amazata: Moderate distribution. Noted from the La Amazata Mine. (Precious opal).

La Mazata: Moderate distribution. Most noted from the La Estancia Mine. Massive. (Precious opal).

La Silleta: Moderate distribution. Massive. (Precious opal)

Magdalena: Moderately distributed, 9km southeast. Most noted from the La Unica and Tepucanapa mines. Massive. (Precious opal).

San Andrés: Moderate distribution, west. Most noted from the La Mara Mine. Massive. (Precious opal).

San Simón: Moderate distribution, 2 km west. Most noted from the La Chela and San Simón mines. Massive. (Precious opal).

Municipio de San Cristóbal de la Barranca

El Salvador: Moderate distribution. Most noted from the El Cobano and Los Hornitos mines. Massive. (Precious opal).

MEXICO

Municipio de Chiautla

San Lucas: Moderate distribution on the Cerro de Tlatecahuacan. Massive. (Wood and common opal).

Municipio del Oro

El Oro de Hidalgo: Moderate distribution. Massive. (Common opal).

Venta del Aire: Limited distribution. Massive. (Hyalite).

Municipio de Zumpango

Hacienda de Casa Blanca: Moderate distribution. Massive. (Common opal).

MICHOACAN

Municipio de Contepec

Contepec: Moderate distribution on the Hacienda de San Isidro. Massive. (Precious and common opal).

Municipio de Maravatío

Cerro Aguatino: Moderate distribution. Massive. (Precious and common opal).

NAYARIT

Municipio de Compostela

Compostela: Limited distribution. Massive. (Precious opal).

Municipio de Jala

Jala: Moderate distribution. Massive. (Preciuos opal).

Municipio de Jalisco

La Curva: Moderate distribution, 6 km southeast. Most noted from the Divisadero Mine. Massive. (Precious opal).

Municipio de La Yesca

La Yesca: Moderately abundant. Noted from the Buena Vista Mine. Massive. (Common opal).

OAXACA

Municipio de San Juan Teposcolula

San Juan Teposcolula: Moderate distribution. Massive. (Common opal).

Municipio de Santa María Asunción Tlaxiaco

Santa María Asunción Tlaxiaco: Limited distribution. Massive. (Precious opal).

PUEBLA

Municipio de Naupan

Capila: Moderate distribution. Massive. (Wood opal and tripolite).

QUERETARO

Municipio de Amealco

Amealco: Moderate distribution. Most noted from the La Purísima Mine. Massive. (Precious opal).

Municipio de Colón

Esperanza: Limited distribution.

Municipio de Ezequiel Montés

Ezequiel Montés: Moderate distribution. Most noted from the Guadalupana Mine. Massive. (Precious opal).

Municipio de Cadereyta

Foentesuela: Moderate distribution within the Arroyo de Ramos on the Hacienda de Foentesuela. Massive. (Precious opal).

Municipio de Tequisquiapan

Hacienda de Tequisquiapan: Limited distribution. Massive. (Precious opal).

Hacienda La Llava: Limited distribution. Massive. (Precious opal).

La Trinidad: Moderate distribution. Most noted from the La Carbonea Mine. Massive. (Precious and cherry opal).

SAN LUIS POTOSI

Municipio de Villa de Arriaga

Tepetate: Moderate distribution. Massive. (Hyalite-colorless to smoky yellow). Associated with topaz.

Municipio de Cerritos

Cerro del Tepozán: Limited distribution. Massive. (Hyalite).

Municipio de Mexquitic

Hacienda de Bocas: Moderate distribution within the Sierra de Mequitic. Massive. (Common opal).

Municipio de Santa María del Río

Hacienda de San Pedro: Limited distribution. Massive. (Hyalite).

Puerto Blanco: Limited distribution. Massive. (Hyalite).

Municipio de Villa de Reyes

Cerro de la Enramada: Limited distribution. Massive. (Precious opal).

TLAXCALA

Municipio de Tlaxcala

Cerro de los Silicates: Moderate distribution. Massive. (Wood opal).

ZACATECAS

Municipio de Villanueva

Hacienda de La Tayahua: Moderate distribution. Massive. (Common opal).

Municipio de Zacatecas

Zacatecas: Limited distribution. Massive. (Common opal).

ORDONEZITE

$ZnSb_2O_6$. A rare mineral found in few lead/antimony deposits.

GUANAJUATO

Municipio de Santa Catarina

Santa Catarina: Limited distribution on the Cerro de Las Fajas. Noted from the Santín Mine. [This mineral was discovered here and was named for Ezequiel Ordónez, a Mexican geologist/mineralogist and former director of the Instituto de Geología. This is the type location for this mineral.]

SONORA

Municipio de Naco

La Morita: Limited distribution, 27 km southwest.

ORPIMENT

As_2S_3. An uncommon secondary mineral formed by the low temperature alteration of other arsenic minerals.

HIDALGO

Municipio de Zimapán

La Bonanza: Limited distribution.

Zimapán: Limited distribution.

QUERETARO

Municipio de Cadereyta

El Doctor: Limited distribution. Most noted from the San Juan Nepomuceno Mine. Associated with realgar.

SAN LUIS POTOSI

Municipio de Cerro de San Pedro

Cerro de San Pedro: Limited distribution.

Municipio de Guadalcázar

Guadalcázar: Limited distribution. Noted from the La Soledad Mine.

ORTHITE. See ALLANITE

ORTHOCLASE

$KAlSi_3O_8$. A common rock forming mineral and member of the feldspar group found in igneous, metamorphic, and sedimentary rocks.

BAJA CALIFORNIA

Municipio de Ensenada

Rosa de Castilla: Moderate distribution. Associated with elbaite, various garnets, danburite, and andesine.

Sierra San Pedro Martir: Widespread throughout the entire mountain range.

CHIHUAHUA

Municipio de Aquiles Serdán

Francisco Portillo: Limited distribution (deep intrussive bodies). Noted from the Buena Tierra Mine. Associated with plagoclase, chalcopyrite, pyrite, galena, and sphalerite.

San Antonio el Grande: Limited distribution. Noted from the San Antonio Mine. Associated with quartz.

Municipio de Saucillo

Naica: Limited distribution.

DURANGO

Municipio de Coneto de Comonfort

Sierra de San Francisco: Widespread.

Municipio de Cuencamé

Velardeña: Limited distribution.

Municipio de Durango

Cerro de Los Remedios: Limited distribution.

La Luz: Moderately distributed within the Rancho Arperos on the Cerro La Estancia (adularia).

Santa Rosa: Limited distribution within the Sierra de Santa Rosa (sanidine).

Municipio de Santa Catarina

Santa Catarina: Limited distribution. Noted from the Santín Mine.

HIDALGO

Municipio de Mineral del Monte

Mineral del Monte: Limited distribution (adularia). Noted from the San Marcial Mine.

Municipio de Pachuca

Pachuca de Soto: Limited distribution.

Municipio de Zimapán

Zimapán: Limited distribution.

MEXICO

Municipio de Toluca

Cerro de la Teresona: Limited distribution.

Municipio de Zacualpan

Zacualpan: Moderate distribution.

MICHOACAN

Municipio de Angangueo

Mineral de Angangueo: Moderate distribution (adularia).

Municipio de Tlalpujahua

Tlalpujahua: Moderate distribution (adularia).

Municipio de Zinapécuaro

Ueareo: Limited distribution.

OAXACA

Municipio de San Francisco Telixtlahuaca

San Francisco Telixtlahuaca: Moderate distribution. Most noted from the Santa Ana Mine. Associated with albite, beryl, tourmaline, and spodumene.

ZACATECAS

Municipio de Concepción del Oro

Concepción del Oro: Limited distribution. Associated quartz and hematite.

Municipio de Fresnillo

Fresnillo de González Echeverría: Limited distribution (adularia).

OURAYITE

$Ag_{25}Pb_{30}Bi_{41}S_{104}$. A rare mineral found in silver/lead deposits.

SONORA

Municipio de Pitiquito

Pitiquito: Limited distribution.

P

PALYGORSKITE

$(Mg,Al)_2Si_4O_{10}(OH)\cdot 4H_2O$. An uncommon clay mineral found in desert soils and as a product of hydrothermal activities.

CHIHUAHUA

Municipio de Saucillo

Naica: Moderate distribution. Massive. Associated with calcite, quartz, sphalerite, galena, and fluorite.

ZACATECAS

Municipio de Concepción del Oro

Bonanza: Moderate distribution. Noted from the Refugio Mine.

PARADAMITE

$Zn_2(AsO_4)(OH)$. A rare secondary mineral found in zinc deposits rich in arsenic.

DURANGO

Municipio de Mapimí

Mapimí: Limited distribution. Noted from the Ojuela Mine. Crystals to 3 cm. Associated with

mimetite, adamite, legrandite, and goethite. [This mineral was discovered here at the Ojuela Mine, which is the type location.]

PARAGUANAJUATITE

$Bi_2(Se,S)_3$. A rare mineral found in very few base metal deposits.

GUANAJUATO

Municipio de Guanajuato

Santa Rosa: Limited distribution within the Sierra de Santa Rosa, on the Rancho Calvillo. Most noted from the Santa Caterina and La Indústria mines. Crystals to 3 cm. Associated with guanajuatite and scheelite.

PARAMENDOZAVILITE

$NaA1_4 Fe_7(PO_4)_5(PMo_{12}O_{40})(OH)_{16}\cdot 56H_2O$. A rare secondary mineral found in the oxidized zone of base metal deposits.

SONORA

Municipio de Cumpus

Cumobabi: Limited distribution, 27 km southwest. San Judas Mine. Yellow masses. Associated with muscovite. [This mineral was discovered here within the La Verde mining district in the mid 1980s. This is the type location for this mineral.]

PARASYMPLESITE

$Fe_3(AsO_4)_2 \cdot 8H_2O$. A rare secondary mineral formed by the alteration of symplesite.

DURANGO

Municipio de Mapimí

Mapimí: Limited distribution. Noted from the Ojuela Mine. Associated with koettigite, symplesite, gypsum (selenite), and goethite.

PARATELLURITE

TeO_2. A rare mineral found in tellurium deposits.

SONORA

Municipio de Cananea

Cancanea: Limited distribution.

Municipio de Moctezuma

Moctezuma: Limited distribution. Noted from the Moctezuma Mine. Associated with tellurite, klockmannite, and Zemannite.

PARAVAUXITE

$FeAl_2(PO_4)_2(OH)_2 \cdot 8H_2O$. A rare mineral found in few mineral deposits.

CHIHUAHUA

Municipio de Aquiles Serdán

Francisco Portillo: Limited distribution.

PARAWOLLASTONITE

$CaSiO_3$. An uncommon mineral found in contact metamorphic environments.

SONORA

Municipio de Hermosillo

Hermosillo: Limited distribution. Associated with wollastonite.

PARGASITE

$NaCa_2(Mg,Fe)_4Al(Si_6Al_2)O_{22}(OH)_2$. An uncommon mineral and member of the amphibole group that forms in metamorphic environments.

GUANAJUATO

Municipio de San Luis de la Paz

Mineral de Pozos: Limited distribution.

PEARCEITE

$Ag_{16}As_2S_{11}$. An uncommon mineral found in low- to moderate-temperature silver veins.

An 8 cm crystal of pearcite from the Las Chipas Mine, Arizpe, Sonora, in the Romero Mineralogical Museum, Tehuacan, Puebla.

BAJA CALIFORNIA SUR

Municipio de Mulegé

Santa Rosalía: Limited distribution. Noted from the Amelia Mine. Associated with boleite.

COAHUILA

Municipio de Sierra Mojada

Sierra Mojada: Limited distribution. Noted from the Veta Rica Mine. Associated with proustite, acanthite, silver, barite, and erythrite.

GUANAJUATO

Municipio de Guanajuato

La Luz: Limited distribution. Noted from the San Carlos Mine. Crystals to 1 cm. Associated with aguilarite.

PUEBLA

Municipio de Zacatlán

Zacatlán: Limited distribution.

SONORA

Municipio de Arizpe

Arizpe: Limited distribution on the Veta Las Chispas. Noted from the Las Chispas Mine. Crystals to 5 cm.

PECTOLITE

$NaCa_2Si_3O_8(OH)$. An uncommon secondary mineral found in cavities in basalt.

DURANGO

Municipio de Mapimí

Mapimí: Limited distribution. Noted from the Ojuela Mine.

PENTLANDITE

$(Fe,Ni)_9S_8$. An uncommon mineral and member of the pentlandite group found in nickel deposits.

SINALOA

Municipio de Badiraguato

Alisitos: Moderate distribution. Noted from the Gloria Mine.

PERCYLITE

$PbCuCl_2(OH)_2$. A rare mineral and member of the boleite group. It is a questionable mineral species.

BAJA CALIFORNIA SUR

Municipio de Mulegé

Santa Rosalía: Limited distribution. Noted from the Amelia Mine. Associated with boleite.

SONORA

Municipio de Arizpe

Arizpe: Limited distribution. Noted from the Las Chispas Mine. Associated with gold. [This location has been listed in previous literature as the San Lorenzo or Pedrazzini mines, which are correctly known as the Las Chispas mine (see acanthite for the mine's history). This is the type location for this mineral.]

PERIDOT. See FORSTERITE

PETZITE

Ag_3AuTe_2. A rare mineral found in base metal deposits.

CHIHUAHUA

Municipio de Moris

Sahuayacán: Limited distribution. Santo Niño and Sahuayacán mines. Associated with galena, pyrite, and nagyagite.

PHARMACOLITE

$CaHAsO_4 \cdot 2H_2O$. An uncommon secondary mineral found as an alteration product of other primary arsenic-rich minerals.

ZACATECAS

Municipio de Mazapil

Mazapil: Limited distribution. Noted from the Jesús María Mine. Massive white to gray fibers.

PHARMACOSIDERITE

$KFe_4(AsO_4)_3(OH)_4 \cdot 6-7H_2O$. An uncommon secondary mineral formed by hydrothermal solutions or as an alteration product of other arsenic minerals.

DURANGO

Municipio de Mapimí

Mapimí: Limited distribution. Noted from the Ojuela Mine.

SONORA

Municipio de Trincheras

La Mur: Limited distribution. Noted from the Los Animas Mine. Crystals to 2 mm. Associated with carmenite. [This location is west and north of Benjamin Hill.]

ZACATECAS

Municipio de Fresnillo

Fresnillo de Gonzalez Echeverría: Limited distribution. Noted from the San Ricardo Mine. [This location is better known as Fresnillo.]

PHENAKITE

Be_2SiO_4. An uncommon mineral found principally in granitic pegmatites.

DURANGO

Municipio de Durango

Victoria de Durango: Limited distribution. Noted from the Cerro de Mercado Mine. [This location is better known simply as Durango, the capital city of the State of Durango.]

PHLOGOPITE

$KMg_3Si_3AlO_{10}(F,OH)_2$. A common mineral and member of the mica group that forms in contact metamorphic or metasomatic environments.

DURANGO

Municipio de Mapimí

Mapimí: Limited distribution. Noted from the Ojuela Mine.

OAXACA

Municipio de Santo Domingo Tehuantepec

Santo Domingo Tehuantepec: Limited distribution.

SONORA

Municipio de Cananea

Cananea: Limited distribution.

PHOSGENITE

$Pb_2(CO_3)Cl_2$. A rare secondary mineral formed by the oxidation of primary lead minerals.

BAJA CALIFORNIA SUR

Municipio de Mulegé

Santa Rosalía: Limited distribution. Noted from the Amelia Mine. Crystals to 5 cm. [This mineral is found as gray to white crystals to 5 cm in length and about 8-10 mm in diameter. It is presently found occasionaly on the dump of the Amelia Mine.]

SAN LUIS POTOSI

Municipio de Catorce

Catorce: Limited distribution. Associated with chlorargyrite and wulfenite.

Municipio de Guadalcázar

Guadalcázar: Limited distribution within the Sierra de San Cristóbal. Associated with wulfenite.

PICKERINGITE

$MgAl_2(SO_4)_4 \cdot 22H_2O$. An uncommon secondary mineral formed as a product of weathering and also as a by-product of oxidation of pyrite-rich ore bodies.

DURANGO

Municipio de Mapimí

Mapimí: Limited distribution. Noted from the Ojeula Mine.

PITCHBLENDE. See **URANINITE**

PLAGIOCLASE is a series.

PLAGIONITE

$Pb_5Sb_8S_{17}$. An uncommon mineral and member of the plagionite group found in the sulfide ore zone of lead/antinommy deposits.

DURANGO

Municipio de Canatlán

Tejamén: Limited distribution. Noted from the La Gloria Mine.

HIDALGO

Municipio de Cardonal

Cardonal: Limited distribution.

PLATINUM

Pt. A rare native element found in ultramafic rocks, in quartz veins, and in blacksands as placer.

HIDALGO

Location unknown

Limited distribution. Massive.

TLAXCALA

Municipio de Tlaxacala

Limited distribution.

PLATTNERITE

PbO_2. An uncommon mineral formed under extreme oxidizing environments.

CHIHUAHUA

Municipio de Aquiles Serdán

San Antonio el Grande: Limited distribution. Noted from the San Antonio Mine's tin chimney, 4th level. Associated with vanadinite, wulfenite, mimetite, and calcite.

Municipio de Hidalgo del Parral

Hidalgo del Parral: Limited distribution.

DURANGO

Municipio de Mapimí

Mapimí: Limited distribution. Most noted from the Ojuela Mine. Black acicular crystals to 5 mm. Associated with goethite, hemimorphite, calcite, smithsonite, aurichalcite, rosasite, hydrozincite, malachite, murdockite and alpha lead oxide.

SAN LUIS POTOSI

Municipio de Cerro de San Pedro

Cerro de San Pedro: Moderate distribution. Associated with goethite and mimetite.

PLUMBOJAROSITE

$PbFe_6(SO_4)_4(OH)_{12}$. An uncommon secondary mineral and member of the alunite group.

CHIHUAHUA

Municipio de Ahumada

Los Lamentos: Moderately distributed. Noted from the Erupción-Ahumada Mine.

Municipio de Aquiles Serdán

Francisco Portillo: Moderate distribution. Noted from the El Potosi Mine's oxide zone. Associated with calcite and quartz.

San Antonio el Grande: Moderate distribution. Noted from the San Antonio Mine.

Municipio de Namiquipa

Namiquipa: Limited distribution. Noted from the American Vein Mine.

Municipio de Saucillo

Naica: Limited distribution.

COAHUILA

Municipio de Sierra Mojada

Atalaya: Limited distribution.

DURANGO

Municipio de Cuencamé

Velardeña: Moderate distribution. Noted from the Santa María Mine.

Municipio de Mapimí

Mapimí: Limited distribution. Noted from the Ojuela Mine's 13th level, San Diego stope. Yellowish brown masses. Associated with bindheimite, mimetite, and wulfenite.

HIDALGO

Municipio de Zimapán

Zimapán: Widespread. Most noted from the Lomo de Toro Mine. [It is found at the Lomo del Toro in concentric layers that the miners call "eyes of St. Peter."]

POLYBASITE

$(Ag,Cu)_{16}Sb_2S_{11}$. An uncommom mineral formed in low- to moderate-temperature ore veins.

AGUASCALIENTES

Municipio de Asientos

Asientos: Moderate distribution. Noted from the Santa Francisca Mine.

CHIHUAHUA

Municipio de Ascención

Sabinal: Moderate distribution. Most noted from the Adventurerea Mine. Associated with pyrargyrite.

An 8 cm × 6 cm cluster of polybasite crystals from the Las Chipas Mine, Arizpe, Sonora, in the Pinch Mineralogical Museum, Rochester, New York.

Municipio de Batopilas

Batopilas: Moderately distributed.

Municipio de Cusihuiriáchic

Cusihuiriáchic: Moderate distribution. Noted from the Princesa Mine.

Municipio de Guadalupe y Calvo

Guadalupe y Calvo: Moderate distribution. Most noted from the Santa Lucía Mine.

Municipio de Maguarichic

Maguarichic: Limited distribution.

Municipio de Morelos

Zapuri: Moderate distribution. Noted from the San Martín Mine.

Municipio de Urique

Urique: Limited distribution. Noted from the El Rosario Mine.

DURANGO

Municipio de Canelas

Birimoa: Moderate distribution.

Municipio de Guanaceví

Guanaceví: Moderate distribution. Most noted from the San Gil, Rosario, and Pelayo mines.

Municipio del Oro

Santa María del Oro: Limited distribution. Noted from the El Salvador Mine.

Municipio de Pueblo Nuevo

El Salto: Moderate distribution. Crystals to 2 cm. Associated with Pyrargyrite.

Municipio de San Dimas

Tayoltita: Moderate distribution. Most noted from the La Libertad Mine. [This location is often referred to as San Dimas and Guarissamey, which are small pueblos around Tayoltita.]

Municipio de San Pedro del Gallo

Peñoles: Limited distribution. Noted from the San Rafael Mine.

Municipio de Topia

Topia: Moderate distribution. Most noted from the San José and Santa Eduwigis mines.

GUANAJUATO

Municipio de Guanajuato

Guanajuato: Widespread. Most noted from the Valenciana, El Nopal, Cubo, Peregrina, La Aldana, and Sirena mines. Crystals to 4 cm. Associated with calcite, quartz, and pyrargyrite.

La Luz: Widespread. Most noted from the La Luz, Santa Rita (Bolañitos), and San Pedro mines. Crystals to 3 cm. Associated with chalcopyrite, quartz, and calcite.

Municipio de San Felipe

San Felipe: Limited distribution. Noted from the Providencia Mine.

Municipio de San Luis de la
Paz

Mineral de Pozos: Limited
distribution. Noted from the
Augustias Mine.

GUERRERO

Municipio de Taxco

Taxco de Alcarón: Limited
distribution. Noted from the San
Agustín Mine. Associated with
azurite.

HIDALGO

Municipio de Mineral del
Chico

Mineral del Chico: Limited
distribution. Noted from the San
Marcial Mine.

Municipio de Mineral del
Monte

Mineral del Monte: Widespread.
Most noted from the Escobar,
Cabrera, and Carretera mines.

Municipio de Pachuca

Pachuca de Soto: Widespread. Most
noted from the Barron and El
Rosario mines.

MEXICO

Municipio de Tejupilco

Juloapán: Moderate distribution.
Most noted from the Esperanza and
La Magdalena mines.

Tejupilco de Hidalgo: Limited
distribution. Associated with
stephanite.

Municipio de Zacualpan

Zacualpan: Widespread. Most noted
from the Carboncillo, Cuchara,
Santa Inés, Alacran, and San
Adrian mines.

MICHOACAN

Municipio de Angangueo

Mineral de Angangueo: Moderate
distribution. Most noted from the
El Carmen and Carrillos mines.

Municipio de Tlalpujahua

Tlalpujahua: Widespread. Most
noted at the Borda, Antigua, and
Tiro Pinto mines. Associated with
native silver, acanthite, and
pyrargyrite.

OAXACA

Municipio de Natividad

Natividad: Limited distribution.

Municipio de Ocotlán de
Juárez

Ocotlán de Juárez: Moderate
distribution.

PUEBLA

Municipio de Tetela de
Ocampo

Tetela del Oro: Limited distribution. Noted from the Covadonga
Mine.

SAN LUIS POTOSI

Municipio de Catorce

Catorce: Widespread. Most noted
from the Bombillo, Concepción,
and Refugio mines.

Municipio de Guadalcázar

Guadalcázar: Moderate distribution.

Municipio de Moctezuma

Hacienda de San Antonio Rul:
Limited distribution.

Municipio de Villa de Ramos

Ramos: Moderate distribution.
Noted from the Cocinera Mine.
Associated with acanthite,
galena, and chalcopyrite.

SINALOA

Municipio de Cosalá

Cosalá: Moderate distribution on
the Veta de Cienega. Noted from
the Nuestra Senora Mine.

SONORA

Municipio de Arizpe

Arizpe: Moderate distribution.
Most noted from the Las Chipas
Mine. Crystals to 6 cm. Associated with acanthite, stephanite,
chlorargyrite, and fluorite.
[Crystals of polybasite have been
found here up to 6 cm in diameter
and 2 cm thick. In 1908, on the
2nd level of the Las Chispas
Mine, a 29.5 kilogram (65 pounds)
crystal mass of polybasite was
uncovered in the Las Chipsas
vein. The mine owner, Mr.
Pedrazzini, had it shipped to the
Columbia School of Mines for its
Egleston Collection, but unfortunately it arrived in two
pieces! The Egleston Collection
belongs now to the American
Museum of Natural Histroy in New
York. The mine producing these
fine specimens, the Las Chipas,
has been called in some history
books as the Pedrazzini mine,
however, this was the name of the
mining company that operated the
mine.]

Municipio de Sahuaripa

Mulatos: Moderate distribution.
Most noted from the San Andrés,
Valle de Tecupeto, and Del Arco
mines.

ZACATECAS

Municipio de Fresnillo

Fresnillo de González Echeverría:
Moderately distributed throughout

the Cerro de Proaño. Crystals to
4 cm. Associated with pyrargyrite and quartz. [This location
is best known simply as Fresnillo.]

Municipio de Mazapil

Mazapil: Limited distribution.
Noted from the San Martín Mine.

Municipio de Pinos

Pinos: Widespread. Most noted
from the Cata, Vaca, El Pinto,
Purísima, Fresnillo, Matatusa,
and Salomon mines.

Municipio de Zacatecas

Zacatecas: Moderate distribution.
Most noted from the Guadalupe, La
Carnicería, and El Bote mines.
Crystals to 2 cm. Associated with
acanthite, galena, and chalco-
pyrite.

PORTLANDITE

Ca(OH)$_2$. An uncommon mineral
found in metamorphic and sedimentary environments.

COAHUILA

Municipio de Morelos

Morelos: Moderately distributed
throughout the Cerro de la
Corona.

MICHOACAN

Municipio de Zitácuaro

Suspuato: Limited distribution
within Cerro Mazahua.

MORELOS

Municipio de Cuernavaca

Cerro de La Coronita: Moderate
distribution.

POUGHITE

$Fe_2(TeO_3)_2(SO_4) \cdot 3H_2O$. A rare mineral found in tellurium deposits.

SONORA

Municipio de Moctezuma

Moctezuma: Limited distribution. Noted from the Moctezuma Mine. Associated with zemannite.

POWELLITE

$CaMoO_4$. An uncommon secondary mineral found in tungsten rich deposits.

SONORA

Municipio de Huépac

Huepac: Moderately distributed. Noted from the Rosa María Mine. Crystals to 5 cm. Associated with scheelite and pyrite.

Municipio de Santa Cruz

Santa Cruz: Moderately distributed. Noted from Guadalupana Mine. Crystals to 3 cm. Associated with scheelite.

POYARKOVITE

Hg_3ClO. A rare mineral found in mercury deposits.

DURANGO

Municipio de Canatlán

Elena: Limited distribution.

PREHNITE

$Ca_2Al_2Si_3O_{10}(OH)_2$. An uncommon mineral found in mafic volcanic rocks and in contact metamorphosed limestones.

BAJA CALIFORNIA

Municipio de Ensenada

Juárez: Moderate distribution. Associated with zoisite.

Municipio de Tecate

Los Gavilanes: Moderate distribution. Noted from the El Formento Mine. Associated with epidote.

GUANAJUATO

Municipio de Guanajuato

Guanajuato: Moderately distributed within the Veta Madre.

HIDALGO

Municipio de Pachuca

Pachuca de Soto: Limited distribution.

SAN LUIS POTOSI

Municipio de Charcas

Charcas: Limited distribution. Associated with datolite.

Municipio de Villa de la Paz

La Paz: Moderate distribution.

PRICEITE

$Ca_4B_{10}O_{19} \cdot 7H_2O$. A rare mineral found in a few ore deposits.

ZACATECAS

Municipio de Sombrerete

Sombrerete: Limited distribution.

PROCHLORITE. See **CLINOCHORE**

PROSOPITE

$CaAl_2(F,OH)_8$. An uncommon mineral found in tin veins and in pegmatites.

ZACATECAS

Municipio de Mazapil

Mazapil: Limited distribution. Noted from the Santa Rosa Mine. Pale blue botryoidal masses.

PROUDITE

$Cu_{0-1}Pb_{7.5}Bi_{9.3-9.7}(S,Se)_{22}$. A rare mineral found in base metal deposits.

CHIHUAHUA

Municipio de Janos

Janos: Limited distribution. Associated with sphalerite.

PROUSTITE

Ag_3AsS_3. An uncommon mineral forming in the late stage of development of low-temperature hydrothermal veins within silver deposits.

AGUASCALIENTES

Municipio de Asientos

Asientos: Limited distribution.

CHIHUAUA

Municipio de Aquiles Serdán

Francisco Portillo: Limited distribution. Noted from the Mina Vieja. Associated with native silver and pyrargyrite. [This mineral was found here in the early years of mining in the Santa Eulalia mining district.]

San Antonio el Grande: Limited distribution. Noted from the San Antonio Mine.

Municipio de Batopilas

Batopilas: Moderate distribution. Crystals to 2 cm. Associated with calcite and acanthite.

Jesús María: Limited distribution. Most noted from the Yedros Mine. Associated with acanthite and pyrargyrite.

Municipio de Cusihuiriáchic

Cusihuiriáchic: Moderate distribution. Noted from the Princesa Mine.

Municipio de Guazapares

Guazapares: Moderate distribution. Most noted from the San Juan de Dios Mine.

Municipio de Saucillo

Naica: Moderate distribution within the Sierra de Naica.

COAHUILA

Municipio de Sierra Mojada

Sierra Mojada: Moderately distributed. Most noted from the Veta Rica and Buenaventura mines. Crystals to 2 cm. Associated with native silver, acanthite, erythrite, and barite.

DURANGO

Municipio de Canelas

Brimoa: Limited distribution.

Municipio de Indé

Indé: Moderate distribution. Most noted from the Las Coloradas, El Agua, Garabatos, La Cruz, Los Machos, Restauradora, San José de la Cumbre, and Soledad mines. Crystals to 2 cm.

Municipio de San Dimas

San Dimas: Moderate distribution.

Municipio de Topia

Topia: Moderate distribution.
Associated with quartz and
calcite.

GUANAJUATO

Municipio de Guanajuato

Guanajuato: Moderately distrib-
uted throughout the Veta Madre.
Most noted from the Valenciana,
El Cubo, Esperanza, and the Santa
Rita mines. Crystals to 2 cm.
Associated with calcite, quartz,
and pyrargyrite.

La Luz: Moderate distribution.
Most noted from the El Rosario,
La Luz, Santa Lucía, Ovejera,
Melladito, and San Cayetano
mines.

GUERRERO

Municipio de Coyuca de
Catalán

Tepantitlán: Widespread. Most
noted from the Descubridora,
Zopilote, La Cruz, La Luz, and
Mina Grande mines.

Municipio de Taxco:

Taxco de Alarcón: Widespread.
Most noted from the Espíritu
Santo, Chocotitlan, Milagro,
Santa Isabel, and Santa Teresa
mines. Crystals to 2 cm. Associated with native silver and
pyrargyrite.

HIDALGO

Municipio de Mineral del
Chico

Mineral del Chico: Moderately
distributed. Noted from the San
Marcial and El Pabellón mines.

Municipio de Pachuca

Pachuca de Soto: Moderately
distributed.

Municipio de Real del Monte

Real del Monte: Moderately
distributed.

MEXICO

Municipio de Temascaltepec

Temascaltepec de González:
Widespread. Most noted from the
La Luz, Caracas, Mina Grande,
Mina de Agua, Ocotillos, Preciosa
Sangre, Quebradilla, Reyes,
Rosario, and Veta Rica mines.
Crystals to 2 cm.

Municipio de Zacualpan

Zacualpan: Moderately distributed.

MICHOACAN

Municipio de Angangueo

Mineral de Angangueo: Moderate
distribution. Noted from the
Santa Lucía and El Carmen mines.

Municipio de Tlalpujahua

Tlalpujahua: Moderately distrib-
uted throughout the Veta Coronas.
Noted from the Dos Estrellas
Mine.

MORELOS

Municipio de Tlaquiltenango

Huautla: Limited distribution.

NAYARIT

Municipio de Ruiz

El Zopilote: Moderate distribution. Most noted from the
Ahuacatlán, Jesús María, Concep-

ción, La Yesca, Los Tajos, San José, and Tenamache mines.

OAXACA

Municipio de Ocotlan de Morelos

Ocotlan de Morelos: Moderate distribution.

PUEBLA

Municipio de Tetela de Ocampo

Tetela del Oro: Limited distribution. Noted from the Covadonga Mine.

SAN LUIS POTOSI

Municipio de Villa de Ramos

Ramos: Moderate distribution.

ZACATECAS

Municipio de Concepción del Oro

Aranzazú: Moderate distribution. Noted from the Aranzazú Mine.

Municipio de Fresnillo

Fresnillo de Gonzáles Echeverría: Moderate distribution throughout the Cerro de Proaño. Crystals to 2 cm. Associated with pyrargyrite and quartz. [In the early 1980s, a new silver vein was discovered within the famous Cerro de Proaño. The Veta de Santo could be modern day Mexico's richest silver vein; rich not only in the amount of silver ore, but also in the number of crystallized specimens. Blood red proustite crystals to 2 cm have been found. This location is better known as Fresnillo.]

Municipio de Sombrerete

Sombrerete: Widespread. Most noted from the Veta Negra, El Pabellón, and Tocayos mines.

Crystals to 4 cm. [This small mining district is Mexico and the world's leading producer of proustite. In two major silver stikes alone, almost 5,000 cubic meters of proustite was mined! (For complete history, see the Introduction, Sombrerete.) Mining continues today and proustite is still being mined as one of the principle silver minerals of the district. Crystals of proustite are encountered frequently, but are usually small, less than 10-15 mm in size, but occasionally pockets are uncovered that have produced blood red crystals up to 4 cm in size. But with the modern mining techniques used, most of these crystals are destroyed! The proustite is kept separate from the rest of the ore and used to "sweeten" the silver ore to keep the silver content never less than 250 grams per metric ton (7 ounces/ton) silver. Specimens are occasionaly taken to Zacatecas and sold there as being from Zacatecas or Vetagrande in order to protect the miners. Many of the proustites sold as being from Zacatecas or Vetagrande over the past decade or two have come from this small district and its true has been location lost!]

Municipio de Zacatecas

Zacatecas: Moderate distribution. Noted from the La Plata, Prodigio, and Refugio mines.

PSEUDOBOLEITE

$Pb_5Cu_4Cl_{10}(OH)_8 \cdot 2H_2O$. A rare secondary mineral that forms in extremely minor amounts in the oxidized zone of copper/lead deposits.

BAJA CALIFORNIA SUR

Municipio de Mulegé

Santa Rosalía: Limited distribution within the Cañada de Curuglú. Noted from the Amelia and Curuglú mines. Crystals to

1 cm. Associated with boleite and cumengite.

PSEUDOBROOKITE

Fe_2TiO_5. An uncommon mineral formed as a product of pneumatolytic or fumerolic activities.

DURANGO

Municipio de Durango

Cerro de los Remedios: Moderate distribution. [This location is in the city of Victoria Durango and is now a city park.]

Municipio de Coneto de Comonfort

Sierra de San Francisco: Limited distribution. Noted from the Orosco Mine.

GUANAJUATO

Municipio de San Felipe

San Felipe: Moderate distribution. Most noted from the Santa Barbara Mine.

PSEUDOMALACHITE

$Cu_5(PO_4)_2(OH)_4 \cdot H_2O$. A rare secondary mineral that forms in the oxidized zones of copper deposits.

MEXICO

Municipio del Oro

El Oro de Hidalgo: Limited distribution. Noted from the El Carmen Mine. Associated with bismutite.

SONORA

Municipio de Naco

La Morita: Limited distribution.

PSILOMELANE. See **ROMANECHITE**

PYRARGYRITE

Ag_3SbS_3. An uncommon mineral formed as a late primary or secondary enrichment vein mineral.

AGUASCALIENTES

Municipio de Asientos

Asientos: Moderate distribution. Noted from the Santa Francis mine.

CHIHAUHAU

Municipio de Ascensión

Sabinal: Moderate distribution. Noted from the Adventurerea Mine. Associated with polybasite.

Municipio de Batopilas

Batopilas: Moderate distribution. Most noted from the Santo Niño, San Miguel Giral, Jesús María, Urique, Santa Rosa, San Antonio, and La Unión mines. Crystals to 2 cm. Associated with native silver and acanthite.

Jesús María: Moderate distribution. Noted from the Yedros

A 10 cm × 12 cm cluster of crystals of pyrargyrite from the Los Ninos Vein, Fresnillo, Zacatecas.

Mine. Associated with acanthite and proustite.

Municipio de Guazapares

Guazapares: Moderate distribution.

Municipio de Hidalgo del Parral

Hidalgo del Parral: Moderate distribution. Noted from the El Tajo Mine.

Municipio de Saucillo

Naica: Moderate distribution.

COAHUILA

Municipio de Sierra Mojada

Sierra Mojada: Moderately distributed.

DURANGO

Municipio de Canelas

Canelas: Moderate distribution. Noted from the La Portilla and Birimoa mines.

Municipio de Guanaceví

Guanaceví: Widespread. Most noted from the Santa Ana, Guanasevi, La Cruz, and San José mines.

Municipio de Indé

Indé: Moderate distribution. Noted from the Matracal Mine.

Municipio de Mapimí

Mapimí: Moderate distribution. Noted from the Ojuela Mine. Associated with native silver, pyrite, chalcopyrite, sphalerite, calcite, stibnite, siderite, barite, ankerite, and fluorite.

Municipio de Pueblo Nuevo

El Salto: Moderate distribution. Noted from the Nueva Esperanza Mine. Crystals to 2 cm.

Municipio de San Pedro del Gallo

Peñoles: Moderate distribution. Noted from the Peñoles and San Rafael mines.

Municipio de Topia

Topia: Moderate distribution. Noted from the Contaranas Mine.

GUANAJUATO

Municipio de Guanajuato

Guanajuato: Moderately distributed throughout the Veta Madre. Most noted from the Valenciana, San Juan de Rayas, Sirena, El Nopal, Esperanza, Madrina, and Angustias mines. Crystals to 5 cm. Associated with quartz (rock and amethyst) native silver, polybasite, and calcite.

La Luz: Moderate distribution. Noted from the La Luz, Santa Rita (Bolañitos), Santa Lucía, Santa Clara, La Joya, Purísima, and San Pedro mines. Crystals to 6 cm. Associated with quartz (both rock and amehyst) calcite, and silver. [At the La Luz Mine in the late 1700s, pyrargyrite was reported to have been found with crystals of azurite up to 2 cm in length. Several large crystallized specimens of pyrargrite have been found at La Luz. One weighed just over 11 kilograms (25 pounds) and was made up of 3 cm long crystals. It was presented to Maximilian, Emperor of Mexico in the 1860s.]

Municipio de San Felipe

San Felipe: Moderate distribution. Noted from the La Providencia Mine.

GUERRERO

Municipio de Taxco

Taxco de Alcarón: Moderate distribution. Noted from the Santa Gertrudis, Acatitlán, and

Santa Bárbara mines. Crystals to 2 cm. Associated with quartz. [Several years ago the Santa Barbara Mine produced specimens of pyrargyrite with crystals up to 2 cm long associated with quartz crystals.]

HIDALGO

Municipio de Mineral del Monte

Mineral del Monte: Moderate distribution. Associated with chlorargyrite and acanthite.

Municipio de Pahuca

Pahuca de Soto: Widespread: Most noted from the El Encino, El Cristo, San Cayetano, San Buenaventura, Concordia, El Bordo, El Chico, Maravillas, Pabellon, and Sacramento mines. Crystals to 2 cm. Associated with acanthite and chlorargyrite.

Municipio de Zimapán

Zimápan: Moderate distribution. Noted from the Lomo del Toro Mine.

JALISCO

Municipio de Bolaños

Bolanos: Moderate distribution. Noted from the Santa Fe, Iguana, Verde, Pichardo, Descubridora, and Barranco mines. Associated with acanthite.

MEXICO

Municipio de Temascaltepec

Real de Arriba: Moderate distribution. Noted from the El Rincón Mine.

Temascaltepec de González: Widespread along the Vetas Rica and Negra.

Municipio de Zacualpan

Zacualpan: Widespread. Most noted from the Carboncillo, Durazano, Preciosa Sangre, Reforma, and Santa Eduwigis mines. Crystals to 2 cm.

MICHOACAN

Municipio de Angangueo

Mineral de Angangueo: Moderate distribution. Noted from the Santa Lucía and El Carmen mines.

Municipio de La Huacana

Inguaran: Moderately distributed. Noted from the La Palma Mine.

Municipio de Tlalpujahua

Tlalpujahua: Widespread. Noted from the La Borda, El Carmen, La Luz, San Juan Nepomuceno, Santa Lucía, and Santa Rita mines. Crystals to 1 cm. Associated with native silver, acanthite, and polybasite.

MORELOS

Municipio de Tlaquiltenango

Huautla: Moderate distribution.

OAXACA

Municipio de Natividad

Natividad: Moderate distribution. Noted from the Natividad Mine.

Municipio de Ocotlán de Morelos

Ocotlán de Morelos: Moderately distributed. Most noted from the Benjamín and San Francisquito mines.

Municipio de Temaxcalapan del Progresso

Villa: Moderately distributed. Most noted from the Candelaria, Dolores, San Andrés, and San Joaquín mines.

PUEBLA

Municipio de Tetela de Ocampo

Tetela del Oro: Moderate distribution. Noted from the Covadonga Mine.

SAN LUIS POTOSI

Municipio de Catorce

Catorce: Moderate distribution. Noted from the Concepción, Santa Ana, San Augustín, La Purísima, and San Antonio mines.

Municipio de Charcas

Charcas: Moderate distribution.

Municipio de Villa de Ramos

Ramos: Moderate distribution. Noted from the Cocinera and La Purísima mines. Associated with azurite.

SINALOA

Municipio de Badiraguato

Soyatita: Moderate distribution. Noted from the San José and Yedra mines. Associated with azurite.

Municipio de Concordia

Pánuco: Moderate distribution. Noted from the La Virgen, San Antonia, and San Diego mines.

Municipio de Cosalá

Cosalá: Moderately distributed. Noted from the La Dura Mine.

SONORA

Municipio de Arizpe

Arizpe: Moderate distribution. Noted from the Las Chipas, Dios Padre, El Arco, El Carmen, and Rafa mines. Crystals to 2 cm. Associated with chalcopyrite and native silver.

Municipio de Baviacora

Suaqui: Moderately distributed. Noted from the Santa Ana and Roasrio mines.

ZACATECAS

Municipio de Fresnillo

Fresnillo de González Echeverría: Widespread within the Cerro de Proaño. Crystals to 8 cm. Associated with proustite, polybasite, acanthite, calcite, quartz, galena, sphalerite, chalcopyrite, arsenopyrite, pyrrhotite, and pyrite. [In 1975 the Veta de Santo Niña was discovered at Fresnillo. Crystals of pyrargyrite have been found up to 8 cm in length and 3 cm in diameter. The Veta de Santo Nino is a remarkable discovery. Like other silver discoveries before within Mexico, it is not the great richness, but rather the large volume of low grade ore that makes it important. At the present time, Compañia Fresnillo, the mine's owner and operator, states that there is 500,000 metric tons of proven ore and 1,225,770 metric tons of indicated ore in reserve and the deposit remains open with depth. The average assay of the known and indicated ores of the Santo Niño vein is 45 grams/metric ton gold (1.45 troy ounces/ metric ton), 610 grams/metric ton silver (18.97 troy ounces/metric ton), 0.4% lead, 0.7% zinc, and 0.01% copper. This location is better known simply as Fresnillo.]

Municipio de General
Francisco Murguía

Nieves: Moderate distribution.
Noted from the San José Mine.

Municipio de Mazapil

Noche Buena: Moderate distribution. Noted from the Noceh Buena
Mine.

Municipio de Noria de
Angeles

Noria de Angeles: Widespread.
Most noted from the Candelaris,
La Palma, Matavacas, Azul,
Trinidad #1, Trinidad #2, Santa
Rita, San José, Porvenir, and
Sierra de Peñon Blanco mines.
[This location is listed in
old records as Real de Los
Angeles. The area is currently
being developed into the largest
open pit silver mine in the
world: the Real de Angeles
Mine. Silver from this new
project is contained galena,
tetrahderite, and tennantite. It
is estimated that when the new
open pit mine is in full production it will produce 7,000,000
troy ounces of silver per year
from the lead concentrates. The
silver reserve of this mine is
58,800,000 metric tons of ore
that contains 74 grams per metric
ton of silver (2.3 troy
ounces/metric ton). This places
values of 135,240,000 troy ounces
of silver in the orebody worth in
excess of 1.1 billion in 1980
dollars!]

Municipio de Sombrerete

San Pantaleón de la Noria: Moderate distribution. Noted from the
La Noria Mine. Associated with
wires of native silver.

Sombrerete: Moderate distribution
on the Vetas de La Negra y
Pabellón. Noted from the La
Negra, El Pabellón, and Tocayos
mines. Associated with proustite
and arsenopyrite.

Municipio de Vetagrande

Vetagrande: Widespread. Most
noted from the Purísima, San
Fernando, San Pedro, San Rafel,
San Vicente, Tiburcio, El
Pródigo, Mela Noche, La Gallego,
and San Marcos mines. Crystals to
1 cm. Associated with native
silver.

Zacatecas: Widespread. Noted from
the Quebradillas, La Cantera, El
Rufugio, and Carnicería mines.
Crystals to 2 cm. Associated with
native silver.

PYRITE

FeS_2. A common mineral formed
under wide varieties of conditions.

AGUASCALIENTES

Municipio de Asientos

Asientos: Widespread. Most noted
from the Cinco Señores, Descubridora, La Merced, Santa Elena, and
Santa Francisca mines. Crystals.

Municipio de Tepezalá

Tepezalá: Widespread. Crystals.

BAJA CALIFORNIA

Municipio de Ensenada

Isla de Cedros: Moderate distribution. Noted from the Máscara de
Hierro mine. Crystals to 1 cm.
Associated with hematite,
chalcocite, and gold.

San Fernando: Widespread. Most
noted from the San Fernando del
Cobre, Adela, Chalcocita, Copper
Queen, Calumet, and Rosario
mines. Crystals to 1 cm. Associated with bornite, chalcocite,
chalcopyrite, cuprite, and
hematite.

Real del Castillo: Widespread.

Rosa de Castilla: Moderate distribution. Noted from the El Fenómeno Mine. Associated with chalcopyrite, pyrrhotite, and calcite.

BAJA CALIFORNIA SUR

Municipio de La Paz

Cacachilas: Widespread.

Punta Aguaja: Moderate distribution.

Municipio de Mulegé

Santa Rosalía: Widespread.

Municipio de San Antonio

San Antonio: Widespread. Crystals to 1 cm.

Triunfo: Widespread. Crystals to 1 cm.

CHIHUAHUA

Municipio de Aquiles Serdán

Francisco Portillo: Widespread. Most noted from the El Potosí and Buena Tierra mines. Crystals to 2 cm. Associated with galena, sphalerite, calcite, and pyrrhotite.

San Antonio el Grande: Widespread. Noted from the San

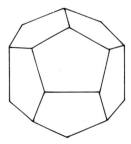

A 2 cm crystal of pyrite from the San Antonio Mine, San Antonio el Grande, Chihuahua. (Modified by Shawna Panczner from Sinkankas, 1964)

Antonio Mine. Crystals to 4 cm. Associated with calcite, quartz, galena, sphalerite, pyrrhotite, and fluorite.

Municipio de Batopilas

Batopilas: Moderate distribution.

Municipio de Casas Grandes

San Pedro Corralitos: Widespread. Most noted from the Congreso-León, Candlelaria, San Nicolás-Cobriza, and San Pedro mines. Crystals to 1 cm. Associated with calcite.

Municipio de Chihuahua

Labor de Terrazas: Moderate distribution. Noted from the Rio Tinto Mine.

Municipio de Camargo

La Negra: Moderate distribution. Noted from the La Negra Mine. Associated with magnetite, hematite, and barite.

La Perla: Moderate distribution. Noted from the La Perla Mine. Associated with barite, magnetite, and hematite.

Municipio de Cusihuiári-achic

Cusihuiariáchic: Widespread. Most noted from the La Renina, Canadelaria, Santo Niño, Santa Marina, San Miguel, El Burro, Mexicanna, San Antonio, San Bartolo, and San Francisco mines.

Municipio de Guadalupe y Calvo

Guadalupe y Calvo: Moderate distribution. Noted from the El Rosario Mine.

Municipio de Hidalgo del Parral

Hidalgo del Parral: Widespread. Most noted from the Coyote, La Inmensidad, and San Juanico Mines.

Municipio de Santa Bárbara

Santa Bárbara: Widespread. Most noted from the El Tajo, La Prieta, Mecatena, Santo Tomás and Tecolotes mines.

Municipio de Saucillo

Naica: Widespread throughout the Sierra de Naica. Noted from the Naica Mine. Crystals to 2 cm. Associated with fluorite, galena, sphalerite, quartz, and calcite. [This mineral has been found here with a beautiful iridescent coating on calcite.]

Municipio de Urique

Urique: Widespread. Most noted from the Blanca, Guerra al Tirano, Patrona, Piedras Verdes, Realito, San Antonio, Santa Rita, Rosario, and Truyo mines.

Municipio de Uruáchic

Uruáchic: Moderate distribution. Noted from the San Martín Mine.

COAHUILA

Municipio de Arteaga

Arteaga: Moderate distribution. Noted from the San Antonio de las Alianzas mine.

Municipio de Candela

Pánuco: Widespread. Noted from the Pánuco Copper Mine. Crystals to 15 cm. (Pyritohedral).

DURANGO

Municipio de Cuencamé

Velardeña: Widespread. Most noted from the Guardarraya, Velardeña, Terneras, and Santa María mines.

Municipio de Guanaceví

Guanaceví: Widespread on the Veta Esperanza. Most noted from the Probidad, Animas, Paleros, Carmen, La Fortuna, Soto, San Juan de los Llanitos, Barradón, Chica, El Nuevo Porvenir, La Cruz, La Luz, La Unión, Santa Ana, Manzanilla, Republicana, Dos Repúblicas, San Francisco, El Salto, San Gonzalo, Santa Elodia, San Joaquín, San Luis, San Marcos, Santa Rosa, Sirena, Soledad, and Sembranena mines.

Municipio de Indé

Indé: Widespread. Most noted from the Caballo, Colorado, Demasía, El Agua, Cajuelo, Garabatos, La Cruz, Los Machos, Portillo, Restauración, Soledad, Tablas, Tres Varones, La Unión, Unión y Constancia, Continuación, and Matracal mines.

Municipio de Mapimí

Mapimí: Moderately distributed. Noted from the Ojuela Mine. Crystals to 4 cm. [The more common forms of pyrite crystals from here are pyritohedrons, cubes, octahedrons, and diploids.]

Municipio del Oro

Santa María del Oro: Widespread. Noted from the La Reina, Esperanza, Novedad, La Tuna, La Princesa, and Santa Ana mines.

Megistral: Widespread. Noted from the El Lustre, La Cocinera, La Colorada, Guadalupe de la Montaña, and La Predilecta mines.

Municipio de San Dimas

Tayoltita: Widespread. Most noted from the La Arana, Soledad, Santo Tomás, Concepción, Las Cuatro Estrellas, El Ojito, and San Miguel mines.

Municipio de San Pedro del Gallo

Peñoles: Widespread. Noted from the Jesús María, La Cruz, Potosí, Providencia, La Abundancia, and San Rafael mines.

Municipio de Topia

Topia: Widespread. Most noted from the Palmira, El Toro, Magistral, Otatal, Otela, San

Felipe, San José, El Refugio, Providencia, El Picacho, Purisíma, and La Fe mines.

GUANAJUATO

Municipio de Dolores Hidalgo

San Antonio de la Mina: Moderate distribution. Noted from the Ave de Gracia Mine.

Municipio de Guanajuato

Guanajuato: Moderate distribution. Noted from the Sirena Mine. Associated with quartz.

La Luz: Moderate distribution.

Municipio de San Felipe

San Felipe: Moderately distributed. Noted from the Providencia Mine.

Municipio de San Luis de la Paz

Mineral de Pozos: Widespread. Most noted from the Santa Brígida, La Joya, and San Joaquín mines.

GUERRERO

Municipio de Arcelia

Campo Morado: Moderate distribution. Noted from the El Naranjo Mine.

Municipio de Atlixtac

Petatlán: Moderate distribution. Noted from the Copper King Mine.

Municipio de Huitzuco

Huitzuco de los Figueroa: Moderate distribution. Noted from the La Cruz Mine. Associated with gypsum and sulfur.

Municipio de Taxco

Taxco de Alcarón: Widespread within the Vetas de La Argentita and La Marquesae. Most noted at the Trinidad, Babilonia, Chocotitlán, El Rosario, El Cedral, Catrina, La Perla, San Ignacio de la Borda, Varones, La Luz, El Pedregal, San Antonio, Jesús, and Guerrero mines. Crystals to 2 cm. Associated with marcasite and quartz. [At the Guerrero Mine, pyrite has been found as an iridescent coating on large smoky quartz crystals and as a replacement of pyrrhotite crystals.]

Tehuilotepec: Moderate distribution.

HIDALGO

Municipio de Cardonal

Cardonal: Moderate distribution. Noted from the La Punta Mine.

Municipio de Mineral del Chico

Mineral del Chico: Moderate distribution.

Municipio de Mineral del Monte

Mineral del Monte: Moderate distribution.

Municipio de Pachuca

Pachuca de Soto: Widespread.

Municipio de Zimapán

Zimapán: Widespread.

JALISCO

Municipio de Bolaños

Bolaños: Moderate distribution. Noted from the San José, Los Negritos, and Santa Fe mines.

Municipio de Chiquilistlán

Jalapa: Moderate distribution. Noted from the La Mexicana Mine.

Municipio de Pihuamo

Pihuamo: Moderately distributed.

Municipio de Tequila

Hacienda San Martín la Guijar-
rena: Moderate distribution.

MEXICO

Municipio del Oro

El Oro del Hidalgo: Moderate
distribution.

Municipio de Sultepec

Sultepec de Pedro Ascencio de
Alquisiras: Widespread. Most
noted from the San Pedro,
Concepción, Santa Gertrudis, San
Juan Bautista, La Cruz, El
Rufugio, La Luz, Providencia, and
El Zapote mines.

Municipio de Tejupilco de
Hidalgo

Juluapán: Moderate distribution.
Noted from the Esperanaza Mine.

Municipio de Temascaltepec

Temascaltepec de Gonzalez:
Moderately distributed. Noted
from the Magdalena Mine.

Municipio de Zacualpan

Zacualpan: Widespread throughout
Cerro Alto. Most noted from the
San Adrián, Carboncillo, San
Ignacio, San Antonio, and Alacrán
mines.

MICHOACAN

Municipio de Acuitzio

Etucuaro: Moderate distribution
on the Rancho de los Horconcitos
and Hacienda de San Diego
Milpillas.

Municipio de Angangueo

Mineral de Angangueo: Widespread
throughout the Veta de Santa
Margarita. Noted from the San
Cayetano, El Carmen, Concordia,
Catingo, and Santa Gertrudis
mines.

Municipio de La Huacana

Inguaran: Widespread.

Municipio de Tlalpujahua

Tlalpujahua: Moderate distribution. Noted from the Dos Estrellas Mine.

NAYARIT

Municipio de La Yesca

La Yesca: Moderately distributed
throughout the Veta de Santa
María del Oro. Noted from the El
Mirador Mine.

Municipio de Santa María del
Oro

Acuitlapilco: Widespread. Most
noted from the Anonas, Guaje, La
Cuchara, La Noria, La Soledad,
Manzana, Santa Catarina, and
Valenciana.

NUEVO LEON

Municipio de Cerralvo

Cerralvo: Moderate distribution.
Noted from the El Porvenir,
Purísima, and Reforma mines.

Municipio de Lampazos de
Naranjo

Lampazos de Naranjo: Moderate
distribution. Noted from the Flor
de Peña Mine.

OAXACA

Municipio de Natividad

Natividad: Moderate distribution.
Noted from the Natividad Mine.

Municipio de San Francisco
Telixtlahuaca

Rancho San Francisco Telixtlah-
ucaca: Limited distribution.
Noted from the El Chavo and Los
Ocotes mines.

Municipio de San Jeronimo
Taviche

San Jeronimo Taviche: Moderately
distributed. Most noted from the
Consuelo, La Soledad, Tajestio,
Benjamín, La Cruz, and San Martín
de los Cansecos mines.

PUEBLO

Municipio de Tetela de
Ocampo

Tetela de Ocampo: Moderate
distribution. Noted from the
Providencia Mine.

QUERETARO

Municipio de Cadereyta

El Doctor: Moderate distribution.
Noted from the San Juan Nepomuceno Mine.

Municipio de Tolimán

Río Blanco: Limited distribution.

SAN LUIS POTOSI

Municipio de Catorce

Catorce: Widespread. Most noted
from the Concepción, Bravos,
Guadalupana, Guerrero, San
Guillermo, Morelos, San Pablo,
and San Carlos mines.

Municipio de Charcas

Charcas: Widespread throughout
the Veta Rica. Most noted from
the Tiro General, Potosí, La
Recompensa, La Bufa, and Morelos
mines. Crystals to 2 cm. Associated with galena and sphalerite.

Municipio de Guadalcázar

Guadalcazár: Moderately distributed. Noted from the San Rafael
and San Esteban mines.

Municipio de Villa de la
Paz

La Paz: Moderate distribution.
Noted from the Trinidad Mine.

SINALOA

Municipio de Badiraguato

Soyatita: Widespread. Most noted
from the Candelaria, Guadalupe,
El Salto, San Antonio, San
Vicente, Santa Brígida, and Tres
Reyes mines.

Municipio de Concordia

Pánuco: Widespread. Most noted
from the Cata Rica, El Toro, Gran
Capitán, La Virgen, Palo Blanco,
Porvenir, Restauración, San
Antonio, and San Diego mines.

Municipio de Cosalá

Cosalá: Widespread. Noted from
the Nuestra Señora, El Refugio,
and La Culebra mines.

SONORA

Municipio de Alamos

Alamos: Widespread. Most noted
from the Veta Grande, Espíritu
Santo, and La Dura mines.

Municipio de Bacanora

Bacanora: Moderate distribution.
Noted from the El Sacramento and
Mulaos mines.

La Trinidad: Widespread. Noted
from the La Trinidad, Cabeza de
Plata, Dos Padre, Espíritu Santo,
Jaladera, Noche Buena, and San
Francisco mines.

Municipio de Cananea

Cananea: Widespread. Crystals to
2 cm. Associated with quartz,
sphalerite, and rhodochrosite.

Municipio de Huépac

Huépac: Widespread. Noted from
the Rosa María Mine. Crystals to
40 cm. Associated with scheelite
and powellite. [In the early
1970s a pocket was uncovered
that, according to the owners,
was about 15 m long, 3 m wide,
and 3 m high and was lined
with brilliant highly modified
octahedral crystals of pyrite.

Several of the crystals measured
off the pyritic matrix 40 cm!
In the clay on the floor of the
pocket were a few doubly terminated pyrite octrahedrons that
measured to 20 cm on an edge.
One of these massive crystals
weighed just over 100 kilos (220
pounds). Much of the pyrite from
this location is considered
metallurgical grade, which is
extremely pure and very compact
and can be cut and polished.]

Municipio de Nacozari de García

Nacozari de García: Widespread.
Noted from the Pilares and Santo
Domingo mines. Crystals to 5 cm.
Associated with quartz, sphalerite, and rhodochrosite. [Crystals
at the Pilares Mine reach only
2 cm, but at the Santo Domingo
Mine, highly modified crystals of
pyrite have been found up to
5 cm in size.]

Municicpio de Onavas

La Barranca: Moderate distribution. Noted from the Santa Fe,
Tarahumara, and Primavera mines.

VERACRUZ

Municipio de Las Minas

Zomelahuacan: Widespread. Most
noted from the Rosario and San
Aneselmo mines.

ZACATECAS

Municipio de Chalchihuites

Chalchihuites: Widespread. Noted
from the Sante Fe, Chilchihuites,
Colorada, Refugio, Santa Albina,
San Juan, and Trinidad mines.
Crystals to 1 cm.

Municipio de Concepción del Oro

Bonanza: Moderate distribution.
Noted from the Bonanza Mine.
Brilliant crystals to 5 cm.
Associated with andradite.
[Crystals from this location are
found as cubes, octahedrons,

pyritohedrons, and modifications
of them.]

Concepción del Oro: Widespread.
Noted from the Cata Arroyo Mine.

Municipio de Fresnillo

Fresnillo de González Echeverría: Widespread throughout the
Cerro de Proaño. Crystals to
1 cm.

Municipio de General Francisco Murguía

Nieves: Widespread. Most noted
from the San Francisco and Santa
Catarina mines. Crystals to 2 cm.
Associated with jamesonite,
stibnite, and barite.

Municipio de Mazapil

Noche Buena: widespread. Most
noted from the Noche Buena and
Santa Rosa mines. Crystals to
5 cm. Associated with galena,
sphalerite, quartz, bournonite,
jamesonite, and bourlangerite.
[Crystals from here are found as
cubes, octahedrons, pyritohedrons, and modifications of
these forms. At the Noche Buena
mine the pyrite has been found as
fine replacements of pyrrhotite
crystals.]

Municipio de Noria de Angeles

Noria de Angeles: Widespread.
Most noted from the Cubierta,
Anexas, and Cata Urista mines.
Associated with galena, sphaler-
ite, tetrahedrite, and tennant-
ite. [This location was known as
Real de Angeles.]

Municipio de Pinos

Pinos: Widespread. Noted from the
Cata Rica, Estrellas, Compana, El
Pinto, La Blanca, Matavacas,
Pabellón, Puerto de San José, San
Antonio, San Francisco, San Juan,
Trinidad #1, and Trinidad #2
mines.

Municipio de Sombrerete

San Martin: Moderately distributed. Noted from the San Martín

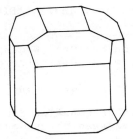

An 8 cm crystal of pyrite from Concepción del Oro, Zacatecas. (Modified by Shawna Panczner from Sinkankas, 1964)

Mine. Associated with chalco-pyrite, bornite, and covellite.

San Pantaleón de la Noria: Moderate distribution. Noted from the La Noria mine. Associa-ted with stibnite, chalcopyrite, and covellite.

Sombrerete: Moderate distribu-tion. Noted from the Tocayos Mine. Crystals to 3 cm. Associ-ated with quartz.

Municipio de Vetagrande

Vetagrande: Widespread. Noted from the Macias, La Calándria, La Tenacidad, and Guadalupe mines. Crystals to 2 cm.

Municipio de Zacatecas

Zacatecas: Widespread. Crystals to 2 cm. Associated with quartz and calcite.

PYROCHROITE

Mn(OH)$_2$. An uncommon member of the brucite group. Forms as a low-temperature hydrothermal mineral.

Location data is unavailable.

PYROLUSITE

MnO$_2$. An common mineral and member of the rutile group that forms in oxidizing conditions.

CHIHUAHUA

Municipio de Allende

Talamantes: Moderate distribu-tion. Massive. Associated with rhodonite and manganite.

Municipio de Aquiles Serdán

Francisco Portillo: Limited distribution. Noted from the El Potosí Mine.

San Antonio el Grande: Limited distribution. Noted from the San Antonio Mine.

Municipio de Cusihuiriáchic

Cusihuiriáchic: Moderate distri-bution.

Municipio de Saucillo

Naica: Limited distribution.

DURANGO

Municipio de Mapimí

Mapimí: Moderate distribution. Noted from the San Pedro and Ojuela mines. Massive. Associated with hausmannite, manganite, cryptomelane, and goethite.

GUANAJUATO

Municipio de San Luis de la Paz

Mineral de Pozos: Moderately distributed. Noted from the La Antigua and Septentrion mines.

HIDALGO

Municipio de Pachuca

Pachuca de Soto: Widespread. Most noted from the Rosario, San Pedro, Jacal, Candado, Porvenir, La Corteza, Santa Gertrudis, Concordia, Amistad, Potosi, Maravillas, Pabellón, San Buenaventura, Sacramento, San Cayetano, El Bordo, El Cristo, Guadalupe, Calicanto, and Encino mines.

NAYARIT

Municipio de La Yesca

La Yesca: Limited distribution. Noted from the La Colorada Mine.

SAN LUIS POTOSI

Municipio de Catorce

Catorce: Limited distribution. Noted from the Santa Ana Mine.

PYROMORPHITE

$Pb_5(PO_4)_3Cl$. An uncommon mineral and member of the apatite group that forms in the oxidized zone of lead deposits.

BAJA CALIFORNIA SUR

Municipio de San Antonio

El Triunfo: Moderately distributed. Noted from the Mina Rica, Gobernadora, Fortuna, and San José mines.

San Antonio: Moderate distribution. Noted from the San Antonio Mine.

Municipio de Mulegé

Santa Rosalía: Limited distribution within the Cañada de Curuglú. Noted from the Amelia and Curuglú mines. Yellow botryoidal mass.

CHIHUAHUA

Municipio de Aquiles Serdán

San Antonio el Grande: Limited distribution. Noted from the San Antonio Mine's tin chimney. Yellow botryodial masses. Associated with wulfenite and mimetite.

Municipio de Ciudad Camargo

Sierra de Las Encinillas: Moderate distribution.

Municipio de Cusihuiriáchic

Cusihuiriáchic: Limited distribution.

Municipio de Guazapares

San José: Moderate distribution. Noted from the San Luis, Monterde, and Calabacillas mines. Bright green crystals to 2 cm. Associated with goethite. [This mineral was so popular with the local miners that, in the cemetery of San José, one of the tombstones was made from a rock slab that was full of small vugs of bright green crystals of pyromorphite. A few years ago this unique tombstone was stolen from the cemetery.
In the late 1970s many of these crystals were recovered when several of the pillars in one of the stopes in the upper oxidized ore zone of the San Luis Mine were mined out for their gold content. This location has been listed as being in the state of Sonora, which is incorrect.]

DURANGO

Municipio de Mapimí

Mapimí: Moderate distribution. Noted from the Ojuela and San José mines. Pale yellow to green crystals to 8 mm. Associated with mimetite, wulfenite, cerussite anglesite, goethite, bindheimite, and plumbojarosite.

HIDALGO

Municipio de Pachuca

Pachuca de Soto: Limited distribution. Noted from the Manzano Mine.

Municipio de Zimapán

Zimapán: Moderately distributed. Noted from the Guadalupe and Dolores mines. Green crystals to 10 mm. Associated with goethite and mimetite. [The best crystals were found at the Dolores Mine.]

SAN LUIS POTOSI

Municipio de Guadalcázar

Guadalcázar: Moderate distribution within the Cerro de San Cristóbal. Noted from the Promontorio Mine.

Municipio de Villa de La Paz

La Paz: Moderate distribution. Noted from the La Paz Mine.

SONORA

Municipio de Cucurpe

Cerro Prieto: Limited distribution. Noted from the San Francisco Mine's 11th level. Crystals to 1 cm. Associated with wulfenite.

ZACATECAS

Municipio de Mazapil

Mazapil: Moderate distribution. Noted from the Los Galemes and Santa Rosa mines. Yellow crystals to 15 mm. Associated mimetite, wulfenite, and goethite. [The best crystal specimens came from the Santa Rosa Mine.]

PYROPE

$Mg_3Al_2(SiO_4)_3$. A common mineral and member of the garnet group that forms in ultrabasic rocks.

BAJA CALIFORNIA SUR

Municipio de La Paz

Punta Arena: Limited distribution.

HIDALGO

Municipio de Zimapán

Zimapán: Moderately distributed.

VERACRUZ

Municipio de Las Vigas

Piedras Parado: Moderately distributed. Crystals to 2 cm. Associated with quartz (amethyst).

PYROPHYLLITE

$Al_2Si_4O_{10}(OH)_2$. An uncommon mineral forming from the hydrothermal alteration of feldspars and found in metamorphic rocks and hydrothermal veins.

MICHOACAN

Municipio de Angangueo

Mineral de Angangueo: Moderately distributed. Noted from the Veta de Carrillos Mine.

Municipio de Madera

Villa Madera: Limited distribution. Gemmy red crystals to 1 cm.

PYROXENE is a group name.

PYRRHOTITE

$Fe_{1-.17x}S$. A common mineral found in high temperature mineral veins as an early formed mineral, in some igneous rocks, in contact metamorphic deposits and, at times, in sediments.

BAJA CALIFORNIA

Municipio de Ensenada

Alamo: Moderately distributed. Noted from the Abbeline Mine. Associated with galena and quartz.

Rosa de Castilla: Limited distribution. Noted from the El Fenómeno Mine. Associated with pyrite, arsenopyrite, and chalcopyrite.

CHIHUAHUA

Municipio de Aquiles Serdán

Francisco Portillo: Moderate distribution. Noted from the El Potosí Mine. Crystals to 15 cm. Associated with galena, pyrite, sphalerite, quartz, and calcite.

San Antonio el Grande: Moderate distribution. Noted from the San Antonio Mine. Crystals to 5 cm. Associated with galena and sphalerite.

Municipio de Saucillo

Naica: Moderate distribution. Crystals to 2 cm. Associated with pyrite, galena, and sphalerite.

Municipio de Urique

Piedras Verdes: Limited distribution.

DURANGO

Municipio de Cuencamé

Velardeña: Moderate distribution. Most noted from the Velardeña and Socavón Hay mines.

Municipio de Guanaceví

Guanaceví: Limited distribution. Noted from the Probidad Mine.

Municipio de Mapimí

Mapimí: Limited distribution. Noted from the Monterrey and Ojuela mines. Associated with diopside. [Large diopside crystals have been found with this mineral at the Monterrey Mine.]

GUANAJUATO

Municipio de Guanajuato

Guanajuato: Moderate distribution.

Municipio de Dolores Hidalgo

San Antonio de las Minas: Moderately distributed. Noted from the Ave de Gracia Mine.

GUERRERO

Municipio de Atlixtac

Petatlán: Moderate distribution. Noted from the Copper King Mine.

HIDALGO

Municipio de Zimapán

Zimapán: Widespread. Most noted from the Los Balcones, San Guillermo, Concordia, Miguel Hidalgo, El Afloramiento, Chiquihuites, Tecomates, and El Chacuaco Viejo mines. Associated with galena, sphalerite, pyrite, chalcopyrite, and arsenopyrite.

MICHOACAN

Municipio de Zitácuaro

Susupuato: Limited distribution on Cerro Mazahua.

OAXACA

Municipio de San Miguel
Peras

San Miguel Paras: Moderate
distribution on the Veta Monserrate. Noted from the Zavaleta
Mine.

SONORA

Municipio de Cananea

Cananea: Moderate distribution.

TAMAULIPAS

Municipio de Villagrán

San José: Moderate distribution.

ZACATECAS

Municipio de Mazapil

Noche Buena: Moderately distributed. Noted from the Noche Buena
Mine. Wafer thin crystals to
4 cm in diameter. Associated with
pyrite, jamesonite, and bourlangerite.

Q

QUARTZ

SiO_2. A common mineral, found in all three rock type environments.

BAJA CALIFORNIA

Municipio de Ensenada

El Marmal: Moderate distribution. (Moss agate and jasper). Massive.

Pino Solo: Moderately distributed. Associated with dravite (inclusions).

Rancho Viejo: Moderate distribution, 15 km east of El Alamo. Noted from the Socorro Mine. Crystals to 5 cm. Associated with elbaite.

El Rodeo: Moderate distribution. Crystals. Associated with epidote (inclusions).

BAJA CALIFORNIA SUR

Municipio de Mulegé

Santa Rosalía: Limited distribution. (Amethyst and chalcedony). Noted from the Carmen Mine. Crystals to 2 cm.

CHIHUAHUA

Municipio de Ahumada

Ahumada: Moderate distribution. (Agate = "crazylace" and "flame").

Gallego: Widespread throughout the Sierra del Gallego, Rancho Gallego. (Agate). Nodules to 32 kilos. [The agate is found in the Rancho El Agate andesite.]

Rancho Coyamito: Moderate distribution. (Agate).

Sueco: Widespread, Rancho Sueco. (Agate). [About 80% of the nodules found from this area of the Rancho El Agate andesite were geodes.]

Municipio de Aquiles Serdán

Francisco Portillo: Widespread. (Amethyst, rock crystal and smoky quartz). Crystals to 8 cm. [The El Potosi Mine has produced water-clear quartz containing inclusions of hematite. The amethyst from here are mainly crystal points but reach large proportions with clusters weighing up to 50 kilos (112 pounds).]

San Antonio el Grande: Widespread. Noted from the San

Antonio Mine's skarn zone. Crystals to 4 cm.

Municipio de Chihuahua

Ejido Esperanza: Moderately distributed, 5 km southeast. (Agate, amethyst, smoky quartz and rock crystal). Noted from the "claims" of La Otra Estrella, La Animosa, El San Antonio, La Morenita, La Paty, and El Mesteno. Associated with birnessite, calcite, goethite, gypsum, hematite, kaolinite, opal, pyrolusite, ramsdellite, rancieite, and todorokite. [The agate from here is found in the geode tuff formation. The nodules are mainly hollow and have become known as "coconut geodes". It is estimated that since 1962 over 34,000 kilos (75,000 pounds) of nodules per year have been mined from these claims!]

Ojo de Laguna: Widespread on Rancho Borunda. (Agate). [The agate from here is known as "Luguna agate" and is found within the Rancho El Agate andesite formation.]

Rancho Agua Nueva: Moderate distribution within the Sierra del Gallego. (Agate).

Municipio de Carmargo

La Negra: Moderate distribution. Noted from the La Negra Mine. Crystals to 2 cm. Associated with hematite, magnetite, and calcite.

La Perla: Moderate distribution. Noted from the La Perla Mine. Crystals to 2 cm. Associated with hematite, magnetite, and calcite.

Villa de Zaragoza: Moderate distribution. (Agate).

Municipio de Coyame

Coyame: Moderate distribution. (Agate).

Municipio de la Ascención

Sabinal: Moderate distribution. Noted from the Adventurera Mine.

Crystals to 3 cm. Associated with hematite (included). [The quartz with the hematite is referred to locally as strawberry quartz. Quartz has been also found replacing crystals of barite and calcite.]

Municipio de Nuevo Casas Grandes

Nuevo Casas Grandes: Widespread on Rancho Apache. (Agate). [The agate has been referred to as "apache agate".]

Municipio de Saucillo

Naica: Widespread throughout the Sierra de Naica. (Rock crystal and amethyst). Crystals to 8 cm. Associated with fluorite, galena, sphalerite, pyrite, and calcite.

Municipio de Uruáchic

Uruachic: Moderately distributed. Crystals to 4 cm. Associated with hematite (heavily included). [This quartz has been referred to as "stawberry quartz". The crystals are extremely etched and eroded.]

DURANGO

Municipio de Durango

Victoria de Durango: Moderate distribution. (Rock crystal). Noted from the Cerro de Mercado Mine. Crystals to 3 cm. Associated with fluorapatite.

Municipio de Mapimí

Mapimí: Moderate distribution. (Rock crystal). Noted from the Ojuela Mine. Crystals to 1 cm.

Municipio de Mezquital

Mezquital: Moderate distribution. (Petrified wood).

Municipio de Rodeo

Rodeo: Moderately distributed. (Agate and jasper).

Municipio de Chilapa

Ollas de Chilapa: Moderate distribution. (Jasper and agate). Massive.

GUANAJUATO

Municipio de Guanajuato

Guanajuato: Widespread throughout the Veta Madre. (Rock crystal and amethyst). Most noted from the Valenciana, Sirena, and Peregrina mines. Crystals to 4 cm. Associated with calcite, acanthite, adularia, polybasite, and hydroxyapophyllite. [Amethyst from the mines of the Veta Madre usually have a dark base and lighter tips. Many of the pockets uncovered within the mines are lined with these 2-cm long crystal druses and have produced large sheets of coarse amethyst.]

La Luz: Widespread. (Rock crystal and amethyst). Noted from the Jesús María, Santa Rita (Bolañitos) and San Lorenzo de Tehuilotepec mines. Crystals to 2 cm. Associated with calcite.

Santa Rosa: Moderate distribtion. (amethyst). Noted from the Calvillo Mine.

GUERRERO

Municipio de Taxco

Barranca de Los Ocotes: Limited distribution. (Amethyst).

Taxco de Alcaron: Widespread. (rock crystal, smoky and amethyst). Most noted from the Guerrero Mine. Crystals to 12 cm. Associated with pyrite, marcasite, and acanthite. [In the mid 1970s, large sheets up to almost 2 m in size of smoky quartz with crystals up to 12 cm in length and 5 cm in diameter were found within the sulfide ores from the Guerrero Mine. These crystals are often partially coated with pyrite and marcasite. The larger crystals of smoky quartz have been used for carving by several of the local artisans in Taxco. Amethyst of extremely dark color and crystal size is also found here. Crystals have been found up to 8 cm in length and 4 cm in diameter, but normally are just sheets of large pointed terminations.]

Municipio de Zumpango del Rio

Amatitlán: Widespread. (amethyst). Most noted from the Santa Margarita, La Valencania, and Rayas mines. Crystals to 30 cm. Associated with hematite and manganese oxide. [This site is located in an extremely remote area of Guerrero with jungle vegetation. The local Indians are good workers, but do not trust the outsiders. This location has been referred to in past literature as Rio Balas, a river which flows near the site. There are two places that one may leave from to reach the amethyst mines; Amatitlán can be reached by a very bad road and Balsas by train. From these points, there still is a 2-3 hour ride by mule via Amatitlán or a 6-7 hour ride from Balsas to reach the mines. The amethyst field is northeast of Amatitlan, which is the closest pueblo. The mines are in reality trenches dug following the amethyst veins at or just below the surface. These trenches are dug to a depth of 8-10 meters and as wide as necessary to mine the amethys. On occasion, they become tunnels rather than trenches. The miners are looking for cutting-grade amethyst and specimens are secondary. The very top grade amethyst looks very much like the Siberian amethyst, even with the red fire when faceted. The amethyst is found in veins that may run for many meters before pinching off. The pockets containing crystals are found where the vein has opened up allowing crystals to develop. The pockets are found throughout the vein usually without much warning. All the mining is done by hand and is extremely difficult and time consuming. The crystal pockets

near the surface are filled with
a reddish mud and at times a
manganese oxide is coating the
crystals. At depth pockets
are encountered with usually only
the manganese coatings on some
of the crystals. Crystal size
varies from very small, less than
1 cm, to over 25 cm in length and
8 cm in diameter! Most of the
crystals are finger sized and
taper downwards with the widest
point nearer the base. They
usually have brilliant crystals
faces. If they are coated, there
is a chance that the coating may
have etched the crystal faces.
Some pockets often produce
crystals showing good phantoms
near the base of the crystals.
The pockets can vary from very
pale to extremely dark amethyst
color. Because the best material
has its deepest color at the base
of the crystals, the miners
determine the crystals gem grade
and value by removing the
crystals from matrix and holding
it up to the sunlight.
The amethyst is taken into México
City or Querétaro for cutting.
Specimens make their way out by
occasional buyers finding
specimens at these cutting
centers or by risking their lives
going into the mining area.
According to the older miners,
many of which have been doing
this all their lives, only a few
"Norte Americanos" have ever been
taken back into the actual mining
area. It can take many years to
gain the miners' confidence
before they will allow outsiders
into the mining area.]

HIDALGO

Municipio de Huichapán

Haciende de Yextho: Limited
distribution near the Salitera
Mine. (Petrified Wood). Massive.

Municipio de Mineral del Chico

Mineral del Chico: Widespread.

Municipio de Mineral del Monte

Mineral del Monte: Widespread.
(Amethyst, chacedony, jasper, and
rock crystal). Most noted from
the Dolores Mine. Associated with
calcite and native silver.

Municipio de Pachuca

Pachuca de Soto: Widespread.
(Amethyst, rock crystal, chalcedony and jasper).

Municipio de Zacualtipán

Zacualtipán: Moderate distribution. (Jasper). Massive.[This
location is southwest of Zacualtipán.]

JALISCO

Municipio de Bolaños

Bolaños: Moderate distribution.
(Amethyst and rock crystal).

Municipio de Zapopan

Calvillo: Widespread. (Chalcedony
and Agate). [This area has been a
major producer of fire agate with
nodules up to 6 cm in diameter.
During peak production periods,
the mines of this area have produced over 10,000 pieces of rough
fire agate per week!]

MICHOACAN

Municipio de Tlalpujahua

Tlalpujahua: Moderate distribu-
tion. (Amethyst and jasper).
Noted from the Espíritu Santo
Mine.

MORELOS

Municipio de Ayala

Ayala: Moderate distribution.
(Green Jasper). Massive.

NAYARIT

Municipio de Ruiz

El Zopilote: Moderate distribution. (Amethyst).

Municipio de Tepic

Juanacaxtla: Moderate distribution in the Sierra de Juanacaxtla. (Green Jasper). Massive.

SAN LUIS POTOSI

Municipio de Charcas

Charcas: Moderate distribution. (Rock crystal and citrine). Noted from the San Bartolo Mine. Crystals to 3 cm. Associated with damburite, datolite, pyrite, and sphalerite. [Quartz is found here as replacements of danburite crystals and as crystals exhibiting the hexagonal bipyramids of the high temperature form of Beta quartz. These high-temperature crystals are less than 1 cm in length. Amethyst has also been found here as crystals up to 5 cm long and associated with datolite.]

Municipio de Villa de Arriaga

Tepetates: Moderate distribution. (Citrine). Massive nodules to 5 cm. Associated with topaz. [This material is gemmy and of a medium to deep yellow color.]

Municipio de Santa María del Oro

El Tula: Widespread at Cerro de Tula. (Chalcedony and agate). [This area produces extremely fine fire agate.]

Municipio de Villa de Rayas

Villa de Rayas: Moderate distribution. (Jasper).

SINALOA

Municipio de Rosario

La Rustra: Moderate distribution. Associated with bismuthinite.

Municipio de Sinaloa de Leyva

Sinaloa de Leyva: Moderate distribution. (Amethyst).

SONORA

Municipio de Arizpe

Tetuachic: Moderately distributed.(amethyst). Associated with hematite (inclusions). Crystals to 6 cm. [This material is known locally as strawberry quartz because the hematite inclusions give the stubby crystals a strawberry red color. These pyramidal crystals reach to 6 cm n length and 3 cm in diameter. Clusters of these crystals to 15 cm in diameter have been found in course druses.]

Municipio de Bacerac

Bacerac: Moderate distribution. (Rock crystal). Crystals to 6 cm.

Municipio de Santa Cruz

Santa Cruz: Widespread throughout the Cerro de Chavito. (smoky quartz and amethyst). Noted from the Guadulapana Mine. Crystals to 40 cm. Associated with scheelite and dravite. [The crystals are found up to 40 cm in length and 18 cm in diameter. They are tapered crystals that are etched and occasionally have a secondary capping on the tip of the terminations.]

TLAXCALA

Municipio de Tlaxcala

Sierra de Las Silicates: Moderate distribution. (Petrified wood).

Massive. [This location is near Tlaxcala de Xicohtencal.]

Municipio de Las Minas

Zomelahuacán: Widespread. (rock crystal). Crystals to 6 cm. [The crystals from here are of the high-temperature morphology of flattened crystals. They form in clusters 3-6 cm long and are doubly terminated.]

Municipio de Profesor Rafael Ramirez

Piedras Parado: Widespread. (Amethyst and rock crystal). Crystals to 10 cm. Associated with several types of garnet varieties. [Recently the municipio and its governmental seat under went a name change from Las Vigas to Profesor Rafael Ramirez. The seams of quartz are found in the local volcanic rocks of rugged and remote lush canyons near this small pueblo (Piedras Parado) closest to the quartz areas.
The quartz seams often open up into pockets lined with crystals of amethyst and rock crystal. The crystal are found from just over 1 cm to 10 cm in length and occur in clusters of brilliant crystals. Because the several hours needed to walk or ride on the burros to the highway at Profesor Rafael Ramirez (Las Vigas), extremely large matrix specimens are usually broken down to keep the weight at a minimum. But a few outstanding pieces up to almost a meter in diameter have been brought out. The crystals are often found with a coating that can be removed by an acid wash, but this will cause

the rock matrix to change color from greenish to grayish.]

ZACATECAS

Municipio de Concepción del Oro

Aranzazú: Moderately distributed. (Rock crystal). Noted from the El Cobre Mine. Crystals to 8 cm. Associated with malachite, azurite, hematite, and scorodite. [Jap twin crystals have been found here coated with hematite.]

Municipio de Sombrerete

Sombrerete: Moderate distribution. (Rock crystal). Crystals to 4 cm. Associated with acanthite.

Municipio de Vetagrande

Vetagrande: Widespread. (Amethyst and rock crystal). Crystals to 2 cm.

Municipio de Zacatecas

Zacatecas: Widespread. (amethyst and rock crystal). Crystals to 2 cm. Associated with galena, acanthite, native silver, and pyrargyrite.

QUETZALCOATLITE

$Zn_8Cu_4(TeO_3)_3(OH)_{18}$. A rare mineral found in tellurium-rich deposits.

SONORA

Municipio de Moctezuma

Moctezuma: Limited distribution. Noted from the Bomallita Mine. [This is the type location for this mineral.]

R

RAMMELSBERGITE

$NiAs_2$. An uncommon member of the loellingite group found in nickel and cobalt deposits.

CHIHUAHUA

Municipio de Batopilas

Batopilas: Limited distribution.

Municipio de La Ascencion

Sabinal: Limited distribution. Noted at the Adventurera Mine.

RAMSDELLITE

MnO_2. A common mineral found in oxidzed zone of manganese deposits.

BAJA CALIFORNIA

Municipio de Tecate

Los Gavilanes: Limited distribution. Noted from the El Fenomeno Mine.

BAJA CALFIFORNIA SUR

Municipio de Mulegé

San Lucifer: Moderate distribution. Noted from the San Lucifer Mine.

CHIHUAHUA

Municipio de Ahumada

Ejido Esperanza: Limited distribution. Associated with quartz.

DURANGO

Municipio de Mapimí

Mapimí: Limited distribution. Noted from the Ojuela and San Pedro mines.

RANCIEITE

$(Ca,Mn)Mn_4O_9 \cdot 3H_2O$. An uncommon mineral found in manganese deposits.

CHIHUAHUA

Municipio de Ahumada

Ejido Esperanza: Limited distribution. Associated with quartz.

REALGAR

AsS. An uncommom mineral found in base metal deposits, in volcanic, fumerols, and in hot springs.

CHIHUAHUA

Municipio de Aquiles Serdán

Francisco Portillo: Limited distribution. Noted from the El Potosí Mine. Crystals to 4 mm. Associated with galena.

San Antonio el Grande: Limited distribution. Noted from the San Antonio Mine. Associated with quartz and calcite.

DURANGO

Municipio de Cuencamé

Velardeña: Limited distribution. Noted from the Velardeña Mine.

Municipio de Mapimí

Mapimí: Limited distribution. Noted from the Ojuela Mine.

HIDALGO

Municipio de Jacala

San Nicolás: Limited distribution. Noted from the Encino Largo Mine.

Municipio de Zimapán

Bonanza: Limited distribution.

Zimapan: Limited distribution. Noted from the Lomo del Toro Mine. Associated with galena.

QUERETARO

Municipio de Cadereyta

El Doctor: Limited distribution. Noted from the San Juan Nepomucno Mine. Associated with orpiment.

ZACATECAS

Municipio de Mazapil

Noche Buena: Limited distribution. Noted from the Noche Buena Mine.

RHODOCHROSITE

$MnCO_3$. An uncommon secondary mineral and member of the calcite group formed in hydrothermal veins, manganese deposits, and pegmatites.

CHIHUAHUA

Municipio de Aquiles Serdán

Francisco Portillo: Moderately distributed. Noted from the El Potosí Mine's silicate orebody. Crystals to 8 cm. Associated with galena, sphalerite, fluorite, huebnerite, and kutnohorite. [This location has produced fine rose red colored crystals up to 8 cm in length. The crystal morphology appears to change with the size of the crystals. In the smaller crystals, rhombohedrons are common, but as crystal size increases, the rhombs modify and change to the rarer scalenohedrons. Perhaps the most commonly found rhodochrosite crystals found at the El Potosí Mine are the botryoidal druses of rhombic crystals. The crystal faces are usually very shiny and the color of the rhodocrosite varies from a faint pink to a deep rose red through shades of pinkish-orange to pinkish-brown.]

San Antonio el Grande: Limited distribution. Noted from the San Antonio Mine. Crystals to 2 cm.

Municipio de Cusihuiriáchic

Cusihuiriáchic: Moderate distribution. Noted from the Princesa Mine.

Municipio de Saucillo

Naica: Limited distribution. Crystals to 3 cm. Associated with galena, fluorite, calcite, and quartz. [This mineral is normally a very fine pale pink druse, but rhombic crystals up to 3 cm in size have been found.]

DURANGO

Municipio de Mapimí

Mapimí: Limited distribution. Noted from the San Pedro Mine. Associated with other manganese minerals.

GUANAJUATO

Municipio de Guanajuato

Guanajuato: Limited distribution. Noted from the Valenciana and San Pedro mines.

La Luz: Moderate distribution. Pale pink crystals to 8 mm. San Lorenzo de Tehuilotepec Mine. Associated with amethyst.

HIDALGO

Municipio de Actopán

Hacienda de La Estancia: Limited distribution. Noted at the La Negra Mine.

Municipio de Mineral del Monte

Mineral del Monte: Moderate distribution. Most noted from the El Manzano, Escobar, Dolores, and Cabrera mines.

Municipio de Molango

Otongo: Moderate distribution. Noted from the Nonoalco and Tetzintla mines. Associated with kutnahorite.

Municipio de Pachuca

Pachuca de Soto: Moderate distribution. Noted from the Rosario, Santa Inés, Entrometida, Guatimotzin, and El Pabellón mines.

JALISCO

Municipio de Mascota

San Sebastián: Limited distribution. Noted from the El Refugio Mine.

MEXICO

Municipio de Zacualpán

Zacualpán: Limited distribution. Noted from the Alacran and Blanca mines.

MICHOACAN

Municipio de Angangueo:

Mineral de Angangueo: Moderate distribution on the Vetas Carrillos and Santa Lucia. Noted from the Santa Barbara, Carrillos, Carmen, and San Cristóbal mines.

PUEBLA

Municipio de Tepeyahualco

Tepeyahualco: Limited distribution. Noted from the La Preciosa Mine.

Municipio de Tetela de Ocampo

Tetela de Ocampo: Moderate distribution. Noted from the Espejeras, Cristo de Covadonga, El Peral, and Providencia mine.

SAN LUIS POTOSI

Municipio de Catorce

Catorce: Limited distribution. Noted from the San Augustín Mine.

SINALOA

Municipio de Rosario

Rosario: Moderate distribution. Noted from the Guadalupe Mine.

SONORA

Municipio de Cananea

Cananea: Moderate distribution. Crystals to 4 cm. Associated with quartz, pyrite, and sphalerite. [The crystals are in the rhombic

form and are a medium pink color
with frosted faces. This material
was found in the older underground mines and in the open pit
mine that formed as the mine
expanded.]

Municipio de Imuris

Sierra Azul: Limited distribution. Noted from the Espirito and
Santa Gertrudes mines. Massive.

Municipio de Nacozari de
Garcia

Nacozari de Garcia: Moderate
distribution. Crystrals to
2 cm. Associated with pyurite,
sphalerite, and quartz.

ZACATECAS

Municipio de Mazapil

Noche Buena: Moderate distribution. Noted from the Noche Buena
Mine. Associated with pyrite.

RHODONITE

$(Mn,Fe,Mg,Ca)SiO_3$. An uncommon
mineral found in metasomatic
manganese deposits,pegmatites
associated with manganese
ore bodies and contact metamorphic environments.

CHIHUAHUA

Municipio de Allende

Talamantes: Moderate distribution. Associated with pyrolusite
and manganite.

Municipio de Aquiles Serdán

Francisco Portillo: Limited
distribution. Noted from the El
Potosí Mine. Associated with
calcite and quartz.

San Antonio el Grande: Limited
distribution. Noted from the San
Antonio Mine.

GUANAJUATO

Municipio de Guanajuato

Guanajuato: Limited distribution.

GUERRERO

Municipio de Taxco

Tehuilotepec: Limited distribution. Noted from the San Lorenzo
Mine.

HIDALGO

Municipio de Mineral del
Monte

Mineral del Monte: Moderate
distribution on the veta de Santa
Brigida. Noted from the San
Ignacio, El Manzano, Velasco,
Santa Inés, Carretera, and
Dificultad mines.

Municipio de Pachuca

Pachuca de Soto: Moderate
distribution. Noted from the
Guatimotzin, San Rafael, Rosario,
Zorra, San Francisco, El Cristo,
and Barron mines.

JALISCO

Municipio de Mascota

San Sebastián: Limited distribution. Noted from the El Refugio
Mine.

MEXICO

Municipio de Zacualpán

Zacualpán: Moderate distribution.
Noted from the El Alacran Mine.

MICHOACAN

Municipio de Angangueo

Mineral de Angangueo: Moderately
distributed on the Vetas de Santa
Lucia and Carrillos.

SAN LUIS POTOSI

Municipio de Catorce

Catorce: Limited distribution. Noted from the Santa Ana Mine.

SONORA

Municipio de Huépac

Huépac: Moderate distribution. Massive. [Massive specimens have been found to just over 100 kilos in size.]

ROMANECHITE

$BaMnMn_8O_{16}(OH)_4$. An common mineral that replaced the mineral psilomelane.

BAJA CALIFORNIA SUR

Municipio de Mulegé

Santa Rosalía: Limited distribution. Noted from the Amelia Mine.

Municipio de San Antonio

El Triunfo: Moderate distribution.

CHIHUAHUA

Municipio de Cusihuiriáchic

Cusihuiriáchic: Limited distribution.

Municipio de Hidalgo del Parral

Hidalgo del Parral: Limited distribution.

DURANGO

Municipio de Guanaceví

Guanaceví: Limited distribution. Noted from the Manzanilla Mine.

Municipio de Mapimí

Mapimí: Moderate distribution. Noted from the Ojuela and San Pedro mines. Massive botryoidal crusts. Associated with cryptomelane and pyrolusite.

GUANAJUATO

Municipio de Guanajuato

Guanajuato: Moderate distribution.

HIDALGO

Municipio de Mineral del Monte

Mineral del Monte: Moderate distribution. Noted from the Jesús María, Santa Brígida, and Resquicio mines.

Municipio de Pachuca

Pachuca de Soto: Moderate distribution. Most noted from the El Bordo, El Cristo, and Encino mines.

MEXICO

Municipio de Zacualpán

Zacualpán: Limited distribution. Noted from the La Blanca Mine.

SONORA

Municipio de Navojoa

Navojoa: Limited distribution.

ZACATECAS

Municipio de Fresnillo

Fresnillo de Gonzalez Echeverría: Moderate distribution within the Cerro de Proaño.

ROSASITE

$(Cu,Zn)_2(CO_3)(OH)_2$. An uncommon secondary mineral found in the oxidized zones of copper-lead-zinc deposits.

CHIHUAHUA

Municipio de Aquiles Serdán

San Antonio el Grande: Moderate distribution. Noted from the San Antonio Mine's 1st to 8th levels. Botryodial crusts. Associated with hemimorphite, calcite, aurichalcite, and goethite.

DURANGO

Municipio de Mapimí

Mapimí: Limited distribution. Noted from the Ojuela and San Juan mines. Botryoidal crusts. Associated with malachite, aurichalcite, hemimorphite, and goethite.

RUBELLITE. See ELBAITE

RUSTUMITE

$Ca_{10}(Si_2O_7)_2(SiO_4)Cl_2(OH)_2$. A rare mineral found in contact mineral deposits.

QUERETARO

Municipio de Cadereyta

Maconi: Limited distribution. Noted from the La Negra Mine. Associated with spurrite.

RUTILE

TiO_2. A common mineral and member of the rutile group formed in metamorphic and sedimentary environments.

BAJA CALIFORNIA

Municipio de Encenada

Real de Castillo: Limited distribution.

DURANGO

Municipio de Cuencamé

Velardeña: Limited distribution.

OAXACA

Municipio de San Miguel Peras

San Miguel Peras: Limited distribution.

S

SAFFLORITE

$CoAs_2$. An uncommon mineral and member of the loellingite group and is found in hydrothermal veins.

CHIHUAHUA

Municipio de Batoplis

Batoplis: Limited distribution.

Municipio de La Ascención

Sabinal: Limited distribution. Noted from the Adventurera Mine.

SAL AMMONIAC

NH_4Cl. An uncommom mineral found in association with active volcanism.

MICHOACAN

Municipio de Uruapán

Volcan de Paricutín: Limited distribution.

SANIDINE

$(K,Na)AlSi_3O_8$. An uncommon mineral and member of the feldspar group found in alkali and felsic volcanic rocks.

DURANGO

Municipio de Coneto de Comonfort

America: Limited distribution within the Sierra de San Francisco. Noted from the Varocites Mine. Associated with quartz.

SAPONITE

$(Ca/2,Na)_{0.33}(Mg,Fe)_3(Si,Al)_4 O_{10}(OH)_2 \cdot 4H_2O)$. An uncommon mineral and member of the smectite group.

GUANAJUATO

Municipio de Guanajuato

Guanajuato: Moderate distribution.

SARTORITE

$PbAs_2S_4$. A rare secondary mineral found in lead deposits.

CHIHUAHUA

Municipio de Neuvo Casas Grandes

San Pedro Corralitos: Limited distribution on Cerro de Corralitos.

Municipio de Uruáchic

Las Animas: Limited distribution.

SAUCONITE

$Na_{0.33}Zn_3(Si,Al)_4O_{10}(OH)_2 \cdot 4H_2O$.
A common mineral and member of
the smectite group.

DURANGO

Municipio de Mapimí

Mapimí: Limited distribution near
the Socavón shaft. Noted from the
Ojuela Mine. Massive. Associated
with hemimorphite.

SCAPOLITE is a group name,
not a specific mineral.

SCHEELITE

$CaWO_4$. An common mineral found
in high-temperature environments
such as metamorphic contact
zones, in pegmatites, and in
high-temperature quartz veins.

BAJA CALIFORNIA

Municipio de Tecate

Laguna Hansen: Moderate distribution. Noted from the Olivia Mine.

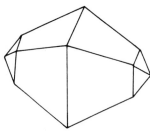

An 8 mm scorodite crystal from the Ojuela
Mine, Mapimi, Durango. (Modified by Shawna
Panczner from Sinkankas, 1964)

Los Gavilanes: Widespread. Noted
from the La Raza and El Audaz
mines.

Rosa de Castilla: Moderately
distributed. Noted from the El
Fenómeno Mine. Associated with
clinozoisite, axinite, and
grossular.

CHIHUAHUA

Municicpio de Ahumada

Banco Lucero: Limited distribution.

Municipio de Coyame

San Eduardo: Moderate distribution.

Municipio de Guadalupe y
Calvo

Guadalupe y Calvo: Limited
distribution. Noted from the La
Flaca Mine.

Municipio de Saucillo

Naica: Limited distribution.

GUERRERO

Municipio de Atoyac de
Alvarez

Atoyac de Alvarez: Moderate
distribution. Noted from the
Lostres Brazos Mine.

HIDALGO

Municipio de Zimapán

Zimapán: Moderate distribution.
Noted from the Concordia Mine.
Crystals to 1 cm.

SINALOA

Municipio de Choix

Choix: Limited distribution.
Noted from the Magistral Mine.

SONORA

Municipio de Alamos

La Salada: Moderate distribution. Noted from the San Alberto Mine.

Municipio de Bacanora

Bacanora: Moderate distribution.

Municipio de Baviacora

Washington: Widespread.

Municipio de Hermosillo

Hermosillo: Widespread. Noted from the Santa Ana, La Cruz, La Providencia, and El Espejo mines.

Las Norias: Moderate distribution, 8 km southwest.

Rancho de Las Víboras: Moderately distributed. Most noted from the El Camino, La Luz Azul, and La Leonora mines.

Villa Seris: Widespread on Cerros Marta, Tepcoripa, and Luján. Most noted from the Beatriz, Carnaval, San Juan de Dios, María Luisa, Cinco de Mayo, and Tungsteno mines. Crystals to 3 cm. Associated with calcite, epidote, quartz, powellite, and pyrite. [The Cinco de Mayo Mine produced large kidney-shaped masses of scheelite up to 6 kilos from the north shaft, 8th level. Scheelite from all these mines has been found occasionally as white crystals up to 3 cm in length.]

Municipio de Guaymas

Rancho Nochebuena: Widespread throughout the Sierra de Nochebuena. Noted from the Virgen de Guadalupe Mine. Crystals to 4 cm.

Municipio de Húepac

Húepac: Widespread. Most noted from the Rosa Maria Mine. Yellow crystals to 4 cm. Associated with pyrite and powellite. [White crystals up to 3 cm have been found on large pyrite crystals.]

Municipio de Rosario

Nacimiento: Widespread, 10 km west. Noted from the El Trueno Mine.

Municipio de Sahuaripa

Sahuaripa: Widespread, 4 km west. Noted from the Coker Mine. Widespread, 14 km north. Noted from the Cadena de Cobre Mine.

Municipio de San Javier

Tecoripa: Widespread, 15-20 km northwest. Noted from the La Paz and El Cobre mines. White crystals to 2 cm. Associated with chalcopyrite, pyrite, molybdenite, chalcocite, and powellite. [Often the crystals from here are gemmy enough to allow the cutting of large gemstones.]

Municipio de San Miguel de Horacasitas

El Picacho: Widespread, 4 km north of Pozo de Toyos and 25 km northeast of Hermosillo. Noted from the El Nublado, El Picacho, and La Florencia mines.

Municipio de Santa Cruz

Santa Cruz: Widespread throughout the Cerro de Chavito. Noted from the Guadalupana Mine. Golden yellow crystals to 6 cm. Associated with quartz, dravite, and powellite.

Municipio de Soyopa

Rancho Telcolote: Moderate distribution, 20 km north of Tonichi. Noted from the Telcolote Mine. Crystals.

Municipio de Tonichi

El Encinal: Moderate distribution, 30 km east. Noted from the Lydia and the Veta Rey mines. Associated with molybdenite.

Municipio de Ures

Ures: Moderate distribution. Noted from the Lobo Mine.

Colorless crystals to 3 cm in size. Associated with wulfenite.

Municipio de Villa Pesqueira

La Venada: Widespread. Crystals to 2 cm. [The ore is taken a few kilometers southwest to Mazatan for processing.]

Municipio de Yécora

La Dura: Widespread, 5 km east. Noted at the El Tungsteno Mine. White crystals to 2 cm. Associated with quartz and feldspar.

Santa Ana: Widespread, 1 km north. Noted from the San Julian (San Carlos) and Puerto de Buenavista mines.

San Nicolás: Widespread. Noted from the La Cruz and El Búfalo mines. Associated with pyrite, chalcopyrite, molybdenite, chlorite, and cuprotungstite.

ZACATECAS

Municipio de Villanueva

Villanueva: Moderate distribution.

SCHMITTERITE

$(UO_2)TeO_3$. A rare mineral found in tellurium deposits.

SONORA

Municipio de Moctezuma

Moctezuma: Limited distribution. Noted from the San Miguel and Bambolla mines. Associated with mackayite. [The San Miguel Mine is the type location for this mineral, which was named after Eduardo Schmitter, past mineralogist for Instituto de Geologia de México.]

SCHORL

$NaFe_3Al_6(BO_3)_3Si_6O_{18}(OH)_4$. A common mineral and member of the tourmaline group found in granites, granitic pegmatites, schists, and gneisses. This mineral is presently under study and possibly could be divided into dravite and uvite.

BAJA CALIFORNIA

Municipio de Ensenada

Pino Solo: Moderate distribution within pegmatites.

SONORA

Municipio de Santa Cruz

Santa Cruz: Moderate distribution on the Cerro de Chavito. [Recent studies on this mineral from this location has shown it to be mostly dravite and uvite.]

SCHORLOMITE

$Ca_3(Fe,Ti)_2(Si,Ti)_3O_{12}$. An uncommon member of the garnet group found in metamorphite rocks.

VERA CRUZ

Municipio de Professor Rafael Rumeriz

Piedras Parada: Limited distribution. Associated with andradite, grossular, quartz, and calcite.

SCHREIBERSITE

Found in meteorites at a few locations within México.

SCHWARTZEMBERGITE

$Pb_6(IO_3)_2Cl_4O_2(OH)_2$. A rare secondary mineral found in few metal deposits.

SONORA

Municipio de Arizpe

Arizpe: Limited distribution. Noted from the Las Chipas Mine. Associated with percylite.

SCOLECITE

$CaAl_2Si_3O_{10} \cdot 3H_2O$. A member of the zeolite group found in volcanic and metamorphic rocks that have been formed hydrothermally.

JALISCO

Municipio de San Cristóbal de la Barranca

San Cristóbal de la Barranca: Moderately distributed throughout the Barranca de Río Grande. Crystals. Associated with mesolite.

SAN LUIS POTOSI

Municipio de Charcas

Charcas: Limited distribution. Crystals to 4 cm. Associated with calcite and datolite.

SCORODITE

$FeAsO_4 \cdot 2H_2O$. A relatively common mineral forming in oxidizing conditions and rarely as a primary hydrothermal mineral.

DURANGO

Municipio de Cuencamé

Velardeña: Limited distribution. Noted from the Santa María Springs prospect.

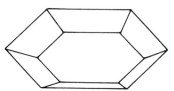

An 8 cm sheelite crystal from the Guadalupana Mine, Santa Cruz, Sonora. (Modified by Shawna Panczner from Sinkankas, 1964)

Municipio de Mapimí

Mapimí: Limited distribution. Noted from the Ojuela Mine. Crystals to 1 cm. Associated with goethite, carminite, dusserite, arseniosiderite, and arsenopyrite.

HIDALGO

Municipio de Zimapán

Zimapán: Limited distribution. Noted from the Miguel Hidalgo Mine. Crystals to 15 mm.

SAN LUIS POTOSI

Municipio de Villa de la Paz

La Paz: Limited distribution. Noted from the Dolores Mine.

SONORA

Municipio de Alamos

San Bernardo: Limited distribution. Noted from the Sara Alicia and San Bernardo mines. Crystals to 3 mm. Associated with quartz.

ZACATECAS

Municipio de Concepción del Oro

Aranzazú: Limited distribution. Noted from the El Cobre Mine. Crystals to 3 cm. Associated with quartz. [This location has produced the finest crystals of scolecite from Mexico.]

Municipio de Mazapil

Mazapil: Limited distribution.

Noche Buena: Limited distribution. Noted from the Noche Buena Mine.

SELENIUM

Se. A rare secondary mineral (element) found in selenium rich deposits.

GUANAJUATO

Municipio de Guanajuato

Guanajuato: Limited distribution.

SONORA

Municipio de Moctezuma

Moctezuma: Limited distribution. Noted from the Moctezuma Mine.

SENARMONTITE

Sb_2O_3. A secondary mineral formed from the oxidation of stibnite.

OAXACA

Municipio de San Juan Mixtepec-Juxtlahuaca

Tejocotes: Limited distribution. Noted from the Yucunani Mine.

QUERETARO

Municipio de Cadereyta

El Doctor: Limited distribution.

Municipio de Tolimán

Soyatal: Limited distribution.

SONORA

Municipio de Caborca

El Antimonio: Limited distribution.

SEPIOLITE

$Mg_4Si_6O_{15}(OH)_2 \cdot 6H_2O$. A common mineral formed by alteration of magnesium rocks in metamorphic conditions or in hydrothermal veins.

DURANGO

Municipio de Durango

Victoria de Durango: Limited distribution. Noted from the Cerro Mercado Mine. Associated with fluoapatite and hematite. [This location is better known as Durango.]

ZACATECAS

Municipio de Concepción del Oro

Bonanza: Limited distribution. Noted from the Refugio Mine.

SERPENTINE is a group name and not a specific mineral.

SERPIERITE

$Ca(Cu,Zn)_4(SO_4)_2(OH)_6 \cdot 3H_2O$. An uncommon secondary mineral found in base metal deposits.

SONORA

Municipio de Cabórica

Heróica Carbórica: Limited distribution. Noted from the Otona Mine.

SIDERITE

$FeCO_3$. A common mineral and member of the calcite group found both in sedimentary rocks and in hydrothermal ore veins.

AGUASCALIENTES

Municipio de Tepezalá

Tepezalá: Widespread throughout the Vetas de Amarilla and Nopal.

CHIHUAHUA

Municipio de Aquiles Serdán

Fransicso Portillo: Widespread. Noted from the El Potosí Mine. Crystals to 1 cm. Associated with quartz, rhodochrosite, and pyrrhotite.

San Antonio el Grande: Moderately distributed. Noted from the San Antonio Mine's 14th level. Crystals to 1 cm. Associated with galena, sphalerite, and quartz.

Municipio de Saucillo

Naica: Moderate distribution. Crystals to 1 cm.

COAHUILA

Municipio de Múzquiz

Melchor Múzquiz: Widespread. Noted from the Cedral, Rosario, and San Francisco mines.

Municipio de Sierra Mojada

Sierra Mojada: Moderate distribution. Noted from the San Jose Mine.

DURANGO

Municipio de Mapimí

Mapimí: Moderate distribution. Noted from the Ojuela Mine. Tan crystals to 10 mm. Associated with quartz, calcite, smithsonite, hemimorphite, and goethite.

GUANAJUATO

Municipio de Guanajuato

Guanajuato: Moderately distributed throughout the Veta Madre.

HIDALGO

Municipio de Pachuca

Pachuca de Soto: Widespread.

Municipio de Mineral del Monte

Mineral del Monte: Widespread.

JALISCO

Municipio de Pihuamo

Pihuamo: Moderate distribution.

MICHOACAN

Municipio de Tlalpujahua

Tlalpujahua: Moderate distribution on the Cerro del Campo del Gallo.

NUEVO LEON

Municipio de Lampazos de Naranjo

Lampazos de Naranjo: Moderate distribution. Noted from the Flor de Peña Mine.

SONORA

Municipio de Onavas

Santa Ana: Moderate distribution. Noted from the Santa Ana Mine.

ZACATECAS

Municipio de Fresnillo

Fresnillo de González Echeverría: Moderate distribution.

SILLENITE

Bi_2O_3. An uncommom mineral found very few metal deposits deposits.

DURANGO

Municipio de Durango

Limited distribution. [From an undisclosed location within this municipio.]

SILLIMANITE

Al_2SiO_5. A rather common mineral found in metamorphic rocks.

PUEBLA

Municipio de Acatlán

Acatlán: Limited distribution.

SILVER

Ag. A rather common mineral found in the upper zones of silver deposits and in zones of sulfide enrichment of copper deposits.

AGUALSCALIENTES

Municipio de Tepezalá

Tepezalá: Limited distribution. Noted from the San Matias Mine.

BAJA CALIFORNIA SUR

Municipio de La Paz

Cacachilas: Limited distribution.

Municipio de Mulegé

Santa Rosalía: Moderate distribution.

Municipio de San Antonio

El Triurfo: Widespread. Most noted from the Animas, Anima Sola, Bebelama, Chivato, Casualidad, Jesús María, Nacimiento, San Cayetano, and Santa Teresa mines.

CHIHUAHUA

Municipio de Aquiles Serdán

Francisco Portillo: Widespread. Most noted from the Mina Vieja. Wires. Associated with pyrargyrite, proustite, and acanthite. [In the early years of mining within the Santa Eulalia Mining district, this mineral was found within most of the district's mines.]

San Antonio el Grande: Widespread. Noted from the San Antonio Mine. Associated with acanthite.

Municipio de Ascensión

Sabinal: Moderate distribution. Noted from the Adventurerea Mine.

Municipio de Batopilas

Batopilas: Widespread. Most noted from the Camuchin, Descubridora, Giral, La Unión, Los Tajos, Pastrana y Anexas, Roncesalles, San Agustín, San Nestor, Santa Martina, Santa Rosa, Santo Niño, Todos Santos, La Naveda, Ballinas, El Carmen, San Pedro, Martinez, San Animas, San Antonio, La Cata, Purísima, San Antonio de los Tachos, San Miguel, and New Nevada mines. Crystals to 5 cm. Associated with calcite, gold, quartz, acanthite, bromargyrite, chlorargyrite, polybasite, proustite, and pyrargyrite. [The Spanish discovered silver here in 1632 along the banks of the Río Batopilas in the rugged Baranaca de Batopilas. Little is known about the early years of operations within the district because a fire in 1740 destroyed the archives and most of its records. From the records that are available, the discovery was made when the Río Batopilas had receded after a flood and exposed native silver in some of the rocks at the river edge. Because the silver looked like freshly fallen snow, the first mine was named the La Nevada Mine. Rich native silver specimens from the developing mine were sent to the Spanish viceroy, Don Diego López Paucheco, in México City. The viceroy, in turn, sent several of these fine crystallized specimens to the King of Spain. The King quickly ordered the Viceroy to collect data and develop the region for the Spanish Crown. From records available in Spain, from the time the Real de San Pedro de Batopilas was officially denounced to 1790, $10,000,000.00 had been officially taxed from the district's mines! The exact figure probably was much higher because much of the rich native silver ore was stolen or taken out as "contraband" and, therefore, taxed by the Crown.

In the late 1700s, Don Rafael Alonzo de Pastrana, owner of several of the larger silver mines in the district, invited the Bishop of Durango to visit. Upon the bishop's arrival, he found the street paved with silver bars from the church to the house where he was to reside. The bishop spoke with the mine owner about his great show of vanity and his need to help the poor. From this point on, Pastrana did what the bishop had instructed and distributed much of his wealth to the poor of the region.

In 1792, Angel Bustamente arrived from Spain and began development of the El Carmen Mine and in a few short months he discovered extremely large masses of native silver. Special scales had to be brought from the blacksmith shop because several of the pieces were too large for the scales at the mine. The largest of these specimens weighed 225 k (496 lbs). Bustamente was lavish with his new found wealth and gave much to Batopilas and its poor. The King of Spain conferred upon Bustamente the title of Marquis de Batopilas and Grandee of Spain for his donation of silver to the Spanish treasury. Bustamente also owned and operated the rich silver mines of San Antonio, San Pedro, La Cata, Martinez, and Purísima.

In 1861, John Robinson, an American, bought the San Miguel and San Pedro mines. In 1879, another American, Alexander Shephered signed a contract for the purchase of the San Miguel. Several years later, he bought Robinson's other mine, the San Pedro. Sheperd also acquired the Ballinas, Descubridora, Martinez, San Antonio de los Tachos, and San Animas mines. The mining continued until it was stopped during the revolution of the early 1900s. Mining began again in the early 1920s and by the end of the 1930s the mines began to close. Some mining continues today, but on a limited basis.

Native silver has been found here in large masses and in the early 1800s, von Humbolt reported the finding of masses of pure silver "papas de plata" up to over 200 kilograms (441 lbs). Pockets of crystallized native silver were often uncovered; the crystals would vary from large spiked crystals to fine arborescent crystal sprays to hairlike wires so small they were almost invisible. The pockets were often so large and filled with native silver, that they had to be cut with hacksaws in order to remove them. At the Santo Niño Mine, fine specimens of native silver

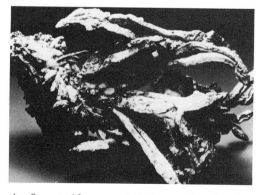

An 8 cm × 10 cm crystal cluster of native silver from Batopilas, Chihuahua in the William Larson Collection, Fallbrook, California.

with crystals of pyrargyrite and proustite were mined in the late 1890s. Crystals of silver have been found up to about 5 cm in length, but usually are less than 1 cm long and 2-3 mm in diameter. They form in arborescent herringbone crystal clusters, for which Batopilas is famous. Wires of native silver have been found over 15 cm in length and over 5 cm in diameter.
Batopilas is located within the Andres del Río Mining District and has had over ninety mines operating during peak periods of production. But the district has had more than 250 mines operating at some period of time during the district's 353-year history. The true amount of production will never be known, but, from existing records over $250 million has been recovered from the district.]

Municipio de Chihuahua

Labor de Terrazas: Moderate distribution. Noted from the San Francisco mine.

Municipio de Cusihuiriáchic

Cusihuiriáchic: Moderate distribution. Noted from the Princesa Mine. Associated with gold.

Municipio de Hidalgo del Parral

Hidalgo del Parral: Moderate distribution. Most noted from the El Refugio and San Francisco mines.

Municipio de Morelos

Zopural: Moderate distribution. Noted from the Independencia Mine. Associated with stephanite and acanthite.

Municipio de Ocampo

Ocampo: Moderately distributed. Noted from the Santa Juliana Mine. Associated with gold.

Rancho Jesús del Monte: Moderate distribution. Most noted from the Green Gold and Santa Eduwigis mines.

Municipio de Urique

Zapuri: Widespread. Most noted from the San Martín, San Nestor, San Rafael, Tajos, and Santa Rosalia mines.

Urique: Widespread. Noted from the El Rosario and Blanca mines.

COAHUILA

Municipio de Jiménez

Santa Rosa: Moderate distribution. Santa Gertrudis Mine.

Municipio de Matamoros

Jimulco: Limited distribution.

Municipio de San Francisco del Oro

San Francisco del Oro: Widespread. Most noted from the San Francisco del Oro and San Patricio mines.

Municipio de Santa Bárbara

Santa Bárbara: Widespread. Noted from the El Tajo Mine.

Municipio de Saucillo

Naica: Moderate distribution.

Municipio de Sierra Mojada

Sierra Mojada: Widespread. Most noted from the Fortuna and Veta Rica mines. [Wires of native silver were associated with acanthite, proustite, barite, and erythrite from the Veta Rica Mine.]

DURANGO

Municipio de Guanaceví

Guanaceví: Widespread. Most noted from the San Gil, La Arianena, and El Rosario mines.

Municipio de Indé

Indé: Widespread. Noted from the El Agua, La Cruz, and Los Machos mines.

Municipio de Mapimí

Mapimí: Moderately distributed. Noted from the San Juan and Ojuela mines. Wires and plates to 1 cm. Associated with cerussite, anglesite, and mimetite.

Municipio de El Oro

Magistral: Moderate distribution. Porvenir Mine. Associated with gold, chalcocite, and chalcopyrite.

Santa María del Oro: Moderate distribution. Associated with gold.

Municipio de San Dimas

Tayoltita: Widespread. Most noted from the Aranzazú, Carrillo, Descubridora, San Jose, Guadalupe, Los Moertos, Maximiliano, Soledad, and Tayoltita mines. Associated with gold and electrum. [This old Spanish mining area has been listed in previous literature as either San Dimas or Gavilanes.]

Municipio de San Pedro del Gallo

Peñoles: Moderate distribution. San Rafael Mine.

Municipio de Pánuco de Coronado

Pánuco de Coronado: Moderately distributed. Noted from the Avino Mine. Associated with gold, galena, and chalcopyrite. [This location in old literature was known as San José de Avino and dates back into the late 1500s.]

Municipio de Topia

Pueblo Nuevo: Moderately distributed. Wires to 2 cm. Associated with pyrargyrite.

Topia: Moderate distribution. Santa Eduwigis Mine.

GUANAJUATO

Municipio de Guanajuato

Guanajuato: Widespread on the Veta Madre. Most noted from the Cata, Nopal, and Valenciana mines. Crystals and wires. Associated with calcite and acanthite.

La Luz: Widespread. Noted from the Santa Rita (Bolañitos) Mine. Associated with gold.

Municipio de San Felipe

Mineral de la Providencia: Moderate distribution. Providencia Mine.

Municipio de San Luis de la Paz

Mineral de Pozos: Widespread. Most noted from the Santa Brigida, San Rafael, Cinco Señores, Progreso, Angustias, and Trinidad mines.

GUERRERO

Municipio de Arcelia

Campo Morado: Moderate distribution. La Reforma Mine.

Municipio de Taxco

Taxco de Alarcón: Widespread. Most noted from the Analco, Bermeja, Chocotitlán, Espíritu Santo, Mora, Golondrinas, Milagro, Pedregal, San Lorenzo, Trinidad, San Pedro y San Pablo, Santa Catrina, Santa María, Guerrero, Zapote, and Zumpancualmil mines. Crystals to 1 cm. Associated with azurite, acanthite, proustite, calcite, and selenite. [Native silver has been found in the district's first three periods of vein mineralization. The third period, the richest, is noted for its spheres composed of silver, acanthite, and proustite. These spheres were

found to almost 14 cm in diameter. The zone of oxidation
within the district is found down
to a depth of less than 100
m below the surface, with the
richest zone being between 20 m
and 60 m deep. When mining was
conducted within this zone
reports of finding delicate
branches of arborescent crystallized native silver enclosed in
crystals of gypsum (selenite)
were common the selenite contained bubbles with crystals of
native silver inside the bubble.
It was also not uncommon in this
zone for crystals of azurite up
to 1 cm or more in size to be
found associated with crystallized native silver. This
location is better known simply
as Taxco.]

Municipio de Tetipac

Tetipac: Moderate distribution.
Xitinga Mine.

HIDALGO

Municipio de Mineral del Monte

Mineral del Monte: Widespread.
Most noted from the Cabrera,
Dolores, Escobar, La Reyna, San
Ignacio, San Pedro y San Pablo,
and El Resquicio mines. Crystals.
[This location is also known as
Real del Monte.]

Municipio de Pachuca

Pachuca de Soto: Widespread. Most
noted from the Camelia, Dolores,
El Encino, El Bordo, El Cristo,
El Lobo, Entrometida, Guatimozín,
Moctezuma, San Cristóbal,
Rosario, San Antonio, San Miguel,
San Nicanor, San Rafael, Santa
Gertrudis, La Sorpresa, and Xacal
mines. Crystals and wires. Associated with calcite, selenite,
and acanthite. [At the El Encino
Mine, crystals of native silver
were found associated with the
water clear crystals of gypsum
(selenite). This locations is
perhaps better known as Pachuca.]

Municipio de Zimapán

Zimapán: Moderate distribution.
Nuestra Señora Mine.

JALISCO

Municipio de Bolaños

Bolaños: Moderately distributed.
Most noted from the Santa Fe, San
José, Iguana, Pichardo, San
Cayetano, and Verde mines.
Crystals. Associated with calcite
and cerussite.

Municipio de Etzatlán

Etzatlán: Moderate distribution.
La Mazata Mine.

Municipio de Tequila

San Pedro Analco: Widespread.
Most noted from the El Tajo, El
Bramador and La Cocina mines.

MEXICO

Municipio del Oro

El Oro de Hidalgo: Moderately
distributed within the Vetas de
San Rafael and Chihuahua. Most
noted from the Esperanza, El Oro,
Dos Estrellas, Nolan, Mexico,
Consuelo, Dolores, El Buen
Suseco, El Carmen, Piedad,
Rosario, San Antonio, San Juan
Nepomuceno, San Miguel, San Jose,
Coronado, and Santa Rita mines.
Associated with gold and quartz.
[This location is best known
simply as El Oro and is within
the Rancho del Oro Mining
District.]

Municipio de Sultepec

Saltepec de Pedro Ascencio de
Alquisiras: Moderate distribution. San Juan Mine.

Municipio de Temascaltepec

Temascaltepec de Gonzalez:
Widespread throughout the Veta
Rica. Most noted from the Animas,

Asuncion, Barranquita, Cruz
Blanca, Guitarra, La Luz, La
Negra, Loma Pelada, Mina Grande,
Ocotillos, Paula, Preciosa
Sangre, Quebradillas, Reyes,
Rosario, San Jose, Santa Rita,
Sierra, and Soledad mines. [This
location is best known as
Temascaltepec.]

Municipio de Zacualpan

San Miguel: Moderate distribution. Tlaxpampa Mine.

Zacualpan: Widespread. Most noted
from the Chucara, Alacran,
Camote, Canal, Carboncillo, El
Toro, Esperanza, Guadalupe, Las
Animas, Marsellesa, Santa Ana,
Socavon, Dios nos Guie, and
Dolores Mines. Crystals and
wires. Associated with calcite
and quartz.

MICHOACAN

Municipio de Angangueo

Mineral de Angangueo: Widespread.
Most noted from the Carmen,
Catingon, Carrillos, El Tigre
and San Luis mines. Crystals and
wires. [This location is best
known and more often referred to
as Angangue.]

Municipio de Tlalpujahua

Tlalpujahua: Widespread. Most
noted from the Las Animas, Los
Zapateros, Laborda, San Estevan,
Coloradilla, Los Olivos, Capulin,
Concepcion, Santos Martires,
Ocotes Carmen, and Coronas mines.
Crystals and wires. Associated
with acanthite, pyrargyrite,
polybasite, calcite, and quartz.
[This location was formerly known
as San Pedro y San Pablo and is
located in the San Pedro y
San Pablo de Tlalpujahua Mining
District.]

MORELOS

Municipio de Tlaquiltenango

Huautla: Moderate distribution.
Noted from the San Anastasio and
Santa Ana mines.

NAYARIT

Municipio de Compostela

Compostela: Moderate distribution. Concepcion Mine.

Municipio de La Yesca

La Yesca: Moderately distributed.
Noted from the Aurora and Florida
mines.

OAXACA

Municipio de Natividad

Natividad: Widespread. Natividad
Mine. Wires and crystals. Associated with calcite, acanthite, and
quartz. [In the early 1980s
several pockets containing silver
wires up to 6 mm in diameter and
6 cm in length were uncovered.]

Municipio de San Jerónimo Taviche

San Jerónimo Taviche: Moderate
distribution. Noted from the San
Francisquito and San Martin de
los Cansecos mines.

PUEBLA

Municipio de Tetela de Ocampo

Tetela de Ocampo: Moderate
distribution. Most noted from the
Covadonga, Esperanza, and
Providencia mines.

QUERETARO

Municipio de Amoles

Pinal de Amoles: Moderate
distribution.

Municipio de Cadereyta

Las Aguas: Moderate distribution.
Noted from the Las Azulitas, La
Luz, Progreso, San José de Los
Amigos, and Santa Inés mines.

Municipio de Ezequiel Montés

Bernal: Moderate distribution.

Municipio de Tolimán

Río Blanco: Moderate distribution. Guadalupe Mine.

SAN LUIS POTOSI

Municipio de Catorce

Catorce: Widespread. Most noted from the Concepción, El Rufugio, and San Agustín mines.

Municipio de Charcas

Charcas: Moderately distributed.

Municipio de Guadalcázar

Guadalcázar: Moderate distribution.

Municipio de Villa de Ramos

Ramos: Moderate distribution. Cocinera Mine.

SINALOA

Municipio de Badiraguato

Soyatita: Moderate distribution. Noted from the San José and San Luis mines.

Municipio de Concordia

Pánuco: Moderate distribution. Noted from the La Virgen, San Antonio, and San Diego mines.

Municipio de Cosalá

Cosalá: Moderately distributed. Nuestra Senora Mine.

Municipio del Fuerte

Realto: Widespread. Most noted from the Animas, Echadas, Guadalupe, La Colorada, La Huerta, La Verde, Jesús María del Oro, Minas del Padre, Monserrate, San Antonio, San José, Santa Catarina, Todos Santos, Vacas, and Veta Grande mines. Crystals and wires. Associated with calcite and quartz.

Municipio de Sinaloa

Rancho de la Joya: Moderate distribution.

SONORA

Municipio de Alamos

Alamos: Widespread. Most noted from the Veta Grande Mine. Wires to 6 cm. Associated with acanthite, chalcocite, chalcopyrite, quartz, and calcite. [Production from these mines made Alamos the silver capital of the world in the 1700s.]

Municipio de Arizpe

Arizpe: Moderate distribtion. Las Chipas Mine. Associated with acanthite and polybasite. [Native silver has been found as a replacement of polybasite crystals.]

Municipio de Cananea

Cananea: Moderate distribution.

Municipio de Nogales

Ranchería de Arissona (Arizonac): Widespread. Las Bolas or Las Planchas de Plata Mine. Massive. [This is the site of the famous discovery of Planchas de Plata. It was discovered in 1736 by a Yaqui Indian miner, Antonio Siraumea, who found chunks of native silver lying on the ground. The location quickly turned of Real de Arissona mining district. The native silver was found in plate or slabs and balls weighing upwards of 1,590 kilos (3500 lbs)! Spanish authorities tried in vain to collect the Royal Fifth (tax) or Quinto and when all efforts failed, the king ordered the mines closed. Fortunately the order came at a point when the mine had just about run out of silver.]

VERACRUZ

Municipio de Las Minas

Zomelahuacán: Moderate distribution.

ZACATECAS

Municipio de Fresnillo

Fresnillo de González Echeverría: Moderately distributed throughout the Cerro de Proano. Wires and crystals. Associated with quartz.

Municipio de General Francisco Murguía

Nieves: Moderate distribution. Santa Catarina Mine.

Municipio de Guadalupe

Guadalupe: Moderate distribution. Los Campos Mine.

Municipio de Mazapil

Mazapil: Moderate distribution. La Nueva Mine.

Santa Rosa: Moderate distribution. Santa Rosa Mine. Associated with gold and acanthite.

Municipio de Pinos

Pinos: Widespread. Noted from the Azul, El Porvenir, Grande, San Bartolo, and Tajo de Ibarra mines. Wires and crystals. Associated with azurite and acanthite.

Municipio de Sombrerete

San Pantaleón de la Noria: Moderate distribution. La Noria Mine. Wires and crystals. Associated with pyrargyrite.

Sombrerete: Moderate distribution. Veta Negra Mine.

Municipio de Vetagrande

Vetagrande: Widespread.

Municipio de Zacatecas

Zacatecas: Widespread. Noted fron the Quebradila, El Bote, Purísima, and Austuriana mines. Wires and crystals. Associated with quartz, stephanite, acanthite, fluorite, and calcite. [At the Austuriana Mine the silver was found associated with stephanite, quartz, and fluorite. Native silver from this district was usually found as wires up to 5-6 mm in diameter and up to 5 cm long.]

SINNERITE

$Cu_6As_4S_9$. An rare mineral found in base metal deposits.

GUANAJUATO

Municipio de Guananjuato

Guanajuato: Limited distribution. Crystals to 1 cm.

SKUTTERUDITE

$CoAs_{2-3}$. An uncommon mineral found in medium-temperature veins in cobalt and nickel deposits.

DURANGO

Municipio de Guanaceví

Guanaceví: Limited distribution.

JALISCO

Municipio de Pihuamo

Pihuamo: Moderate distribution. Esmeralda Mine. Associated with cobaltite and erythrite.

SINALOA

Municipio de Cosalá

Cosalá: Moderate distribution. Assocated with cobaltite and erythrite.

SMITHSONITE

$ZnCO_3$. A rather common secondary mineral and member of the calcite group is found in the oxidized ore zones of zinc deposits in limestone and forms from the alteration of sphalerite.

AGUASCALIENTES

Municipio de Tepezalá

Tepezalá: Moderate distribution.

BAJA CALIFORNIA SUR

Municipio de San Antonio

El Triunfo: Widespread. Noted from the Mina Rica, Gobernadora, San Antonio, Animas, Fortuna, and San Jose mines.

Municipio de Mulegé

Santa Rosalía: Limited distribution (cobaltian smithsonite). Noted from the Ranchería, El Cuarenta, and San Luciano mines. Botryodial masses. Associated with gypsum (selenite).

CHIHUAHUA

Municipio de Aquiles Serdán

Francisco Portillo: Moderate distribution. El Potosí Mine.

San Antonio el Grande: Moderate distribution. Noted from the San Antonio Mine's 8th level. Botryodial masses. Associated with sphalerite, greenockite, and goethite. [In the early 1980s, large masses of botryodial smithsonite were uncovered here ranging in shades of color from gray, green, yellow, orange, and rose red.]

Municipio de Coyame

Coyame: Moderately distributed. Faivre Mine.

Las Plomosas: Moderately distributed. Noted from the Las Cuevitas, El Lago, and Cigarrero mines.

Municipio de Casas Grandes

San Perdro Corralitos: Moderate distribution. El León Mine. [This is the location for the mineral herrerite, which proved not to be smithsonite.]

Municipio de Saucillo

Naica: Moderately distributed. Associated with sphalerite.

COAHUILA

Municipio de Parras

Parras de la Fuente: Widespread throughout the Sierra de Alejo. Noted from the Alfa, Norias de Baján, and Favorita mines. Associated with sphalerite.

DURANGO

Municipio de Mapimí

Mapimí: Widespread. Noted from the Ojuela Mine. Crystal. Associated goethite, calcite, plattnerite, and hemimorphite. [This mineral has been found in three forms in the district: as a replacement of limestone being grayish tan in color, as a gray to white botryodial crust, and, lastly, as a highly modified rhombic pale blue crystal.]

HIDALGO

Municipio de Zimapán

Zimapán: Limited distribution. Yellow green botryodial crusts. Lomo del Toro Mine.

NUEVO LEON

Municipio de General Escobedo

General Escobedo: Moderate distribution. San José Mine.

Municipio de Lampazos de Naranjo

Carrizal: Moderate distribution. Los Angles Mine.

Lampazos de Naranjo: Moderate distribution. Noted from the La Fraternal and El Refugio mines.

Municipio de Monterrey

Cerro de la Mitra: Moderate distribution. Noted from the La Cruz and Soledad mines.

Municipio de Sabinas Hidalgo

Sabinas Hidalgo: Moderately distributed. Noted from the Joya and Sabinense mines.

Municipio de Villaldama

Villaldama: Widespread. Noted from the Viejas, Llanitos, Buenavista, San Francisco, and San Salvador mines.

SAN LUIS POTOSI

Municipio de Catorce

Catorce: Moderate distribution.

SINALOA

Municipio de Choix

Choix: Widespread. Botryoidal. [Large masses of botryoidal smithsonite ranging in shades of gray, blue, pink, and violet. In the early 1970s, several large masses of this mineral were mined at this location.]

SONORA

Municipio de Magdalena

Magdalena de Kino: Moderately distributed.

Municipio de Yécora

Yécora: Moderate distribution.

ZACATECAS

Municipio de Concepción del Oro

Concepción del Oro: Widespread.

Municipio de Mazapil

Mazapil: Moderately distributed. San Eligio Mine.

Municipio de Zacatecas

Zacatecas: Moderate distribution. Noted from the Albarradon, San Francisco, and San Salvodor mines.

SONORAITE

$FeTeO_3(OH) \cdot H_2O$. A rare mineral found in tellurium deposits.

SONORA

Municipio de Moctezuma

Moctezuma: Limited distribution. Moctezuma Mine. [This is the type location for this mineral, which was named for the State of Sonora.]

SPESSARTINE

$Mn_3Al_2(SiO_4)_3$. A less common member of the garnet group formed in skarns, in other manganese-rich metamorphic deposits, and in pegmatites.

BAJA CALIFORNIA SUR

Municipio de La Pas

Cacachiles: Moderate distribution in the Sierra de Cachiles. Gemmy red crystals to 1 cm.

Municipio de Los Cabos

San José del Cabo: Moderate distribution. Crystals are found in the rocks of the harbor.

Municipio de San Antonio

El Triunfo: Limited distribution.
Crystals to 1 cm.

SPHAEROCOBALTITE

$CoCO_3$. A rare secondary mineral
and member of the calcite group
found in the oxide zone of
cobalt, nickel, and other metal
deposits.

BAJA CALIFORNIA SUR

Municipio de Mulegé

Santa Rosalía: Limited distribution. Ranchería, Amelia, and
Curuglú mines. Crystals to 2 mm.
Associated with smithsonite,
brochantite, and remingtonite.

ZACATECAS

Municipio de Concepción del
Oro

Aranzazú: Moderate distribution.
Noted from the El Cobre and
Aranzazu mines. Pale pink
crystals to 4 cm. [The Aranzazu
Mine has produced the finest
specimens of this mineral from
Mexico.]

SPHALERITE

(Zn,Fe)S. A common mineral
formed in a wide range of
environments.

AGUASCALIENTES

Municipio de Asientos

Asientos: Widespread throughout
the Veta de los Pilares. Noted
from the Descubridora, San
Francisco, and Santo Cristo
mines. Crystals to 2 cm.

Municipio de Tepezalá

Tepezalá: Widespread.

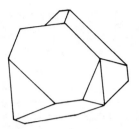

A "spinel"twinned crystal of sphalerite from
the San Antonio Mine, San Antonio el Grande,
Chihuahua. (Modified by Shawna Panczner
from Sinkankas, 1964)

BAJA CALIFORNIA

Municipio de Ensenada

Isla de Cedros: Limited distribution. Máscara de Hierro Mine.

San Fernando: Moderate distribution. Chalcocita Mine.

BAJA CALIFORNIA SUR

Municipio de San Antonio

El Triunfo: Widespread. Most
noted from the Anima Sola,
Comstock, Casualidad, Espinosa,
Fortuna, Gobernadora, Humboldt,
La Soledad, Nacimiento, El
Progreso, La Reforma, Rosario,
Peruana, San Antonio, San
Cayetano, San José, San Pedro,
Santa Lucía, Santa Teresa,
Trinidad, and Tesoro mines.
Crystals. Associated with galena
and pyrite.

CHIHUAHUA

Municipio de Aquiles Serdán

Francisco Portillo: Widespread.
Noted from the Buena Tierra and
El Potosí mines. Crystals to 4
cm. Associated with galena,
pyrite, calcite, pyrrhotite, and
quartz.

San Antonio el Grande: Widespread. San Antonio Mine.
Crystals to 12 cm. Associated
with pyrite, galena, calcite,

quartz, smithsonite, pyrrhotite, and fluroite. [Crystals have been found as simgle and twins, with the larger crystals usually being twinned. In the early 1980s several pockets of rare spinal twinned sphalerite crystals were uncovered up to 12 cm in size.]

Municipio de Batopilas

Batopilas: Limited distribution. Noted from the San Nestor and Los Tajos mines.

Municipio de Cusihuiriáchic

Cusihuiriáchic: Widespread. Noted from the Candelaria, El Madrono, La Durana, La Gloria, Mexicana, San Antonio, and San Bartolo mines.

Municipio de Coyame

Plomosas: Widespread. Noted from the Plomosas, El Lago, and Calera mines. Crystals.

Municipio de Hidalgo del Parral

Hidalgo del Parral: Widespead within the Vetas de Colorado and San Patricio. Noted from the Sierra Plata, El Verde, Terrenates, La Argentina, Palmillo, and El Refugio mines. Crystals. Associated with pyrite, galena, quartz, calcite, and fluorite.

Municipio de Manuel Benavides

San Carlos: Widespread. San Carlos Mine. Associated with galena and pyrite.

Municipio de Nuevo Casas Grandes

San Pedro Corralitos: Widespread. Candelaria Mine. Crystals.

Municipio de San Francisco del Oro

San Francisco del Oro: Widespread within the Vetas de San Francisco del Oro and Victoria. Noted from the San Francisco del Oro and Clarines mines. Associated with galena, pyrite, quartz, calcite, and fluorite. Crystals to 5 cm.

Municipio de Santa Bárbara

Santa Bárbara: Widespread. Noted from the Santa Bárbara, San Diego, Coyote, Segovedad, Bronces, Cobriza, Mina del Agua, San Martín, Alfarena, Tecolotes, Hidalgo, La Rica, and San Albino-Cabrestante mines. Associated with galena, pyrite, quartz, and calcite.

Municipio de Saucillo

Naica: Widespread throughout the Sierra de Naica. Noted from the Naica, Descubridora, El Carmen, El Convento, Los Pajaritos, San Felipe, San Francisquito, San Vicente, and Santa Rita mines. Crystals to 6 cm. Associated with galena, pyrite, quartz, fluorite, and calcite.

Municipio de Temósachic:

Guerrero: Moderate distribution. Calcera Mine.

Municipio de Urique

Urique: Widespread. Piedras Verdes Mine.

DURANGO

Municipio de Canatlán

Canatlán: Widespread. Noted from the La Purisima, La Cruz, and La Providencia mines.

Municipio de Cuencamé

Cuencamé de Ceniceros: Widespread. Noted from the La Chona, La Reina del Cobre, and Tio Samuel mines.

Velardeña: Widespread. Noted from the Velardeña and Socavón Hay mines.

Municipio de Guanacevi

Guanacevi: Widespread.

Municipio de Indé

Inde: Widespread. Noted from the El Agua, Garabatos, and La Union mines.

Municipio de Mapimí

Mapimí: Widespread. Ojuela Mine. Associated with galena, chalcopyrite, pyrite, and arsenopyrite.

Municipio de Nombre de Dios

Vacas: Moderate distribution. Yerbabuena Mine.

Municipio del Oro

Santa María del Oro: Widespread. Noted from the La Reina, La Princesa, Recompensa, and La Predilecta mines. Associated with galena and pyrite.

Municipio de Poanas

Poanas: Moderately distributed throughout the Cerro del Sacrificio. Noted from the El Sacramento, San Carlos, and El Rosario mines.

Municipio de Pueblo Nuevo

Pueblo Nuevo: Widespread. Noted from the San Antonio, La Capa de Oro, and La Espina mines.

Municipio de San Pedro del Gallo

Peñoles: Widespread. Noted from the Jesús María, Potosí, and San Rafael mines.

Municipio de San Dimas

Tayoltita: Widespread. Most noted from the La Amistad, La Gloria, El Gatunal, La Montañesa, Santo Tomas, Guadalupe, Promontorio, San Cayetano, La Araña, San Agustín, and Soledad mines.

Municipio de Topia

Topia: Widespread. Noted from the Magistral, Palmira, Ocotal, Restauradora, Candelaria, Cantarranas, El Toro, La Prieta, San Felipe, San José, Santa Eduwigis, Santa Gertrudis, San Antonio, La Charrera, Victoria, La Perla, La Marquesa, Purísima, El Refugio, La Colorada, Santa Juliana, Santa Paulina, El Picacho, and Purísima mines.

GUANAJUATO

Municipio de Guanajuato

Guanajuato: Moderate distribution. Noted from the Guadalupe, Valenciana, and Mellado mines.

La Luz: Moderately distributed on the Rancho de los Arperos. San Pedro Mine.

Municipio de San Luis de la Paz

Mineral de Pozos: Moderate distribution. La Joya Mine.

GUERRERO

Municipio de Arcelia

Campo Morado: Moderate distribution. El Naranjo and Reforma mines. [This municipio has also been spelled Arselia, which is incorrect.]

Municipio de Taxco

Taxco de Alarcón: Widespread. Noted from the El Pedregal, San Antonio, Caridad, Analco, Calandria, Chocotitlan, Espíritu Santo, San Lorenzo, San Ignacio de la Borda, San Pedro y San Pablo, Santa Catarina, Santa Gertrudis, Santa María, Trinidad, Xocotitlan, Zapote, Zumpancahuil, Jesús, and Guerrero mines. Crystals to 9 cm. Associated with galena, pyrite, and quartz.

Tehuilotepec: Moderate distribution.

Municipio de Quechultenango

Quechultenango: Moderate distribution.

HIDALGO

Municipio de Mineral del Chico

Mineral del Chico: Widespread. Noted from the Arevalo, San Marcial, and San Nicolas mines.

Municipio de Mineral del Monte

Mineral del Monte: Widespread throughout the Veta de Santa Ines. Noted from the Dificultad, Escobar, Moran, Reina, Rincón Grande, Santa Brígida, and San Ignacio mines. [This location is also known as Real del Monte.]

Municipio de Pachuca

Pachuca de Soto: Widespread. Noted from the Entrometida, Jacal, and Moctezuma mines.

Municipio de Zimapán

Bonanza: Moderate distribution. San Martín Mine.

Zimapán: Widespread. Las Estacas, La Salvadora, Lomo del Toro, Balcones, Bernal, Dolores, La Luz, La Palma, San Miquel, San Clemente, San Gabriel, San Juan, San Rafael, and Santo Tomás mines. Associated with galena, pyrite, and chalcopyrite. [Sphalerite of pale yellow color has been found at the Las Estacas. A deep red to brown colored sphalerite has been found at the La Salvadora Mine.]

JALISCO

Municipio de Bolaños

Bolaños: Moderate distribution. Noted from the Santa Fe and San José mines. Associated with fluorite, galena, and quartz.

Municipio de Etzatlán

Etzatlán: Moderate distribution. Regeneración Mine.

Municipio de Sayula

Sierra de Tapalpa: Moderate distribution. La Méxicana Mine.

MEXICO

Municipio de Ixtapan del Oro

Ixtapan del Oro: Moderate distribution.

Municipio de Sultepec

Sultepec de Pedro Ascencio de Alquisiras: Moderate distribution. El Malacate Mine.

Municipio de Temascaltepec

Temascaltepec de Gonzalez: Moderate distribution. Noted from the Veta Chica and Protectora mines.

Municipio de Zacualpan

Zacualpan: Widespread. Most noted from the Carboncillo, Coronas, Durazazno, Esperanza, La Cocina, La Cuchara, Santa Inés, Todos Santos, San Jerónimo, San Antonio, and Santa Isabel mines. Crystals to 2 cm. Associated with galena, pyrite, and quartz.

MICHOACAN

Municipio de Angangueo

Mineral del Angangueo: Widespread. Noted from the El Carmen, Catingon, Concordia, San Cristóbal, Purísima, San Atenogenes, San Severiano, San Hilario, Santa Gertrudis, Santa Lucia, Santa Margarita, La Trinidad, and Carrillos mines. Crystals to 2 cm. Associated with galena, pyrite, and quartz.

Municipio de La Huacana

Inguaran: Widespread. Noted from the San Cristóbal and Inguarian mines. Associated with galena, pyrite, quartz, and calcite.

NAYARIT

Municipio de Amatlán de Cañas

Barranca del Oro: Moderate distribution. Noted from the Rondanera and Cirueloa mines.

Municipio de Campostela

Mitavalles: Moderate distribution. Noted from the Mitavalles and Guadalupe mines.

Municipio de la Yesca

La Yesca: Moderately distributed. Noted from the Los Tajos, La Morada, and La Cabrera mines.

Municipio de Santa María del Oro

Santa María del Oro: Moderate distribution. Noted from the Cuele and Minas de Bolas mines.

OAXACA

Municipio de San Jerónimo Taviche

San Jeronimo Taviche: Moderate distribution. Benjamin Mine.

PUEBLA

Municipio de Hueyapán

Hueyapán: Moderate distribution. Noted from the La Grand Cascada and La Prociosa mines.

Municipio de Tetla del Ocampo

Tetla del Ocampo: Moderate distribution. Noted from the Covadonga, Esperanza, Espejeras, Moises, and Providencia mines.

Municipio de Zacapoaxtla

Zacapoaxtla: Moderately distributed on Cerro de Xochiapulco. Dolores Mine.

QUERETARO

Municipio de Cadereyta

Cadereyta de Montes: Moderately distributed. Noted from the Las Azulitas and Santa Inés mines.

El Doctor: Moderate distribution. Noted from the San Juan Nepomuceno and Trinidad mines.

SAN LUIS POTOSI

Municipio de Catorce

Catorce: Widespread. Noted from the Concepcion, Guadalupito, Gerrero, Medellin, San José, Santa Prisca, Valenciana, Guardarraya, Morelos, San Agustín, San Guillermo, and San Pedro mines.

Municipio de Cerro de San Pedro

Cerro de San Pedro: Moderate distribution. Noted from the San Pedro and Victoria mines. Associated with pyrite, galena, and calcite.

Municipio de Charcas

Charcas: Widespread. Noted from San Sebastián, San Bartolo, La Recompensa, and Tiro General. Crystals to 6 cm. Associated with danburite, datolite, galena, and calcite.

Municipio de Guadalcázar

Guadalcázar: Moderate distribution.

Municipio Villa de la Paz

La Paz: Moderate distribution.

Municipio de Villa de Ramos

Ramos: Moderate distribution. Noted from the La Cocinera, El Patrocinio, and La Purísima mines.

Municipio de Villa de Reyes

Villa de Reyes: Moderate distribution. Providencia Mine.

SINALOA

Municipio de Concordia

Copala: Moderate distribution. Noted from the Cata Rica, El Porvenir, El Toro, Gran Capitan, Restauracion, and San Antonio mines.

Municipio de Cosalá

Cosalá: Moderate distribution. Noted from the Candelaria, Culebra, and Nuestra Senora mines.

SONORA

Municipio de Bacanora

Bacanora: Moderate distribution. Noted from the Sacramento and Trinidad mines.

Municipio de Benjamin Hill

Benjamin Hill: Moderate distribution. Sonora Copper Mine.

Municipio de Cananea

Cananea: Widespread. Noted from the Chivera and Manzanal mines. Crystals to 8 cm. Associated with rhodochrosite, pyrite, and quartz. [At the Chivera Mine, the sphalerite is a golden color and has produced both crystal masses up to 8 cm in diameter and massive pieces that have cut fine gemstones. In the Manzanal Mine, the sphalerite has been found as dark green crystal masses up to 6 cm in diameter. Some of this material has been gemmy enough to cut some excellent gemstones.]

Municipio de Onavas

La Barranca: Moderate distribution. Noted from the El Ahuaje, Santa Fe, El Penasco, Primavera, San Javier, Los Tajos, and La Nahuila mines. Associated with galena, chalcopyrite, pyrite, and quartz.

La Tarahumara: Moderate distribution, south of La Barranca. Noted from the Belem, Cerro Verde, Gualdalupe, Las Animas, Noche Buena, Rosario, Texcalama, and San Javier mines. Associated with galena, chalcopyrite, pyrite, calcite, and quartz.

Municipio de Nacozari de García

Nacozari de García: Widespread. Noted from the Pilares and La Lilly mines. Crystals to 4 cm. Associated with quartz and pyrite.

Municipio de Sahuaripa

Sahuaripa: Moderate distribution. El Mezquite and La Batella San Felipe mines.

ZACATECAS

Municipio de Chalchihuites

Chalchihuites: Moderate Distribution. Lena Mine.

Municipio de Concepción del Oro

Aranzazú: Moderate distribution. Aranzazú Mine. Associated with galena, pyrite, calcite, and quartz.

Concepción del Oro: Moderate distribution. Noted from the Jesús María, Las Animas, Refugio, Quebradilla, Santa Eduwigis, El Tajo, Trinidad, Maranjera, and La Perla mines.

Municipio de Fresnillo

Fresnillo de González Echeverria: Moderate distribution on the Cerro de Proano. [This location is better known as Fresnillo.]

Municipio de General Francisco Murguía

Nieves: Moderate distribution. Santa Catarina Mine.

Municipio de Mazapil

Noche Buena: Moderately distributed. Noche Buena Mine. Associated with pyrite and galena.

Santa Rosa: Moderate distribution. Santa Rosa Mine. Associated with galena.

Municipio de Noria de Angeles

Noria de Angeles: Moderate distribution. Noted from the Cata Rica, Cata Urista, and San Francisco mines.

Municipio de Sombrerete

San Martin: Moderately distributed within the Cerro de Gloria, 18 km west of Sombrerete. San Martín Mine. Associated with chalcopyrite, pyrite, quartz, calcite, pyrrhotite, and fluorite. [This mine, at the present time, is the largest producing mine within Mexico. It was discovered 63 years after Columbus discovered America, with mining operations continuing for over 400 years!]

Sombrerete: Moderate distribution.

Noria de la Pentalón: Moderate distribution. La Noria Mine.

Municipio de Vetagrande

Vetagrande: Moderate distribution. Sierra Nevada Mine.

Municipio de Zacatecas

Zacatecas: Widespread. Noted from the El Bote, La Austuriana, La Luz, No Tiene, La Plata, El Refugio, San Agustín, San Fernando, Zaragoza, and San Juan mines.

SPHENE. See TITANITE

SPINEL

$MgAl_2O_4$. A fairly common mineral formed in thermal and contact metamorphic rocks.

CHIHUAHUA

Municipio de Carmargo

Santa Elena: Limited distribution. Chavira-Olivine Peridot Mine. Associated with fosterite and enstatite.

HIDALGO

Municipio de Zimapán

Zimapán: Moderate distribution.

MICHOACAN

Municipio de Melchor Ocampo del Balsas

Playa Azul: Limited distribution. Crystals up to 12 cm.

NAYARIT

Municipio de Santa María del Oro

Santa María del Oro: Moderate distribution. Massive. [Found as small pebbles in the local streams. The spinel is black in color and has yielded several fine cut stones.]

SAN LUIS POTOSI

Municipio de Villa de Ramos

Ramos: Moderate distribution.

SPIROFFITE

$(Mn,Zn)_2Te_3O_8$. A rare mineral found in tellurium deposits.

SONORA

Municipio de Moctezuma

Moctezuma: Limited distribution. Moctezuma Mine. Associated with denningite. [This is the type location for this mineral.]

SPODUMENE

$LiAlSi_2O_6$. An uncommon mineral which forms in lithium-bearing pegmatites.

DURANGO

Municipio de Durango

Unknown location

OAXACA

Municipio de San Franciaso Telixtlahuaca

San Francisco Telixtlahuaca: Moderate distribution. Santa Ana Mine. Associated with fergusonite, columbite, orthoclase, albite, and muscovite.

SPURRITE

$Ca_5(SiO_4)_2(CO_3)$. A rare mineral which forms in a metamorphic contact zone with limestone and silicates.

DURANGO

Municipio de Cuencamé

Velardeña: Limited distribution. Noted from the Santa Rita and Velardena mines. [This is the type location for this mineral.]

HIDALGO

Municipio de Zimapán

Zimapán: Limited distribution. Lomo del Toro Mine.

MICHOACAN

Municipio de Zitácuaro

Susupuato: Limited distribution within Cerro Mazahua.

QUERETARO

Municipio de Cadereyta

Maconi: Limited distribution. Mina La Negra. Associated with rustumite.

STANNITE

Cu_2FeSnS_4. An uncommon mineral found in a few base metal deposits.

DURANGO

Municipio de Mapimí

Mapimí: Limited distribution. Monterrey Mine. Associated with chalcopyrite, galena, sphalerite, bismuth, and bismuthinite.

STEPHANITE

Ag_5SbS_4. An uncommon mineral formed in the late stages of vein formation in hydrothermal deposits.

AGUASCALIENTES

Municipio de Asientos

Asientos: Moderate distribution. Noted from the Descubridora, Santa Francisca, and Santo Cristo mines.

BAJA CALIFORNIA SUR

Municipio de San Antonio

Triunfo: Moderate distribution.

A 6 cm × 7 cm cluster of crystals of stephanite from the Las Chipas Mine, Arizpe, Sonora in the collection of the National Museum of Natural History, Ottawa, (#499331. (Photographed by J. Schekkerman, Courtesy of the National Museums of Canada)

CHIHUAHUA

Municipio de Chihuahua

Labor de Terrazas: Limited distribution. Rio Tinto Mine.

Municipio de Cusihuiriáchic

Cusihuiriáchic: Moderate distribution. Santa Marina Mine.

Municipio de Morelos

Zopural: Moderate distribution. Noted from the Independencia Mine. Associated with native silver.

DURANGO

Municipio de Canelas

Birimoa: Limited distribution.

Municipio de Guanaceví

Guanaceví: Moderate distribution. Fanny Mine.

Municipio de San Pedro del Gallo

Peñoles: Moderate distribution. San Rafael Mine.

Municipio de San Dimas

Tayoltita: Widespread. Most noted from the Candelaria and Guadalupe mines. Crystals to 1 cm. Associated with acanthite. [This mining district is noted for its stephanite. The Candelaria Mine is located at the former mining center of San Dimas and the Guadalupe Mine, a few km southwest. Tayoltita is the largest pueblo in this old and still active mining district.]

GUANAJUATO

Municipio de Guanajuato

Guanajuato: Moderately distributed throughout the Veta Madre. Noted from the El Nopal, Rayas, San Carlos, and the Valanciana mines. Crystals up to 4 cm. Associated with acanthite, polybasite, and native silver.

La Luz: Moderate distribution. Noted from the Santa Rita (Bolañitos) and Jesús María mines. Crystals.

Municipio de San Felipe

Mineral de la Providencia: Moderate distribution. Providencia Mine.

GUERRERO

Municipio de Taxco

Taxco de Alarcón: Widespread. Noted from the Actitlán and Apaga Candela mines. Crystals.

HIDALGO

Municipio de Mineral del Monte

Mineral del Monte: Moderate distribution.

Municipio de Pachuca

Pachuca de Soto: Moderate distribution.

Pueblo de Cerezo: Moderate distribution. Noted from the San Rafael, Soledad, and Sorpresa mines.

JALISCO

Municipio de Mascota

San Sebastián: Moderate distribution. Refugio Mine.

MEXICO

Municipio de Sultepec

Sultepec de Pedro Ascencio de Alquisiras: Moderate distribution. Compania Mine. [This location is usually referred to as Sultepec.]

Municipio de Zacualpán

Zacualpán: Moderate distribution. Resguardo Mine.

MICHOACAN

Municipio de Angangueo

Mineral de Angangueo: Moderate distribution within the Veta de San Cristóbal. Purísima Mine.

Municipio de Tlalpujahua

Tlalpujahua: Widespread. Most noted from the San Estevan, Coloradilla, Los Olivos, San Sebastián, Capulín, Concepción, Santos Mártires, Dos Estrellas, and Ocotes mines. Crystals.

NAYARIT

Municipio de Amatlán de Cañas

Barranca del Oro: Moderate distribution. El Zopilote Mine.

OAXACA

Municipio de San Mateo Capulalpán

Capulalpán de Mendez: Moderate distribution. Crystals. Associated with native gold.

SAN LUIS POTOSI

Municipio de Guadalcázar

Guadalcázar: Limited distribution.

SINALOA

Municipio de Mocorito

San José de Gracía: Moderate distribution. Loreto Mine.

SONORA

Municipio de Arizpe

Arizpe: Moderate distribution within the Veta de Las Chispas. Las Chispas Mine. Crystals to 6 cm. Associated with acanthite, pyrargyrite, and native silver. [The mine owner, Mr. Pedrazzini, sent the largest stephenite specimen found to the Columbia University School of Mining's Egleston collection.]

Municipio de Onavas

La Barranca: Moderate distribution. La Libertad Mine.

ZACATECAS

Municipio de Concepción del Oro

Aranzazú: Moderate distribution. Azanzazú Mine.

Municipio de Noria de Angeles

Noria de Angeles: Moderately distributed. Noted from the Cubierta and Anexas mines.

Municipio de Pinos

Pinos: Widespread. Most noted from the Catavana, El Pinto, Fresnillo, Infantita, Matatuza, Salomon, and Sierra del Penon Blanco mines.

Municipio de Sombrerete

San Martín: Moderate distribution. San Martín mine.

Municipio de Vetagrande

Vetagrande: Moderate distribtion. Veta Grande Mine.

Municipio de Zacatecas

Zacatecas: Moderate distribution. Mala Noche Mine.

STERNBERGITE

$AgFe_2S_3$. A rare mineral found in few metal deposits.

HIDALGO

Municipio de Pachuca

Pachuca de Soto: Limited distribution.

STIBARSEN

SbAs. A rare native alloy found in few metal deposits.

SONORA

Municipio de Moctezuma

Río Moctezuma: Limited distribution, 38 km east. Associated with allemontite, stibiconite, antimony, and cervantite.

STIBICONITE

$Sb_3O_6(OH)$. An uncommon secondary mineral formed by the oxidation of other antimony minerals.

CHIHUAHUA

Municipio de Uruáchic

Arechuybo: Limited distribution. Associated with antimony, valentinite, and kermesite.

DURANGO

Municipio de Cuencamé

Cuencamé de Ceniceros: Moderate distribution. Crystals to 2 mm. Associated with calcite. [This location has produced stibiconite replacing crystals of stibnite up to 25 cm in length and 5 cm in diameter. These whitish tan pseudomorphs have been found occasionally with crystals of pinkish tan calcite partially covering the psuedomorphs forming what the miners call clubs.]

Municipio de Mapimí

Mapimí: Limited distribution. Noted from the San Ilario and Ojuela mines. Pseudomorphic crystals. Associated with bindheimite, cervantite, and valentinite.

GUERRERO

Municipio de Huitzuco

Huitzuco de la Figueroa: Limited distribution. La Cruz Mine.

Municipio de Taxco

Taxco de Alarcón: Limited distribution.

QUERETARO

Municipio de Peñamiller

Extovar: Limited distribution.

OAXACA

Municipio de San Juan Mixtepec-Juxtlahuaca

Tejocotes: Moderate distribution. Yucunani Mine.

SAN LUIS POTOSI

Municipio de Catorce

Catorce: Limited distribution. Associated with stibnite.

Wadley: Moderate distribution. San José and Santa Emilia mines. Crystals to 1 mm. [In 1975, miners at the San Jose Mine uncovered a cave lined with pseudomorphs of stibiconite after stibnite. The cave was about 30 m long, 3 m high and 4 m wide and completely lined with crystals! The crystals on the walls and ceiling were small, up to 3 cm in length, but the ones on the floor were larger, reaching a length of 1.5 m. It was impossible to walk through the cave without stepping on and breaking off the crystals. All the larger crystals disintegrated when touched and fell to a powder on the floor. The largest crystals that could be removed intact were about 38 cm in length and 6 cm in diameter. The largest of the crystal clusters that were removed were over 60 cm in diameter.]

Municipio de Guadalcázar

Guadalcázar: Limited distribution.

SONORA

Municipio de Caborca

El Antimonio: Limited distribution. El Antimonio Mine.

Municipio de Hermosillo

Las Prietas: Limited distribution. California Mine.

Municipio de Moctezuma

Río Moctezuma: Limited distribution, 38 km east. Associated with stibarsen.

VERACRUZ

Municipio de Tlacolulán

Tlacolulán: Limited distribution.

STIBNITE

Sb_2S_3. A fairly common mineral that normally forms in low-temperture hydrothermal ore veins.

BAJA CALIFORNIA SUR

Municipio de San Antonio

San Antonio: Moderately distributed. Noted from the Comstock, La Reforma, San Antonio, and San José mines. Associated with jamesonite, bindheimite, galena, and pyrite.

Triunfo: Moderate distribution. Noted from the Animas, Fortuna, Humboldt, Rica, Rosario, San José, and Valenciana mines. Associated with jamesonite, bindheimite, galena, and pyrite.

CHIHUAHUA

Municipio de Aquiles Serdán

San Antonio: Limited distribution. San Antonio Mine. Associated with calcite.

Municipio de Cusihuiriáchic

Cusihuiriáchic: Moderate distribution. Noted from the Candelaria, El Madrono, La Durana, La Gloria, Mexicana, San Antonio, San Bartolo, San Francisco, San Miguel, San Nicolas, San Saturnino, and Santa María mines.

Municipio de San Francisco del Oro

San Francisco del Oro: Limited distribution. San Francisco Mine.

DURANGO

Municipio de Cuencamé

La Tapona: Limited distribution.

Municipio de Mapimí

Mapimí: Moderate distribution. Asterillo and Ojuela mines. Crystals to 1 cm. Associated with galena, boulangerite, calcite, barite, siderite, anrite, quartz,

and jamesonite. At the Asterillo
Mine, the stibnite forms as
complex crystals associated with
calcite, barite, siderite,
ankerite, and quartz.

Municipio de San Pedro del Gallo

Peñoles: Limited distribution.
Jesús María Mine.

GUANAJUATO

Municipio de Guanajuato

Guanajuato: Limited distribution.

GUERRERO

Municipio de Huitzuco

Huitzuco de los Figueroa:
Moderate distribution. Noted from
the La Cruz, Sorpresa, and
Trinidad mines. Crystals to 2 cm.
Associated with gypsum, sulfur,
and livingstonite.

Municipio de Taxco

Taxco de Alarcón: Limited
distribution. La Purísima Mine.

HIDALGO

Municipio de Pachuca

Pachuca de Soto: Limited distribution. La Rica Mine. Crystal
sprays to 2 cm.

Municipio de Zimapán

La Ortiga: Limited distribution
within the Cerro de la Ortiga.

JALISCO

Municipio de Hostotipa-quillo

Hostotipaquillo: Limited distribution.

Municipio de Tapalpa

San José del Amparo: Limited
distribution within the Cerro de
la Silleta.

MEXICO

Municipio de Sultepec

Sultepec de Pedro Ascencio de
Alquisiras: Moderate distribu-
tion. Noted from the Alberto and
San Antonio mines.

Municipio de Temascaltepec

Rancho de Albarranes: Limited
distribution. La Cruz Mine.

Municipio de Zacualpan

Zacualpan: Limited distribution
within the Cerro de Coronas.
Resuardo Mine.

MICHOACAN

Municipio de Angangueo

Mineral de Angangueo: Moderate
distribution. Noted from the
Cresta de San Cristobal and
El Carmen mines.

Municipio de Huetamo

Huetamo de Núñez: Moderate
distribution. Noted from the Las
Animas, Soledad, and Dolores
mines.

Municipio de Zinapécuaro

Ozumatlan: Limited distribution.

MORELOS

Municipio de Jojutla

Huautla: Limited distribution.

OAXACA

Municipio de San Jerónimo Taviche

San Jerónimo Taviche: Limited
distribution.

Municipio de San Juan Mixteoec-Juxtlahuaca

Tejocotes: Widespread. Yucunani
Mine. Crystals to 50 cm. [In the
early 1900s several pockets of

stibnite were found. Crystals up to 50 cm long, 10 cm wide, and 8 cm diameter were mined! Most of the stibnite crystals were removed as sprays or clusters.]

PUEBLA

Municipio de Tlaola

Chicahuaxtla: Limited distribution.

QUERETARO

Municipio de Cedereyta

El Doctor: Limited distribution.

Municipio de Peñamiller

Peñamiller: Moderate distribution. Santa María de Miera Mine.

SAN LUIS POTOSI

Municipio de Catorce

Catorce: Limited distribution. San Antonio Mine. Associated with stibiconite.

Wadley: Limited distribution. San José Mine. Associated with stibiconite.

Municipio de Charcas

Charcas: Limited distribution. Noted from the El Salvador and Orion mines.

Municipio de Guadalcázar

Guadalacázar: Limited distribution.

Municipio de Villa de la Paz

La Paz: Limited distribution. Santa María de la Paz Mine.

SINALOA

Municipio de Concordia

Pánuco: Moderate distribution. Noted from the La Virgen, San Diego, and San Antonio mines.

Municipio de Cosalá

Guadalupe de Los Reyes: Limited distribution. Noted from the Nuestra Señora and El Refugio mines.

SONORA

Municipio de Caborca

El Antimonio: Moderate distribution.

Municipio de Bacanora

Bacanora: Limited distribution. El Sacramento Mine.

Municipio de Onavas

La Barranca: Limited distribution. Belem Mine.

ZACATECAS

Municipio de Concepción del Oro

Concepción del Oro: Moderate distribution.

Municipio de Fresnillo

Plateros: Limited distribution within the Veta de Virgen. Veta Virgen Mine.

Municipio de General Francisco Murguia

Nieves: Moderate distribution. Santa Rita Mine. Crystals to 6 cm. Associated with barite, jamesonite, and pyrite. [In the later 1970s and early 1980s several pockets lined with stibnite were uncovered and mined for.]

Municipio de Noria de Angeles

Noria de Angeles: Moderate distribution. Noted from the Cubierta and Anexas mines.

Municipio de Pinos

Pinos: Limited distribution. San Ramon Mine.

Municipio de Sombrerete

San Martin: Moderate distribution. San Martin's, 7th level,
and 25th of October mines.
Crystals to 15 cm. Associated
with xanthoconite and calcite.
[Stibnite associated with
xanthocaonite has been found only
at the San Martin mine.]

San Pantaleon de la Noria:
Moderate distribution. La Noria
Mine. Crystal sprays to 15 cm in
diameter.

Municipio de Zacatecas

Zacatecas: Limited distribution.
Quebradilla mine.

STILBITE

$NaCa_2Al_5Si_{13}O_{36} \cdot 14H_2O$. A fairly
common mineral and member of the
zeolite group found in lava
cavities and as a late forming
vein mineral.

BAJA CALIFORNIA

Municipio de Tijuana

Rancho Ballistero: Moderate
distribution, 25 km south of
Tijuana. Associated with
analcime.

CHIHUAHUA

Municipio de Aquiles Serdán

Francisco Portillo: Limited
distribution. Associated with
quartz, barite, and calcite.

GUANAJUATO

Municipio de Guanajuato

Guanajuato: Moderately distributed throughout the Veta Madre.
Noted from the Valenciana, Rayas,
and Cata mines. Crystals to 3 cm.

HIDALGO

Municipio de Mineral del
Chico

Mineral del Chico: Limited
distribution.

Municipio de Huasca

San Miguel Regla: Limited
distribution.

JALISCO

Municipio de San Cristóbal
de la Barranca

San Cristóbal de la Barranca:
Limited distribution throughout
the Barranca de Río Grande de
Santiago.

MEXICO

Municipio de Zacuálpan

Zacuálpan: Limited distribution.

SAN LUIS POTOSI

Municipio de Charcas

Charcas: Limited distribution.
Crystals to 2 cm. Associated with
datolite and axinite.

SONORA

Municipio de Alamos

Alamos: Limited distribution.
Crystals to 4 cm.

San Bernardo: Limited distribution. Sara Alica Mine. Crystals
to 5 cm. Associated with harmo-
tome.

Municipio de Moctezuma

California Camp: Limited distribution.

ZACATECAS

Municipio de Fresnillo

Fresnillo de González Echeverria: Limited distribution.

Municipio de Sombrerete

San Martín: Limited distribution.
San Martín Mine's 7th level.
Crystals to 1 cm. Associated with
stibnite.

STOLZITE

$PbWO_4$. A rare secondary mineral
and member of the wulfenite group
formed in the oxidized zone of
lead deposits.

ZACATECAS

Municipio de Concepción del
Oro

Concepción del Oro: Limited
distribution. El Cabrestante
Mine.

STROMEYERITE

$AgCuS$. An uncommon secondary
mineral found in the sulfide-
enrichment zones of copper
deposits.

CHIHUAHUA

Municipio de Chihuahua

Labor de Terrazas: Limited
distribution. Río Tinto Mexicano
Mine.

Municipio de Ocampo

Yoquivo: Limited distribution.
Yoquivo Mine.

COAHUILA

Municipio de Sierra Mojada

Sierra Mojada: Moderate distribution.

DURANGO

Municipio del Oro

Santa María del Oro: Moderate
distribution. Noted from the
Grande and El Promontorio mines.

GUANAJUATO

Municipio de Guanajuato

Guanajuato: Limited distribution.

JALISCO

Municipio de Talpa de
Allende

Bramador: Moderate distribution.
Santa Eduwigis Mine.

SAN LUIS POTOSI

Municipio de Catorce

Catorce: Limited distribution.

ZACATECAS

Municipio de Sombrerete

Sombrerete: Limited distribution.

Municipio de Zacatecas

Zacatecas: Moderate distribution.

STRONTIANITE

$SrCO_3$. An uncommon mineral found
in low-temperature hydrothermal
ore veins and in sedimentary
environments.

COAHUILA

Municipio de Sierra Mojada

Sierra Mojada: Limited distribution. Noted from the Suiza and
Veta Rica mines.

STUETZITE

$Ag_{5-x}Te_3$. A rare mineral found in base metal deposits. This mineral has also been spelled stutzite.

SONORA

Municipio de Moctezuma

Moctezuma: Limited distribution. Bambollita Mine.

SULFUR

S. A common mineral (element) found in a wide variety of conditions: near volcanic activities, in oxidized sulfide zones, fires within mines, reduction of sulfates, oxidation of sulfides in mine dumps, sulfur-reducing bacteria in sedimentary rocks, or in salt domes.

BAJA CALIFORNIA

Municipio de Mexicali

San Felipe: Widespread, 40 km south. Noted from the Las Delicias and San Carlos mines. Crystals to 5 cm. Associated with gypsum and alum.

La Pruerta: Limited distribution within the Sierra de Cucupas. Mina del Promontorio.

BAJA CALIFORNIA SUR

Municipio de Mulegé

Santa Roslaía: Limited distribution.

Volcán de las Tres Vírgenes: Limited distribution. Crystals to 3 cm.

Municipio de San Antonio

San Antonio: Moderate distribution.

Triunfo: Moderate distribution.

CHIAPAS

Municipio de Juárez

Hacienda de Mezalapa: Limited distribution.

Municipio de La Libertad

La Libertad: Limited distribution.

CHIHUAHUA

Municipio de Ahumada

Los Lomentos: Limited distribution. Erupcíon/Ahumada mine, Adumada section. Crystals to 15 mm.

Municipio de Aquiles Serdán

Francisco Portillo: Limited distribution. El Potosí Mine. Associated with calcite and quartz.

San Antonio el Grande: Limited distribution. San Antonio Mine.

Municipio de Carmargo

Sierra de las Encinillas: Limited distribution.

COAHUILA

Municipio de Parras

San Rafael: Limited distribution throughout the Sierra de Mayrans. San Pedro Mine.

Municipio de Sierra Mojada

Sierra Mojada: Limited distribution.

COLIMA

Municipio de Coquimatlán

Hacienda de Magdalena: Limited distribution.

DURANGO

Municipio de Mapimí

Mapimí: Limited distribution. El Vergel Mine. Massive. Associated with celestite and gypsum.

GUANAJUATO

Municicpio de Xichú

Xichú: Limited distribution within the Cerro de los Alamos. Dulces Nombres Mine.

GUERRERO

Municipio de Huitzuco

Huitzuco de la Figueroa: Limited distribution. La Cruz Mine. Associated with gypsum and livingstonite.

Municipio de Zumpango del Río

Huitziltepec: Limited distribution.

HIDALGO

Municipio de Singuilucán

Cuyamaloya: Limited distribution.

Municipio de Zimapán

Zimapán: Limited distribution.

JALISCO

Municipio de Guadalajara

Cerro del Coll: Limited distribution, 15 km west.

Municipio de Sayula

Medias Aguas: Limited distribution within the Cerro de Cabeza de Perro.

MEXICO

Municipio de Amecameca

Volcan de Iztaccihuatl: Moderate distribution.

Municipio de Atlautla

Volcan de Popocatepetl: Moderate distribution.

MICHOCAN

Municipio de Zinapécuaro

Ucareo: Moderate distribution.

OAXACA

Municipio de Ixtlan de Juárez

Ixtlan de Juarez: Limited distribution. Santa Gertrudis Mine.

QUERETARO

Municipio de Cadereyta

La Canada: Limited distribution.

Xilitlilla: Limited distribution.

SAN LUIS POTOSI

Municipio de Charcas

Cerro del Azufre: Moderate distribution.

Municipio de Guadalcázar

Guadalcázar: Moderate distribution. San Antonio de Padua, Minas de Mercurio, and La Trinidad mines.

TAMAULIPAS

Municipio de Burgos

Burgos: Limited distribution.

VERACRUZ

Municipio de Jaltipán de Morelos

Jaltipán de Morelos: Moderate distribution. Crystals to 2 cm.

Municipio de la Perla

Pico de Orizaba: Moderate distribution.

ZACATECAS

Municipio de General Francisco Murguíra

Norias: Limited distribution.

Municipio de Mazapil

Mazapil: Limited distribution. San Pedro Ocampo Mine.

Municipio de Pinos

Pinos: Limited distribution. Noted from the El Refugio, Animas, and Cruz de Mayo mines.

Municipio de Zacatecas

Zacatecas: Limited distribution.

SVABITE

$Ca_5(AsO_4)_3F$. A rare mineral and member of the apatite group found in few metal deposits.

SAN LUIS POTOSI

Municipio de Guadalcázar

Guadalcázar: Limited distribution.

SYLVANITE

$(Au,Ag)_2Te_4$. A rare mineral and member of the krennerite mineral group found in base metal deposits.

JALISCO

Municipio de Mascota

San Sebastián: Limited distribution. El Refugio Mine.

MEXICO

Municipio de Tlatlaya

Los Ocotes: Limited distribution. La Providencia Mine.

MICHOACAN

Municipio de Morelia

Curucupaseo: Limited distribution. El Angel Mine.

SONORA

Municipio de Hermosillo

Hermosillo: Limited distribution.

SYMPLESITE

$Fe_3(AsO_4)_2 \cdot 8H_2O$. A rare mineral found in few metal deposits.

DURANGO

Municipio de Mapimí

Mapimí: Limited distribution. Ojuela Mine. Crystal sprays to 2 cm. Associated with parasymplesite, goethite, and gypsum.

T

TAENITE

Found in meteorites at several locations within Mexico.

TALC

$Mg_3Si_4O_{10}(OH)_2$. A common mineral found in low grade metamorphic rocks and in hydrothermal mineral veins.

BAJA CALIFORNIA

Municipio de Ensenada

San Marcos: Moderate distribution. San Pedro Mine.

CHIHUAHUA

Municipio de Batopilas

Batopilas: Limited distribution. San Gabriel Mine.

Municipio de Guazapares

Guazapares: Limited distribution. San Juan de Dios Mine.

GUANAJUATO

Municipio de Guanajuato

La Luz: Moderate distribution. Noted from the La Luz and San Pedro mines.

Santa Rosa: Moderate distribution.

HIDALGO

Municipio de Jacala

Jalaca: Moderate distribution.

Municipio de Pachuca

Pachuca de Soto: Moderate distribution. Noted from the Moctezuma, Barron, and El Cristo mines.

JALISCO

Municipio de Mascota

Bramador: Limited distribution. Bramador Mine.

NAYARIT

Municipio de Ruiz

El Zopilote: Limited distribution. Restauracion Mine.

SONORA

Municipio de Huépac

Huépac: Moderately distributed.

ZACATECAS

Municipio de Fresnillo

Fresnillo de Gonzalez Echeverria: Moderately distributed throughout the Cerro de Proaño.

TEEPLEITE

$Na_2B(OH)_4Cl$. An uncommon mineral forming in arid playa lakes. Location data is unknown.

TELLURITE

TeO_2. A rare mineral found in tellurium deposits.

SONORA

Municipio de Cananea

Cananea: Limited distribution.

Municipio de Moctezuma

Moctezuma: Limited distribution. Noted from the Moctezuma and San Miguel mines. Associated with tellurium, paratellurite, barite, mackayite, and quartz.

TELLURIUM

Te. A rare mineral (element) found in hydrothermal ore veins in tellurium rich metal deposits.

SONORA

Municipio de Cananea

Cananea: Limited distribution.

Municipio de Moctezuma

Moctezuma: Limited distribution. Noted from the Moctezuma and San Miguel mines. Associated with denningite, spiroffite, mroseite, and zemannite.

TENNANTITE

$(Cu,Fe)_{12}As_4S_{13}$. A common mineral formed in hydrothermal ore veins.

CHIHUAHUA

Municipio de Saucillo

Naica: Widespread. Crystals to 2 cm. Associated with tetrahedrite, chalcopyrite, sphalerite, and pyrite.

DURANGO

Municipio de Mapimí

Mapimí: Widespread. Noted from the Ojuela mine. Associated with galena, sphalerite, and pyrite.

HIDALGO

Municipio de Zimapán

Zimapán: Moderate distribution. Lomo del Toro Mine.

MICHOACAN

Municipio de Susupuato

Hacienda de Oroatin: Moderate distribution within the Cerro de Toamaro.

OAXACA

Municipio del San Jerónimo Taviche

San Jerónimo Taviche: Moderate distribution. San Pedro Taviche Mine.

PUEBLA

Municipio de Teziutlán

Teziutlán: Moderate distribution.

SONORA

Municipio de Cananea

Cananea: Moderately distributed.

Municipio de Nacozari de
García

Nacozari de García: Moderate
distribution.

ZACATECAS

Municipio de Concepción del
Oro

Aranzazú: Moderate distribution. El Cobre Mine. Crystals to
2 cm. Associated with tetrahedrite.

TENORITE

CuO. A common secondary mineral
found in oxidized ore zones of
copper deposits.

AGUASCALIENTES

Municipio de Asientos

Asientos: Widespread.

BAJA CALIFORNIA SUR

Municipio de Mulegé

Santa Rosalía: Moderately
distributed.

DURANGO

Municipio de Cuencamé

Velardeña: Moderately distributed. Velardeña Mine.

Municipio de Guanaceví

Guanacevi: Moderate distribution.
Noted from the Rosario and Veta
Capuzaya mines.

Municipio de Mapimí

Buruguilla: Moderate distribution. Descubridora Mine.

Mapimí: Widespread. Ojuela Mine.
Associated with cuprite.

Municipio del Oro

Santa María del Oro: Moderate
distribution. Noted from the
Candela and Promontorio mines.

Municipio de San Pedro del
Gallo

Peñoles: Moderately distributed.
Potosí Mine.

HIDALGO

Municipio de Cardonal

Cardonal: Moderate distribution.
La Pinta Mine.

Municipio de Mineral del
Monte

Mineral del Monte: Moderate
distribution. Santa Brígida Mine.

MICHOACAN

Municipio de La Huacana

Inguaran: Widespread.

SAN LUIS POTOSI

Municipio de Villa de la
Paz

La Paz: Moderate distribution.
Dolores Mine.

Municipio de Villa de Ramos

Ramos: Moderate distribution.
Cocinera Mine.

SINALOA

Municipio de Cosalá

Cosalá: Moderate distribution.
Noted from the Culebra and
Nuestra Senora mines.

SONORA

Municipio de Onavas

La Barranca: Moderately distributed. Noche Buena Mine.

Municipio de Moctezuma

Moctezuma: Limited distribution. Bambollita Mine.

Municipio de Ures

Ures: Moderate distribution within the Arroyo Blanco. Noted from the La Libertad and Prietita mines.

ZACATECAS

Municipio de Concepción del Oro

Concepción del Oro: Moderate distribution.

TERLINGUAITE

Hg_2ClO. A rare mineral found in mercury deposits.

CHIHUAHUA

Municipio de Guerrero

Pedernales: Limited distribution.

GUERRERO

Municipio de Taxco

Huahuaxtla: Limited distribution. Noted from the Aurora, Esperanza, and San Luis mines. Associated with egelstonite and montroydite.

SAN LUIS POTOSI

Municipio de Guadalcázar

Guadalcázar: Limited distribution.

Municipio de Moctezuma

Moctezuma: Limited distribution. Dulces Nombres Mine.

TETRADYMITE

Bi_2Te_2S. An uncommon mineral found both in contact metamorphic deposits and in gold quartz veins.

SONORA

Municipio de Yécora

Yécora: Limited distribution.

TETRAHEDRITE

$(Cu,Fe)_{12}Sb_4S_{13}$. A common mineral formed in hydrothermal veins.

AGUASCALIENTES

Municipio de Asientos

Asientos: Moderate distribution. Noted from the La Merced and Santa Elena mines. Associated with tennantite.

BAJA CALIFORNIA SUR

Municipio de San Antonio

El Triunfo: Widespread within the Vetas Chivato, San Rafael, Tesoro, and Tesorito. Noted from the Fortuna, Gobernadora, Jesús María, Nacimiento, and San Antonio mines. Associated with sphalerite, galena, and pyrite.

CHIHUAHUA

Municipio de Guadalupe y Calvo

Barranca de Baburaga: Moderate distribution. Orpineda Mine.

Municipio de Guazapares

Uruapa: Moderate distribution. Guadalupe Mine.

Municipio de Manuel Bena-
vides

San Carlos: Moderate distribution. San Carlos Mine.

Municipio de Santa Bárbara

Boquillas: Moderate distribution.
La Unión Mine.

Municipio de Saucillo

Naica: Widespread throughout the
Sierra de Naica. Crystals to
3 cm. Associated with galena,
sphalerite, pyrite, quartz, and
tennantite.

Municipio de Urique

Cerocahui: Moderately distributed. Cerocahui Mine. Massive.
Associated with galena, sphalerite, chalcopyrite, and pyrite.

Urique: Moderate distribution.
Noted from the Avila, El Rosario,
Esperanza, and Mascota mines.

Municipio de Uruáchic

Uruáchic: Moderate distribution.
San Martín Mine.

COLIMA

Municipio de Manzanillo

Miraflores: Moderate distribution.

DURANGO

Municipio de Cuencamé

Velardeña: Moderate distribution.
Noted from the La Adelfa and
Terneras mines.

Municipio de Guanaceví

Guanaceví: Widespread. Noted from
the San Gil, El Verde, La
Mexicanna, Barradón, La Sirena,

Paleros, Sierra Santa, and La
Fortuna mines.

Municipio de Mapimí

Buruguilla: Moderate distribution. Descubridora Mine.

Mapimí: Moderate distribution.
Ojuela Mine. Associated with
galena, sphalerite, chalcopyrite,
enargite, and marcasite.

Municipio del Oro

Santa María del Oro: Moderate
distribution. Lustre Mine.

Municipio de Pánuco de
Coronado

Pánuco de Coronado: Moderate
distribution. Noted from the
Potesina and Rosa María mines.

Municipio de Poanas

Poanas: Widespread throughout the
Cerro del Sacrificio.

Municipio de San Dimas

Tayoltita: Moderately distributed. Juan Manuel Mine.

Municipio de Topia

Topia: Moderate distribution.
Noted from the Marquesa and La
Perla mines.

GUANAJUATO

Municipio de Guanajuato

Limited distribution. Noted from
the Rayas and Esperanza mines.

La Luz: Limited distribution.
Tiro de Arcangeles Mine.

GUERRERO

Municipio de Taxco

Taxco de Alarcón: Moderate
distribution.

HIDALGO

Municipio de Jacala

Rancho de la Pechuga: Moderate
distribution.

Municipio de Mineral del
Chico

Mineral del Chico: Moderate
distribution.

Municipio de Mineral del
Monte

Mineral del Monte: Moderate
distribution.

Municipio de Pachuca

Pachuca de Soto: Moderate
distribution.

Municipio de Zimapán

Zimapán: Moderate distribution.
Noted from the Santa Gorgonia,
Dolores, La Luz, San Juan, San
Judas, San Miguel, San Rafael,
Santo Tomás, Verdosas, Nuestra
Señora, and Los Balcones mines.

JALISCO

Municipio de Bolaños

Bolaños: Moderately distributed.
Noted from the Descubridora,
Iguana, Pichardo, Santa Fe, and
Verde mines.

Municipio de Chiquilistlán

Chiquilistlán: Limited distribution. La Cobriza Mine.

Municipio de Mascota

Natividad: Limited distribution.
La Purísima Mine.

MEXICO

Municipio de Temascaltepec

Temascaltepec de González: Wide-
spread. Noted from the Asución,
Barranquillas, Cruz Blaanca,
Gachupina, La Luz, Loma Pelada,
Mina de Agua, Mina Grande,
Ocotillos, Paula, Quebradilla,
Rosario, San José de la Sierra,
Soledad, Tres Reyes, Veta Negra,
and Veta Rica mines.

Municipio de Zacualpán

Zacualpán: Moderately distributed.

MICHOACAN

Municipio de Angangueo

Mineral de Angangueo: Moderate
distribution. Purísima Mine.

Municipio de la Huacana

Inguaran: Widespread.

Oropeo: Moderate distribution.
Noted from the San Cristobal and
Puerto de Mayapito mines.

Municipio de Santa Clara

Santa Clara: Moderate distribution.

Municipio de Tlalpujahua

Tlalpujahua: Moderate distribution. Noted from the La Luz and
La Borda mines.

NAYARIT

Municipio de Ruiz

El Zopilote: Widespread. Noted
from the Ahuacatlán, Concepción,
Jesús María, La Yesa, Los Tajos,
San José, and Tenamache mines.

PUEBLA

Municipio de Huauchinango

Rancho de Ozomatlán: Limited
distribution.

Municipio de Tetla de
Ocampo

Tetla de Ocampo: Limited distribution. Covadonga Mine.

Municipio de Tepeyahualco

Sierra de Tepeyahualco: Moderate
distribution. La Dificultad Mine.

QUERETARO

Municipio de Amoles

Pinal de Amoles: Moderate distribution.

Municipio de Ezequiel Montes

Bernal: Limited distribution.

Municipio de Tolimán

Tolimán: Moderate distribution.

SAN LUIS POTOSI

Municipio de Catorce

Catorce: Limited distribution.

Municipio de Charcas

Charcas: Limited distribution. Associated with datolite and axinite.

Municipio de Villa de Ramos

Ramos: Moderate distribution.

SINALOA

Municipio de Cosalá

Cosalá: Moderate distribution.

SONORA

Municipio de Altar

Altar: Limited distribution. Cueva Santa Mine.

Municipio de Bacoachi

Bacoachi: Moderate distribution. Noted from the Chiquita, Chonita, Foetuna, Ravicanora, Santa María, and Santo Tomás mines.

Municipio de Cananea

Cananea: Moderate distribution.

Municipio de Cumpas

Cumpas: Moderate distribution. Noted from the San Pedro and Sonora mines.

Municipio de Nacozari de García

Nacozari de García: Widespread.

Municipio de Onavas

La Barranca: Widespread. Noted from the Blem, America, Cerro Verde, La Sierra, Rosario, and Santa Ana mines.

Municipio de San Javier

San Javier: Moderate distribution. Noted from the Animas, El Aguaje, and El Peñasco mines.

Municipio de San Pedro de la Cueva

Tesache: Limited distribution.

ZACATECAS

Municipio Concepción del Oro

Aranzazú: Widespread. Noted from the El Cobre Mine. Crystals to 4 cm. Associated with tennantite and quartz.

Bonanza: Moderate distribution. Bonanza Mine. Crystals to 3 cm. Associated with bournonite, sphalerite, pyrite, and quartz.

Municipio de Mazapil

Noche Buena: Moderate distribution. Noche Buena Mine.

Municipio de Zacatecas

Zacatecas: Moderate distribution.

THAUMASITE

$Ca_3Si(CO_3)(SO_4)(OH)_6 \cdot 12H_2O$. An uncommon secondary mineral found in metal deposits.

MICHOACAN

Municipio de Zitacuaro

Suspuato: Limited distribution within the Cerro Mazahua.

ZACATECAS

Municipio de Concepción del Oro

Concepción del Oro: Limited distribution. Associated with pyrite.

THENARDITE

Na_2SO_4. A rather common mineral in arid lands formed as an ephemeral mineral in playas or dry lakes and on lavas.

CHIHUAHUA

Municipio de Aquiles Serdán

San Antonio el Grande: Limited distribution. San Antonio Mine.

Municipio de Jiménez

Laguna de Jaco: Moderate distribution. Associated with halite and glauberite.

THERMONATRITE

$Na_2CO_3 \cdot H_2O$. An uncommon mineral forms in arid environments from the partial dehydration of other saline minerals. Location data is unavailable.

THOMSONITE

$NaCa_2Al_5Si_5O_{20} \cdot 6H_2O$. A rather common mineral and member of the zeolite group found in cavities in mafic igneous rocks.

CHIHUAHUA

Municipio de Aquiles Serdán

San Antonio el Grande: Limited distribution. San Antonio Mine.

JALISCO

Municipio de Autlán

Sierra de Perote: Limited distribution.

TIEMANNITE

HgSe. A rare member of the sphalerite group, which is found in mercury deposits.

QUERETARO

Municipio de Cadereyta

El Doctor: Limited distribution. El Doctor Mine.

Municipio de Amole

Pinal de Amole: Limited distribution near the Rio Blanco. Associated with malachite.

TILLEYITE

$Ca_5Si_2O_7(CO_3)_2$. This is an uncommon rock forming mineral.

MICHOACAN

Municipio de Zitácuaro

Susupuato: Limited distribution throughout the Cerro Mazahua.

TIN

Sn. A rare mineral (element) found in few metal deposits.

GUANAJUATO

Municipio de Guanajuato

Sierra de Santa Rosa: Limited distribution within the Veta de Santa Catarina.

TITANITE

$CaTiSiO_5$. An common mineral found in skarns, pegmatites, and other igneous rocks. Formally known as sphene.

BAJA CALIFORNIA

Municipio de Ensenada

El Alamo: Moderate distribution, 24 km east in pegmatites.

El Rodeo: Moderate distribution, 8 km east southeast of Rancho Viejo in pegmatites. Crystals to 13 cm. Associated with quartz, chlorite, and epidote. [Crystals have been found up to 13 cm long, 10 cm wide, and 3 cm thick. Many of the titanite crystals were gemmy and yielded cut gemstones of large size.]

El Socorro: Moderate distribution. Las Delicias Mine. Crystals. Associated with elbiate, topaz, aquamarine, and quartz.

La Huerta: Moderate distribution, northeast of Ojos Negros in pegmatites. Light yellow-green tabular crystals to 13 cm. [Crystals are usually less than 4 cm in length, but several

crystals did reach sizes up to 13 cm in length and weighed up to 1,800 grams (4 pounds) each!]

Piño Solo: Moderate distribution, pegmatite southeast of Ensenada. Green to dark brown tabular crystals to 13 cm. Associated with quartz and epidote. [Crystals have been found up to 13 cm long, 10 cm, wide and 3 cm thick. Many of these crystals were gemmy enough to cut large gemstones.]

San Quintin: Modertate distribution, pegmatites 48 km east. Bright green crystals. Associated with epidote, quartz, and laumontite. [This is a chromium-rich titanite.]

DURANGO

Municipio de Mapimí

Mapimí: Limited distribution. Ojuela Mine.

HIDALGO

Municipio de Zimapán

Zimapán: Limited distribution.

MORELOS

Municipio de Ayala

Jalostoc: Limited distribution. San Luis Mine. [This location in older literature is spelled Xalostoc.]

OAXACA

Municipio de la Pe

La Panchita: Limited distribution, pegmatites northwest of Ayoquezco. La Panchita Mine. Associated with actinolite.

SONORA

Municipio de Santa Cruz

Santa Cruz: Limited distribution within pegmatites of Cerro de

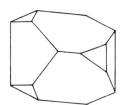

A crystal of titanite from San Quintin, Baja California. (Modified by Shawna Panczner from Sinkankas, 1959)

Chavito. Guadalpana Mine. Associated with dravite and quartz.

TLALOCITE

$(Cu,Zn)_{16}(TeO_3)(TeO_4)_2Cl(OH)_{25} \cdot 27H_2O$. A rare mineral found in copper and tellurium deposits.

SONORA

Municipio de Moctezuma

Moctezuma: Limited distribution. Bambollita (La Oriental) Mine. Associated with azurite and tenorite. [This is the type location for this mineral.]

TLAPALLITE

$H_6(Ca,Pb)_2(Cu,Zn)_3(SO_4)(TeO_3)_4(TeO_6)$. A rare mineral found in copper/tellurium deposits.

SONORA

Municipio de Moctezuma

Moctezuma: Limited distribution. Bambollita (La Oriental) Mine. Associated with azurite, hessite, and tlalocite. [This is the type location for this mineral.]

TOBERMORITE

$Ca_5Si_6O_{16}(OH)_2 \cdot 4H_2O$. A rather uncommon mineral found in hydrothermal veins.

SAN LUIS POTOSI

Municipio de Charcas

Charcas: Limited distribution. Associated with stilbite.

ZACATECAS

Municipio de Mazapil

Noche Buena: Limited distribution. Noche Buena Mine. White spherical masses to 10 cm. Associated with pyrite and hydroxapophyllite.

TODOROKITE

$(Mn,Ca,Mg)Mn_3O_7 \cdot H_2O$. An uncommon mineral found in the oxide ore zones of manganese deposits.

CHIHUAHUA

Municipio de Ahamada

Sierra de Gallego: Limited distribution. Massive. [This mineral is found as a coating in the Coconut geodes.]

TOPAZ

$Al_2SiO_4(F,OH)_2$. A common mineral found in granites, pegmatites, rhyolites, or in the surrounding metamorphic rocks.

A 4 cm long crystal of topaz from La Ventilla, San Luis Potosi. (Modified by Shawna Panczner from Sinkankas, 1959)

BAJA CALIFORNIA

Municipio de Ensenada

Rancho Viejo: Limited distribution, west (Los Pocitos) in Arroyo de Rincón. Las Dilicias Mine. Blue crystals to 1 cm. Associated with elbaite.

Municipio de Tecate

La Jollita: Limited distribution. Blue crystals to 2 cm.

BAJA CALIFORNIA SUR

Municipio de Mulegé

Rosario: Limited distribution. Placeres del Rosario.

Municipio de San Antonio

Las Gallinas: Limited distribution.

CHIHUAHAU

Municipio de Aquiles Serdán

San Antonio el Grande: Limited distribution. San Antonio Mine's tin chimney. Associated with cassiterite, fluorite, quartz, and wolframite.

DURANGO

Municipio de Durango

Cerro de los Remedios: Limited distribution. Remedios Mine.

Municipio de Cento de Comonfort

Sierra de San Francisco: Moderate distribution throughout the arroyos de Culebra and Oroscoebra. Artroyos Mine (at Parrillos). Associated with cassiterite.

GUANAJUATO

Municipio de Guanajuato

Guanajuato: Limited distribution within the Canada de Marfil.

Municipio de León:

Tlachiquera: Limited distribution.

GUERRERO

Municipio de Coyuca de Benitez

Coyuca de Benitez: Moderate distribution.

HIDALGO

Municipio de Metepec

Apulco: Limited distribution on the Cerro del Nado.

SAN LUIS POTOSI

Municipio de Cerritos

Mesa de San José Beunavista: Moderate distribution.

Mesa de Santa Cruz: Moderate distribution.

Topezán: Moderately distributed throughout the Sierra de Combra.

Municipio de Charcas

Charcas: Limited distribution.

Municipio de Villa de Arriaga

Lourdes: Moderate distribution. Crystals to 2 cm. This location has produced fine, sherry colored crystals.

Tepetates: Widespread, 40 km southwest of San Luis Potosí. Crystals to 15 cm. Associated

A 4 cm crystal of topaz from Tepetate, San Luis Potosi. (Modified by Shawna Panczner from Sinkankas, 1959)

with opal (hylite) and quartz. (Citrine and rock crystal). [The topaz are found in the local rhyolitic-trachyte rock, which is mined by hand. Explosives are used sparingly in uncovering the cavities in the rock. These gas-created cavities are often found "dry," containing nothing, or they might be found containing topaz crystals. The crystals from here are stubby to elongate prisms and are terminated by a series of dipyramids to points that are blunted by the basal face. The larger topaz crystals usually have etched faces and are much simpler in crystal form with less clear interiors. The smaller crystals are complex, exhibiting as many as ten different crystal morphologies, and have clean crystal faces with gemmy interiors. The crystals range from micros to 15 cm in length and up to a 1 cm in diameter. The colors of the topaz range from "blancos" colorless, "amarillas" yellowish, "comena" beeswax color, "meil" honey color, to "cafe" a rich reddish brown. This last color, "cafe" appears to be colored at times by swarms of extremely tiny needlelike crystals of rutile. The topaz crystals are mined by individuals or small groups of miners who bring them to Tepetates to be sorted and color graded for the buyers from San Luis Potosí.]

Municipio de Villa de Reyes

La Ventilla: Moderate distribution. Crystals to 12 cm. Associated with opal (hylite), quartz (citrine), and hematite. [The topaz found here are considered the best gem material from the state both in clarity and color.

The crystals and colors are very similar to those from Tepetates area except the topaz colors are deeper and do not fade in the sun.]

ZACATECAS

Municipio de Pinos

Miguel de los Pinos: Limited distribution within the Cerro de la Cruz, 9 km southeast of Pinos.

Pinos: Moderate distribution within the Arroyo de San Juan de Las Herreras, located in the Hacienda de Tepezate.

Municipio de Sombrerete

Sierra de los Organos: Widespread. [Pieces of weathered topaz are mined by ants and are found on the tops of ant hills within the surrounding mountains.]

TORBERNITE

$Cu(UO_2)_2(PO_4)_2 \cdot 8-12H_2O$. An uncommon secondary mineral and member of the autunite group found in the oxidized ore zone of uranium deposits.

SONORA

Municipio de Alamos

Alamos: Limited distribution.

Municipio de Moctezuma

Moctezuma: Limited distribution. El Sapo Mine.

Municipio de Nacozari de García

Nacozari de García: Limited distribution.

Municipio de Soyora

San Antonio de la Huerta: Limited distribution. Crystals to 1 cm. Associated with metatorbernite.

TREMOLITE

$Ca_2(Mg,Fe)_5Si_8O_{22}(OH)_2$. A common mineral and member of the amphibole group occurring in contact metamorphised limestones and dolomites.

BAJA CALIFORNIA SUR

Municipio de la Paz

Isla de Magdalena: Moderate distribution.

CHIPAS

Municipio de Cintalapa

Cintalapa: Limited distribution on the Hacienda de La Razón.

Municipio de Pichucalco

Pichucalco: Moderate distribution. Santa Fe Mine.

CHIHUAHUA

Municipio de Aquiles Serdán

Francisco Portillo: Moderate distribution.

San Antonio el Grande: Limited distribution. San Antonio Mine's skarn zone. Associated with hedenburgite and grossular.

Municipio de Saucillo

Naica: Limited distribution.

DURANGO

Municipio de Mapimí

Mapimí: Limited distribution. Ojuela Mine.

GUANJUATO

Municipio de Guanajuato

Guanajuato: Moderate distribution. Noted from the Santa Clara, Santa Ana, and El Nopal mines.

HIDALGO

Municipio de Mineral del Monte

Mineral del Monte: Limited distribution within the Veta Vizcaina. Noted from the Santa Agueda and Vizcaina mines.

Municipio de Zimapán

Encarnación: Limited distribution within the Cerro del Pilon.

Zimapán: Moderate distribution. Noted from the Las Estacas and Miguel Hidalgo mines.

PUEBLA

Municipio de Acatlán

Acatlán de Osorio: Moderate distribution.

SAN LUIS POTOSI

Municipio de Catorce

Catorce: Limited distribution. Associated with asbestos.

Municipio de Guadalcázar

Guadalcazár: Limited distribution within Cerro de San Cristóbal.

VERACRUZ

Municipio de Las Minas

Las Minas: Moderately distributed. Noted from the La Purísima, Ampliación, and San Valentín mines.

ZACATECAS

Municipio de Concepción del Oro

Concepción del Oro: Moderate distribution.

TRIDYMITE

SiO_2. An uncommon mineral found in felsic volcanic rocks.

DURANGO

Municipio de Durango

Victoria de Durango: Moderate distribution on Cerro de los Remedios. Los Remedios Mine. Associated with fluorite. [This is a city park in the city better known as Durango.]

HIDALGO

Municipio de Pachuca

Cerro de San Cristóbal: Moderate distribution. Crystals to 4 mm. Associated with cristobalite.

TROILITE

FeS. A rare mineral found in few metal deposits and in meteorites.

CHIHUAHUA

Municipio de Hidalgo del Parral

Rancho Chupadero: Limited distribution.

TRONA

$Na_3(CO_3)(HCO_3) \cdot 2H_2O$. A rather common mineral found in dry lakes.

CHIHUAHUA

Municipio de Carmargo

Laguna de Jaco: Moderate distribution.

JALISCO

Municipio de Sayula

Zacoalco: Moderate distribution at Lagunas de la Cuenca de Sayula.

MEXICO

Municipio de Texcoco

Lago de Texcoco: Moderate distribution. Associated with halite.

PUEBLA

Municipio de Tepeaca

Tepeaca: Limited distribution.

TUNGSTITE

$WO_3 \cdot H_2O$. An uncommon mineral found in the oxidation zone of tungsten rich deposits.

BAJA CALIFORNIA

Municipio de Tecate

Rosa de Castillo: Limited distribution. El Fenomeno Mine.

SONORA

Municipio de Alamos

Alamos: Limited distribution.

Municipio de Ures

Ures: Limited distribution. La Cruz Mine.

TURGITE. See HEMATITE

TURQUOISE

$CuAl_6(PO_4)_4(OH)_8 \cdot 5H_2O$. An uncommon secondary mineral formed in arid regions by surface waters acting on aluminous rocks in copper-rich metal deposits.

BAJA CALIFORNIA

Municipio de Ensenada

El Aguajito: Moderate distribution, 4 km south. La Reina(La Turquesa), Vincent, Hermosa, and Preciosa mines. [The turquoise from this area is considered as good gem quality.
Another turquoise mine, by the same name as La Turquesa Mine, is located several km east of El Aguajito and has been operated for turquoise. This is the more famous of the two and the older operation. It too produces good gem grade turquoise.]

El Rosario: Moderate distribution. Evan Turquoise and American Hole mines. [The turquoise from this area is of fair gem quality.]

Laguna Chapala: Limited distribution, 40 km northeast.

Los Arrastos: Moderate distribution, 10 km northwest.

SONORA

Municipio de Cananea

Cananea: Moderate distribution. [The turquoise from here is considered to be extremely pure and considered "chalk," but will treat very well. This copper deposit is one of the world's largest with a 1.85 billion metric ton ore reserve!]

Municipio de Nacozari de García

La Carida: Moderate distribution. La Caridad Mine. Nodules to 38 cm. [The developing open pit copper mine here has already produced large nodules of pure "chalk" turquoise.]

Nacozari de García: Moderate distribution.

Municipio de Onavas

La Barranca: Moderate distribution. La Barranca mine.

ZACATECAS

Municipio de Concepción del Oro

Bonanza: Limited distribution. Bonanza Mine.

Municipio de Mazapil

Mazapil: Moderate distribution. Santa Rosa, Santa Isabel, Todos Santos, and Socován de las Turquesas.

TYUYAMUNITE

$Ca(UO_2)_2(VO_4)_2 \cdot 5-8H_2O$. An uncommon secondary mineral found in limestone uranium and vanadium deposits.

CHIHAUHUA

Municipio de Aquiles Serdán

Francisco Portillo: Limited distribution.

U

UMANGITE

Cu_3Se_2. A rare mineral found in copper/selenium deposits.

DURANGO

Municipio de Mapimí

Mapimí: Limited distribution. Monterrey Mine. Associated with galena, sphalerite, bismuth and emplectite.

URALITE. See AMPHIBOLE

URANINITE

UO_2. An uncommon mineral found in pegmatites, sedimentary rocks, and hydrothermal veins.

CHIHUAHUA

Municipio de Aldama

Puerto del Aire: Moderate distribution. Noted from the La Virgen, Bolanos, La Chiva, La Judía, Puerto de Aire, La Purísima, San Blas, San Martin, and San Rafael mines. Crystals to 3 cm. Associated with gold, uran-ophane, and calcite. [Crystals are found both as cubes and octahedrals and as singles and clusters of crystals. It also has been found massive (pitchblende). Associated with gold.]

Municipio de Chihuahua

Sierra Pena Blanca: Moderate distribution. Associated with carnotite and margaritasite.

OAXACA

Municipio unknown

El Desengaño: Limited distribution. Crystals to 1 cm. Uncommon.

URANOPHANE

$Ca(UO_2)_2Si_2O_7 \cdot 5H_2O$. An uncommon secondary mineral found within uranium deposits.

CHIHUAHUA

Municipio de Aldama

Puerto del Aire: Moderate distribution. La Virgen Mine. Associated with uraninite and calcite.

UVAROVITE

$Ca_3Cr_2(SiO_4)_3$. An uncommon mineral and member of the garnet group found in metamorphic and sedimentary rock environments.

BAJA CALIFORNIA

Municipio de Ensenada

El Alamo: Limited distribution. Crystals to 2 cm.

VERACRUZ

Municipio de Profesor Rafael Ramirez

Piedras Parada: Limited distribution. Crystals to 2 cm.

UVITE

$(Ca,Na)(Mg,Fe)_3Al_5Mg(BO_3)_3Si_6O_{18}(OH,F)_4$. A common mineral and member of the tourmaline group found in all three rock types.

SONORA

Municipio de Santa Cruz

Santa Cruz: Widespread within the Cerro de Chavito. Crystals to 14 cm. Associated with dravite, quartz, and various feldspars.

V

VALENCIANITE. See **ORTHOCLASE** var. adularia

VALENTINITE

Sb_2O_3. An uncommon secondary mineral found in antimony deposits.

BAJA CALIFORNIA SUR

Municipio de San Antonio

El Triunfo: Limited distribution. Noted from the San Jose and Gobernadora mines.

CHIHUAHUA

Municipio de Uruáchic

Arechuybo: Limited distribution.

DURANGO

Municipio de Mapimí

Mapimí: Limited distribution. Ojuela Mine. Associated with stibiconite.

GUERRERO

Municipio de Huitzuco

Huitzuco de los Figueroa: Moderate distribution. Noted from the La Cruz, Sorpresa, Tumbaga, and Victoria mines. Associated with stibnite, livingstonite, and gypsum (selenite).

Municipio de Taxco

Taxco de Alarcón: Limited distribution. San Miguel del Monte Mine.

MEXICO

Municipio de Zacualpan

Zacualpan: Limited distribution. Coronas Mine.

OAXACA

Municipio de San Juan Mixtepex-Juxlahuaca

Tejocotes: Moderate distribution. Yucunani Mine. Associated with stibnite.

Municipio de San Jerónimo Taviche

San Jerónimo Taviche: Limited distribution.

QUERETARO

Municipio de Cadereyta

El Doctor: Moderate distribution. San Juan Nepomuceno Mine.

Municipio de Tolimán

Soyatal: Widespread. Santa María
de Miera and Santo Niño mines.
[This mineral has been found near
here as replacing stibnite.]

SAN LUIS POTOSI

Municipio de Catorce

Catorce: Limited distribution. La
Luz Mine. Associated with
stibnite.

Wadley: Limited distribution. San
José Mine.

Municipio de Charcas

Santa Gertrudis: Limited distribution on the Hacienda de Santa
Gertrudis.

A cluster of vanadinite crystals from San
Carlos, Chihuahua. (Modified by Shawna
Panczner from Dana/Ford, 1966)

ZACATECAS

Municipio de General
Francisco Murguía

Nieves: Limited distribution 8 km
southwest. Crystals to 8 mm.
Santa Rita Mine. Associated with
barite, jamesonite, and stibnite.

VANADINITE

$Pb_5(VO_4)_3Cl$. An uncommon mineral
and member of the apatite group,
which is found in the oxidized
ore zones of lead deposits.

CHIHUAHUA

Municipio de Ahumada

Los Lamentos: Moderate distribution [arsenatian vanadinite].
Erupción/Ahumada Mine, 5th level
to water table. Crystals to 6 cm.
Associated with descloizite,
calcite, wulfenite, and goethite.
[This mineral has been found on
the 7th level as fine brown to
brownish-red hexagonal prisms and
cavernous crystals up to 1 cm in
length. The smaller crystal

prisms are solid while the larger
crystals are usually cavernous.
Large, lustrous dark brown
crystals up to 6 cm long have
also been found on this level of
the mine. A vanadinite variety
endlichite has been found here,
but this mineral has been
reclassified as arsenatian
vanadinite. Several stopes on the
7th level were nothing except
loose spongy masses of arsenatian
vanadinite crystals. Vanadinite
has been found coating and
replacing wulfenite in some areas
of the Erupcion/Ahumada Mine.]

Municipio de Aquiles Serdán

Francisco Portillo: Limited
distribution. Santo Domingo Mine.
San Antonio el Grande: Limited
distribution. San Antonio Mine's
Cock's oreboby and tin chimney.
Yellow to orange red crystals to
4 mm. Associated with wulfenite,
mimetite, plattnerite, and
quartz.

Municipio de Camargo

Villa Rosales: Limited distribution. San Francisquito Mine.

Municipio de Coyame

Cuchillo Parado: Widespread arsenian vanadinite. Noted from the Oruro and La Aurora mines. Crystals to 2 cm. Associated with calcite and desclozite. [The La Aurora Mine has produced fine crystallized specimens of dark brown to a tanish vandinite. The crystals have been the typical hexagonal prisms to 8 mm size, but some have elongated and taper at the tops, reaching lengths up to 2 cm. These elongated crystals are a dark tan color and are what was called endlichite, but is an arsenatian vanadinite.]

Municipio de Manuel Bena- vides

San Carlos: Limited distribution. San Carlos Mine. Crystals to 6 cm. Associated with calcite. [The vanadininte is found as simple red to red-orange single hexagonal hollowed crystal prisms up to 6 cm in length and as complex groups in calcite.]

Municipio de Santa Bárbara

Santa Bárbara: Limited distribution. Associated with wulfenite.

DURANGO

Municipio de Mapimí

Mapimí: Limited distribution. Ojuela Mine. Brownish orange crystals to 8 mm. Associated with wulfenite, calcite, and goethite. [This mineral is found only where the arsenic level is extremely low or nonexistent.]

Municipio del Oro

Santa María del Oro: Limited distribution.

GUERRERO

Municipio de Taxco

Taxco de Alcarón: Limited distribution.

HIDALGO

Municipio de Cardonal

Cardonal: Limited distribution. Soledad Mine.

Municipio de Pachuca

Pachuca de Soto: Moderate distribution. Noted from the San Antonio and El Puerco mines.

Municipio de Zimapán

Zimapán: Limited distribution. Lomo del Toro Mine, San Damian stope. [The reknown mineralogist Andres Manuel del Río discovered in the late 1790s an unknown reddish mineral composed of small hexagonal "barrel" shaped crystals. He concluded from his analysis of this new mineral that it contained lead, but also something else, maybe a new element. He named the mineral, "plomo rojo de Zimapán" and sent ALL his notes, findings and specimens with his good friend, Barron von Humboldt, to Europe for further study. The Barron had been visiting Mexico from 1804 to 1808 and was returning to Europe. Unfortunately, in his transatlantic crossing during rough seas the box containing the paperwork and specimens was lost overboard. When von Humboldt wrote Del Rio of the great misfortune, Del Rio naturally was upset, especially when a few years later in Scandinavia vanadinite was rediscovered! "In 1836 Rio made a sarcastic reference to the episode in a paper that he wrote as a sort of protest against the injustice done to him, in calling the new element after the Scandinavian goddess Riita." Del Rio did not receive the credit for discovering vanadinite, but did receive the credit for discovering the element, vanadium.]

JALSICO

Municipio de Bolaños

Bolaños: Limited distribution. La Igucana Mine.

SAN LUIS POTOSI

Municipio de Catorce

Catorce: Limited distribution. Concepción Mine.

Municipio de Charcas

Charcas: Limited distribution within the Sierra de Coronados.

SINALOA

Municipio de Culiacán

Cerro de la Chiche: Limited distribution near Culiacán Rosales. Noted from the La Aurora and El Chicho mines. Reddish-brown crystals to 5 mm. Associated with descloizite.

SONORA

Municipio de Baviacora

Mazocahui: Limited distribution. Crystals to 2 mm.

ZACATECAS

Municipio de Ojo Caliente

Rancho de Encinillas: Limited distribution.

VARISCITE

$AlPO_4 \cdot 2H_2O$. An uncommon member of the variscite group, which forms by the action of surface waters on rocks containing aluminum.

QUERETARO

Municipio de Tolimán

Soyatal: Limited distribution. Santa María de Mura Mine.

SONORA

Municipio de Nacozari de García

Nacozari de García: Limited distribution.

VAUQUELINITE

$Pb_2Cu(CrO_4)(PO_4)(OH)$. A rare secondary mineral found in the oxidizied ore zone of metal deposits.

JALISCO

Municipio de Exatlán

Exatlán: Limited distribution.

NAYARIT

Municipio de La Yasca

La Yasca: Limited distribution.

VERMICULITE

$(Mg,Fe,Al)_3(Al,Si)_4O_{10}(OH)_2 \cdot 4H_2O$. An uncommon mineral formed from the alteration of other mica minerals and found in contact zones between felsic and mafic or ultra mafic rocks or in soils.

OAXACA

Municipio de San Francisco Telixtlahuaca

San Francisco Telixtlahuaca: Limited distribution. Santa Ana Mine.

VESUVIANITE

$Ca_{10}Mg_2Al_4(SiO_4)_5(Si_2O_7)_2(OH)_4$. A rather common mineral found in contact metamorphic limestones.

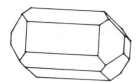

A typical crystal of vesuvianite from the Sierra de La Cruz, Coahuila. (Modified by Shawna Panczner from Sinkankas, 1964)

CHIHUAHUA

Municipio de Aquiles Serdán

San Antonio el Grande: Limited distribution. San Antonio Mine's skarn zones. Associated with grossular, anyhdrite, hedenbergite, and actinolite.

Municipio de Hidalgo del Parral

Hidalgo del Parral: Limited distribution.

COAHUILA

Municipio de Sierra Mojada

Sierra de la Cruz: Widespread. Crystals to 14 cm. Associated with grossular. [This location is often referred to as being in the state of Chihuahua at Lago de Jaco.]

DURANGO

Municipio de Cuencamé

Velardeña: Moderate distribution. Santa Juana Mine.

Municipio de Mapimí

Mapimí: Limited distribution. Ojuela Mine's skarn zones.

HIDALGO

Municipio de Zimapán

La Encarnacion: Limited distribution. San Jose del Oro Mine's skarn zone.

Zimapán: Limited distribution. Noted from the Las Estacas and San Cayetano mines.

MICHOACAN

Municipio de Zitácuaro

Suspuato: Moderate distribution within the Cerro Mazahua.

MORELOS

Municipio de Ayala

Jalostoc: Widespread throughout the Sierra Tlaycac on the Rancho de San Juan. Crystals to 6 cm. Associated with grossular and wollastonite. [This location has also been spelled in the older literature as Xalostoc.]

OAXACA

Municipio de Santo Domingo Zanatepec

Río Putla: Moderate distribution.

SONORA

Municipio de Santa Cruz

Santa Cruz: Limited distribution within the Cerro Chavito. Guadalapana Mine.

VERACRUZ

Municipio de Profesor Rafael Ramirez

Profesor Rafael Ramirez: Limited distribution. Crystals to 6 cm.

VESZELYITE

$(Cu,Zn)_3(PO_4)(OH)_3 \cdot 2H_2O$. A rare secondary mineral found in base metal deposits.

PUEBLA

Municipio de Zacapoaxtla

Zacapoaxtla: Limited distribution. La Esperanza Mine. Crystals to 2 mm.

VIOLARITE

Ni_2FeS_4. A rare mineral and member of the linnaeite group found in nickel deposits.

SINOLOA

Municipio de Alistos

Alistos: Moderate distribution. Gloria Mine.

VIVIANITE

$Fe_3(PO_4)_2 \cdot 8H_2O$. An uncommon secondary mineral and member of the vivianite group found in the weathered portions of base metal deposits and in pegmatites.

CHIHAUHAU

Municipio de Aquiles Serdán

San Antonio el Grande: Limited distribution. San Antonio Mine's 13th level. Crystals to 8 cm. Associated with ludlamite, pyrite, siderite, and galena.

[This mineral has been found on the 13th level stope as fine, water-clear green crystals up to 8 cm in length.]

MEXICO

Municipio de Sultepec

Sultepec de Pedro Ascencio de Alquisira: Limited distribution.

Municipio de Temascaltepec

Temascaltepec de Gonzáles: Limited distribution.

VOLTAITE

$K_2Fe_5Fe_4(SO_4)_{12} \cdot 18H_2O$. An uncommon secondary mineral found in ore deposits.

DURANGO

Municipio de Cuecamé

Velardeña: Limited distribution. Santa Maria mine. Associated with Krausite.

W

WAGNERITE

$(Mg,Fe)_2(PO_4)F$. An uncommon mineral. Forms in quartz veins, which are in clay-rich slates. Specimens exist but location data is not available.

WAVELLITE

$Al_3(PO_4)_2(OH,F)_3 \cdot 5H_2O$. An uncommon mineral formed by low-grade metamorphism or as a product of hydrothermal alterations.

BAJA CALIFORNIA

Municipio de Ensenada

El Rosario: Limited distribution, 58 km south of San Quintín. Evans Turquoise mine.

WEEKSITE

$K_2(UO_2)_2Si_6O_{15} \cdot 4H2O$. A rare mineral found in uranium deposits.

CHIHUAHUA

Municipio de Chihuahua

Sierra Pena Blanca: Moderate distribution. Noted from the Margaritas #1, #2, Nopal #1, #2, and #3 mines. Associated with carnotite, uraninite, and margaritasite.

WEIBULLITE

$Pb_5Bi_8(S,Se)_{17}$. A rare mineral found in metalic deposits.

CHIHUAHUA

Municipo de Janos

Janos: Limited distribution. Associated with sphalerite and calcite.

WERNERITE. See MEIONITE

WHITLOCKITE

$Ca_9(Mg,Fe)H(PO_4)_7$. An uncommon secondary mineral found in mineral deposits.

DURANGO

Municipio de Mapimí

Mapimí: Limited distribution. San Ilario mine. Yellow to yellow brown stalactites.

NUEVO LEON

Municipio de Sabinas
Hidalgo

Mercedes: Limited distribution.

WILLEMITE

Zn_2SiO_4. An uncommon mineral
found in limestones and as a
secondary mineral in the oxidized
portions of zinc deposits.

CHIHUAHUA

Municipio de Ahumada

Los Lamentos: Limited distribution. Erupción/Ahumada mine, 7th
level, Ahumada section. Crystals
to 8 mm. Associated with wulfenite. Brown hexagonal crystals
and white tuffs of acicular
crystals.

Municipio de Aquiles Serdán

San Antonio el Grande: Limited
distribution. San Antonio mine.

JALISCO

Municipio de Bolaños

Bolaños: Moderate distribution.
Noted from the Santa Fe, San
José, and Pichado mines. Associ-
ated with quartz and calcite.

SAN LUIS POTOSI

Municipio de Charcas

Charcas: Limited distribution.

WOLFRAMITE

$(Fe,Mn)WO_4$. An uncommon mineral
found in quartz veins, schists,
and pegmatites.

CHIHUAHUA

Municipio de Aquiles Serdán

Francisco Portillo: Limited
distribution.

San Antonio el Grande: Limited
distribution. San Antonio mine's
tin chimney. Associated with
fluorite, cassiterite, and
quartz.

DURANGO

Municipio de Conto de
Comonfort

Sierra de San Francisco: Limited
distribution.

SAN LUIS POTOSI

Municipio de Guadalcázar

Guadalcázar: Limited distribution.

ZACATECAS

Municipio de Villanueva

Villanueva: Limited distribution.

WOLLASTONITE

$CaSiO_3$. A common mineral and
member of the pyroxene group
found in contact metamorphic
limestones and in mineral
deposits formed by metamorphic
conditions.

BAJA CALIFORNIA

Municipio de Mexicali

San Felipe: Moderate distribution, 8 km north.

CHIAPAS

Municipio de Pichucalco

Pichucalco: Limited distribution. Santa Fe mine.

CHIHUAHUA

Municipio de Aquiles Serdán

Francisco Portillo: Limited distribution. Noted from the Mina Vieja and Purísima Mines.

Municipio de Saucillo

Naica: Moderate distribution within the Sierra de Naica.

DURANGO

Municipio de Mapimí

Mapimí: Limited distribution. Ojuela mine.

Municipio del Oro

Santa Maria del Oro: Moderate distribution. La Celestial mine.

Municipio de Poanas

Cerro de Sacrificio: Moderate distribution.

HIDALGO

Municipio de Zimapán

Zimapán: Moderate distribution.

MICHOACAN

Municipio de Zitácuaro

Susupuato: Moderate distribution within Cerro Mazahua.

MORELOS

Municipio de Ayala

Jalostoc: Moderate distribution. Massive. Associated with grossular and vesuvianite. [This mineral has been cut and polished from this location. This location has been spelled in older literature as Xalostoc.]

SAN LUIS POTOSI

Municipio de Guadalcázar

Guadalcázar: Moderate distribution.

Municipio de Villa de la Paz

La Paz: Limited distribution.

SONORA

Municipio de Hermosillo

Hermosillo: Moderate distribution. Associated with para-wollastonite.

ZACATECAS

Municipio de Concepción del Oro

Concepción del Oro: Moderate distribution.

Municipio de Sombrerete

Sombrerete: Limited distribution.

WULFENITE

$PbMO_4$. An uncommon secondary mineral found in the oxidized ore zones of lead deposits.

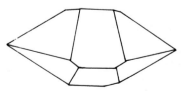

A typical wulfenite crystal from Cuchillo Parado, Chihuahua. (Modified by Shawna Panczner from Williams, 1966)

CHIHUAHUA

Municipio de Ahumada

Los Lamentos: Moderately distributed, 65 km east within the
Sierra de Los Lamentos. Erup-
ción/Ahumada mine. Crystals to 6
cm. Associated with calcite,
vanadinite, descloizite, willemite, hydrozincite, hematite,
goethite, and pyromorphite.
[This location, offically
recorded in 1907, was always an
area of unrest; first with the
Apache Indians playing havoc in
the region, then bandits, and,
last, the revolutionaries. During
the revolution, the revolutionaries under the command of Pancho
Villa in 1918 took the mine
superintendent and company
president hostage and held them
for ransom. After the ransom was
paid, Villa promised to protect
the Erupción Mining Company's
Erupción Mine from further raids
and allowed the mine to begin
operations again. Development
went slowly and soon the Company
began to run short of capitol. In
1919 the newly formed Ahumada
Lead Mining Company took over
operations and changed the name
of the mine to the Ahumada Mine.
All the old offical mine records
referred to the mine as the
Erupcion/Ahumada mine. The older
upper mine workings were referred
to as the "Erupción territory"
and the newly developing lower
workings as the "Ahumada territory." The mine adit was enlarged
and new drifts built. Even a new
niche inside the mine near the
entrance was made for the shrine
for the mine's patron saint,
Santa Cruz.
The mine operated until the 1930s
when it changed to Mexican
ownership, Compania Minera de
Plomo, S.A. and eventually ceased
operations in the early 1940s. It
has been operated on a limited
basis by a miners' cooperative
and at the present time is being
drilled and studied for possible
development and operations
again.
Wulfenite has been found in most
of the mine with the exception of
the old Erupción portion of the
mine. Wulfenite here is considered rare, but was found on
occasion in the deeper portions
of the Erupción workings and had
distinctly different crystal
morphologies. Here the miners
found small pockets of the
bipyramidal or pseudocubic
crystals of wulfenite.
Wulfenite increased in occurrence
with depth within the Erupcion-
Ahumada mine. From the 5th to 7th
levels of the mine, wulfenite is
found as orange to reddish
orange, blocky crystals up to
3 cm in size. From there on
wulfenite has been found down to
the watertable in various crystal
modifications. The color of the
wulfenite from here ranges in
shades of yellow, orange, and
brownish red. Wulfenite from Los
Lamentos is often color zoned
perpendicular to the crystals
"c" axis, thus giving an appearance of being a sandwich. The
size of the crystals has ranged
from just under 1 cm to 6 cm.]

Municipio de Aquiles Serdán

Francisco Portillo: Limited
distribution. Santo Domingo mine.

San Antonio el Grande: Limited
distribution. San Antonio Mine's
tin chimney and 5th level of the

RAMP project. Yellow crystals to
8 mm. Associated with mimetite.

Municipio de Coyame

Cuchillo Parado: Moderate
distribution. Noted from the La
Aurora, Oruro, and La Plomosas
mines. Crystals to 3 cm. Associated with goethite. [The La
Aurora Mine has been the most
prolific producer with two
different crystal morphlogies
being seen. The most common
crystal type is the small reddish
orange to brownish orange
bipyramidal crystals up to 3 cm
in length. The other crystal
type, which is less commonly
seen, is the thick tabular,
reddish orange crystals up to
1 cm thick and 3 cm on an edge.]

Municipio de Manuel Bena-
vides

San Carlos: Limited distribution.
San Carlos Mine. Associated with
calcite and vanadinite.

Municipio de Nuevo Casas
Grandes

San Pedro Corralitos: Limited
distribution within the Sierra
del Capulín. Unnamed prospect,
3 km north of San Pedro Mine.
Yellow crystals to 5cm. Associated with quartz, goethite, and
mimetite. [This location has been
previously given in the literature as the San Pedro Mine. A
major pocket of wulfenite was
uncovered in the early 1970s that
produced paper-thin yellow
crystals.]

Municipio de Santa Bárbara

Santa Bárbara: Limited distribution. Associated with vanadinite.

Municipio de Saucillo

Naica: Limited distribution.

COAHUILA

Municipio de Cuatrociénegas

Cuatrociénegas de Carranza:
Limited distribution. Reforma
Mine.

Municipio de Sierra Mojada

Sierra Mojada: Limited distribution. La Mula Mine.

DURANGO

Municipio de Mapimí

Mapimí: Widely distributed. Noted
from the Ojuela Mine. Crystals to
5 cm. Associated with calcite,
goethite, aurichalcite, cerussite, adamite, bindheimite, hemimorphite, mimetite, duftite,
litharge, and plattnerite. [Two
different crystal morphologies,
tabular and bipyramidal, have
been found at Mapimi. Crystals

have been found up to 5 cm in
size on the tubular morphology
and 2 cm long on the bipyramidal
crystal morphology. The color
ranges from colorless, through
shades of green, yellow, orange,
red, and brown. The small bright
red wulfenite crystals occasionally seen from here have been
identified as containing litharge. The yellow tabular crystals
are on mimetite and botryodial
green duftite. In 1983, a vug was
discovered of brownish orange
bipyramidal wulfenite associated
with mimetite.]

GUERRERO

Municipio de Taxco

Taxco de Alarcón: Limited
distribution.

HIDALGO

Municipio de Zimapán

Zimapan: Limited distribution.
San Pascual, Salon Grande de los
Carbonatos stope, Lomo del Toro,
and Albarradon mines. Associated
with vanadinite.

MORELOS

Municipio de Tlaquilenango

Huautla: Limited distribution.
San Francisco Mine.

SAN LUIS POTOSI

Municipio de Catorce

Catorce: Limited distribution.
Associated with phosgenite and
chlorargyrite.

One of the most common wulfenite crystal
morphology from the San Francisco Mine,
Cucurpe, Sonora. (Modified by Shawna
Panczner from Williams, 1966)

Municipio de Villa de la
Paz

La Paz: Limited distribution.
Santa María de la Paz Mine.

SINOLOA

Municipio de San Blas

Sierrita de San Blas: Limited
distribution.

SONORA

Municipio de Alamos

Alamos: Limited distribution.
Associated with leadhillite and
alamosite.

Municipio de Arizpe

Huelpai: Limited distribution.
Plumosa Mine.

Municipio de Cananea

Cananea: Limited distribution.

Municipio de Cucurpe

Cucurpe: Limited distribution
within the Sierra Prieta. San
Francisco Mine. Crystals to
13 cm. Associated with quartz
(rock crystal, amethyst), mimetite, gold, pyromorphite, galena
cerussite, and barite. [Gold was
discovered in the late 1800s and
the developing mine was named for
the mountain on which it was
located, the Sierra Prieta Mine.
It was operated first for its
gold and later for gold, silver,
and molybdenum. By 1909 the mine
was supplying its mill (100
stamps) with 720 tons of ore per
day. The mine had been developed
by eight levels and a shaft sunk
from the 8th level to develop the
downward extension of the ore
shoots. A level was developed on
the 10th level (1055 ft) and a
pump station at the 11th level
(1100 ft). The bottom of the
shaft then acted as a sump. As
the ore increased with depth, the
gold decreased and the silver and
molybdenum increased. By 1912,

with the beginning of the Mexican
Revolution, the mine closed. In
1934, Anaconda acquired the mine
and began some development work
on the 10th level (1055 ft) and
core drilled the deeper ore zone.
They eventually dropped their
lease and during the 1940s and
1950s the mine was operated by a
series of small groups of
independent mine operators. In
the mid 1980's the mine was again
operated on a small scale for its
gold and silver ore.
The exact date of the change of
the name of the mine is unknown,
but as early as the 1940s the
offical Mexican records list the
mine as the San Francisco Mine,
named after one of the two claims
it is developed upon, the San
Francisco and San Felix claims.
Wulfenite was note as early as
1912 in the mining records and
Anacanda's records in the mid
1930s also mentioned large
pockets of wulfenite in the
mine's lower levels. Even in
logs of their core drills,
wulfenite was mentioned in depth.
Wulfenite at the San Francisco
Mine has been found above the 8th
level in a few of the older gold
stopes as a dark yellow tabular
crystal up to 5 cm wide and
associated with mimetite, gold,
calcite, barite, and a manganese
oxide coating. On the 10th level
pockets of wulfenite and mimetite
have been uncovered in the
silica-rich limestones and
goethite. Not until the 1940s and
1950s, when exploration for
silver took place in this part of
the mine, did the mineral world
first see the beauty of the
wulfenite from the San Francisco
Mine. The color of the wulfenite
ranged from a yellow orange to
orange red, with the crystals
reaching upwards to just over 13
cm wide and 9 cm high! The
crystals were paper thin and
usually transparent. At times
they would show color zoning and
even phantoms. The crystals were
normally tabular with some slight
modifications. The color of the
wulfenite was also reflected in
the mimetite that would coat the
wulfenite and at times replace it
completely. Probably the most

spectacular pocket mined to date was in the early 1970s when a bright orange wulfenite was uncovered associated with spheres of firey orange red mimetite!]

Municipio de Hermosillo

Hermosillo: Limited distribution. California Mine.

Municipio de Húepac

Húepac: Limited distribution.

Municipio de Nogales

Heroica Nogales: Limited distribution. Noted from the Santa Ana and Tumacacori mines.

Municipio de Rayón

Rayón: Limited distribution. Yellow to tan crystals to 2 cm.

Municipio de Sahuaripa

Trinidad: Limited distribution. Colon Mine.

ZACATECAS

Municipio de Concepción del Oro

Concepción del Oro: Limited distribution.

Municipio de Gerneral Pánfilo Natera

General Pánfilo Natera: Limited distribution. Los Azulagues.

[This location was formerly known as El Blanca.]

Municipio de Mazapil

Mazapil: Limited distribution. Santa Rosa Mine. Yellow (colored zoned) crystals to 4 cm. Associated with mimetite and pyromorphite.

Municipio de Zacatecas

Zacatecas: Limited distribution.

WURTZITE

$(Zn,Fe)S$. A rare and unstable mineral found in the sulfide ore zone of zinc deposits.

CHIHUAHUA

Municipio de Guerrero

Guerrero: Limited distribution. La Calera Mine. Aassociated with sphalerite.

DURANGO

Municipio de Canatlán

Tejamen: Moderate distribution. Noted from the Mina la Gloria, Cia Minera Eureka, and Anexas mines.

X

XANTHOCONITE

Ag_3AsS_3. A rare mineral found in sulfide ore zones of silver deposits.

ZACATECAS

Municipio de Sombrerete

San Martín: Limited distribution. San Martín Mine. Crystals to 4 mm. Associated with stibnite.

XENOTIME

YPO_4. A rare mineral found in granitic rocks, pegmatites, and gneisses.

SAN LUIS POTOSI

Municipio de Guadalcázar

Guadalcázar: Limited distribution.

XOCOMECATLITE

$Cu_3TeO_4(OH)_4$. A rare mineral found in copper telluride deposits.

SONORA

Municipio de Moctezuma

Moctezuma: Limited distribution. Bambollita Mine. [This is the type location for this mineral.]

XONOTLITE

$Ca_6Si_6O_{17}(OH)_2$. An uncommon mineral found in contact zones.

HIDALGO

Municipio de Pachuca

Pachuca de Soto: Limited distribution. Guadalupe Mine.

Municipio unknown

Tetla de Xonotla: Limited distribution. [This is the type location for this mineral. It was named for the location.]

PUEBLA

Municipio de Tetla de Ocampo

Tetla de Ocampo: Moderate distribution. Noted from the Providencia, Del Peral, Cristo, Covadonga, and Espejeras mines. Associated with johnnsenite.

Y

YECORAITE

$Fe_3Bi_5(TeO_3)(TeO_4)_2O_9 \cdot nH_2O$.   A rare mineral found in tellurium-rich metal deposits.

SONORA

Municipio de Yécora

Yécora: Limited distribution, west of Yecora. San Martin de Porres Mine. Massive. Associated with quartz, goethite, chalcopy-rite, tetradymite, pyrite, and sphalerite. [This is the type location for this mineral named after the location.]

Z

ZAPATALITE

$Cu_3Al_4(PO_4)_3(OH)_9 \cdot 4H_2O$. A rare secondary mineral found in copper deposits.

SONORA

Municipio de Naco

La Morita: Limited distribution at Cerro Morita.

ZEMANNITE

$(Zn,Fe)_2(TeO_3)_3Na_xH_{2-x} \cdot nH_2O$. A rare mineral found in tellurium deposits.

SONORA

Municipio de Moctezuma

Moctezuma: Limited distribution. Moctezuma mine. Associated with paratellurite.

ZINCROSASITE

$(Zn,Cu)_2(CO_3)(OH)_2$. An uncommon secondary mineral and member of the rosasite mineral group found in the oxidized zone of copper-zinc deposits.

DURANGO

Municipio de Mapimí

Mapimí: Limited distribution. Ojuela mine. Associated with goethite.

ZINKENITE

$Pb_6Sb_{14}S_{27}$. An uncommon mineral found in the sulfide ore zone of lead deposits.

CHIHUAHUA

Municipio de Hidlago del Parral

Hidalgo del Parral: Limited distribution.

HIDALGO

Municipio de Zimapán

Zimapán: Limited distribution. San Francisco mine.

ZIRCON

$ZrSiO_4$. A common mineral found in all three rock types.

DURANGO

Municipio de Coneto de Comonfort

Sierra de San Francisco: Moderate distribution. Associated with cassiterite.

Municipio de Durango

Victoria de Durango: Limited distribution a Cerro de Las Remedios. Los Remedios mine.

Municipio de Mapimí

Mapimí: Limited distribution. Ojuela mine.

GUANAJUATO

Municipio de Santa Catarina

Santa Catarina: Limited distribution. El Santín mine. Crystals to 2 cm.

HIDALGO

Municipio de Zimapán

Zimapán: Moderate distribution.

OAXACA

Municipio de La Pe

La Panchita: Moderate distribution, west of Ayoquiztco. La Panchita Mine. Red crystals to 12 cm. Associated with melonite and augite.

Municipio de San Francisco Telixtlahuaca

Huitzo: Limited distribution. El Muerto mine. Cystals to 3 cm. Associated with fluorite. [Zircon is found here both in deep red and blue colors with some of the crystals being extremely gemmy.]

Municipio de Pluma Hidalgo

Pluma Hidalgo: Limited distribution.

SAN LUIS POTOSI

Municipio de Guadalcázar

Guadalcázar: Moderate distribution.

ZACATECAS

Municipio de Saín Alto

Sierra de Chapultepec: Moderate distribution, 25 km south of Saín Alto.

ZOISITE

$Ca_2Al_3(SiO_4)_3(OH)$. A common mineral and member of the epidote mineral group found in contact and thermal metamorphic environments.

AGUASCALIENTES

Municipio de Tepezalá

Tepezalá: Moderate distribution.

BAJA CALIFORNIA

Municipio de Ensenada

Juarez: Limited distribution. Trace mine.

CHIHUAHUA

Municipio Santa Bárbara

Santa Bárbara: Moderate distribution.

Municipio de Saucillo

Naica: Moderate distribution within the Sierra de Naica.

HIDALGO

Municipio de Zimapán

Zimapán: Limited distribution. San Francisco mine.

OAXACA

Municipio de San Francisco Tellixtlahuaca

Rancho La Carbonera: Limited distribution.

SAN LUIS POTOSI

Municipio de Villa de la Paz

La Paz: Moderate distribution.

SONORA

Municipio de Cananea

Cananea: Limited distribution.

APPENDIX

MEXICAN STATES, COUNTIES (municipios), AND COUNTY SEATS (cabeceras)

Note: All data used in this list are based on 1980 statistics and are arranged alphabetically using the Spanish alphabet.

AGUASCALIENTES

Capital: Aguascalientes
Population: 503,410

State abbreviation: Ags
Area: 5,471 sq km

County	County Seat
Aguascalientes	Aguascalientes
Asientos	Asientos
Calvillo	Calvillo
Cosío	Cosío
Jesús María	Jesús María
Pabellón de Arteaga	Pabellón
Rincón de Ramos	Rincón de Ramos
San José de García	San José de García
Tepezalá	Tepezalá

BAJA CALIFORNIA

Capital: Mexicali
Population: 1,225,436

State abbreviation: BCN
Area: 69,921 sq km

County	County Seat
Ensenada	Ensenada
Mexicali	Mexicali
Tecate	Tecate
Tijuana	Tijuana

BAJA CALIFORNIA SUR

Capital: La Paz
Population: 221,389

State abbreviation: BCS
Area: 73,475 sq km

County	County Seat
Cabos, Los	San José del Cabo
Comondú	Comondú
Mulegé	Santa Rosalía
Paz, La	La Paz

CAMPECHE

Capital: Campeche
Population: 188,004

State abbreviation: Camp
Area: 50,812 sq km

County	County Seat
Calkiní	Calkiní
Campeche	Campeche
Carmen	Ciudad del Carmen
Champotón	Champotón
Hecelchakán	Hecelchakán
Hopelchén	Hopelchén
Palizada	Palizada
Tenabo	Tenabo

COAHUILA de ZARAGOZA

Capital: Saltillo
Population: 1,558,401

State abbreviation: Coah
Area: 149,982 sq km

County	County Seat
Abasolo	Abasolo
Acuña	Villa Acuña
Allende	Allende
Arteaga	Arteaga
Candela	Candela
Castaños	Castaños
Cuatrociénegas	Cuatrociénegas de Carranza
Escobedo	Escobedo
Francisco I. Madero	Francisco I. Madero
Frontera	Villa Frontera
General Cepeda	General Cepeda
Guerrero	Guerrero
Hidalgo	Hidalgo
Jiménez	Jiménez
Juárez	Juárez
Lamadrid	Lamadrid
Matamoros	Matamoros
Monclova	Monclova
Morelos	Morelos

Múzquiz	Melchor Múzquiz
Nadadores	Nadadores
Nava	Nava
Ocampo	Ocampo
Parras	Parras de la Fuente
Piedras Negras	Piedras Negras
Progreso	Pregreso
Ramos Arizpe	Ramos Arizpe
Sabinas	Sabinas
Sacramento	Sacramento
Saltillo	Saltillo
San Buenaventura	San Buenaventura
San Juan de Sabinas	San Juan de Sabinas
San Pedro	San Pedro de las Colonias
Sierra Mojada	Sierra Mojada
Torreón	Torreón
Viesca	Viesca
Villa Unión	Villa Unión
Zaragoza	Zaragoza

COLIMA

Capital:	Colima		*State abbreviation:*	Col
Population:	339,202		*Area:*	5,191 sq km

County	**County Seat**
Armería	Armería
Colima	Colima
Comala	Comala
Coquimatlán	Coquimatlán
Cuauhtémoc	Cuauhtémoc
Ixtlahuacán	Ixtlahuacán
Manzanillo	Manzanillo
Minatitlán	Minatitlán
Tecomán	Tecomán
Villa de Alvarez	Villa de Alvarez

CHIAPAS

Capital:	Tuxtla Gutierrez		*State abbreviation:*	Chis
Population:	2,096,812		*Area:*	74,211 sq km

County	**County Seat**
Acacoyagua	Acacoyagua
Acala	Acala
Acapetagua	Acapetagua
Altamirano	Altamirano
Amatán	Amatán
Amatenango de la Frontera	Amatenango de la Frontera
Amatenango de Valle	Amatenango de Valle
Angel Albino Corzo	Angel Albino Corzo
Arriaga	Arriaga
Bejucal de Ocampo	Bejucal de Ocampo
Bella Vista	Bella Vista

Berriozábal	Berriozábal
Bochil	Bochil
Bosque, El	El Bosque
Cacahoatán	Cacahoatán
Catazajá	Catazajá
Cintalapa	Cintalapa de Figueroa
Coapilla	Coapilla
Comitán de Domínguez	Comitán de Domínguez
Concordia, La	La Concordia
Copainalá	Copainalá
Chalchihuitán	Chalchihuitán
Chamula	Chamula
Chanal	Chanal
Chapultenango	Chapultenango
Chenalhó	Chenalhó
Chiapa de Corzo	Chiapa de Corzo
Chiapilla	Chiapilla
Chicoasén	Chicoasén
Chicomuselo	Chicomuselo
Chilón	Chilón
Esuintla	Escuintla
Francisco León	Francisco León
Frontera Comalapa	Frontera Comalapa
Frontera Hidalgo	Frontera Hidalgo
Grandeza, La	La Grandeza
Huehuetán	Huehuetán
Huistán	Huistán
Huitiupan	Huitiupan
Huixtla	Huixtla
Independencia, La	La Independencia
Ishuatán	Ishuatán
Ixtacomitán	Ixtacomitán
Ixtapa	Ixtapa
Ixtapangajoya	Ixtapangajoya
Jiquilpilas	Jiquilpilas
Jitotol	Jitotol
Juárez	Juárez
Larrainzar	Larrainzar
Libertad, La	La Libertad
Mapastepec	Mapastepec
Margaritas, Las	Las Margaritas
Mazapa de Medero	Mazapa
Mazatán	Mazatán
Metapa	Metapa de Domínguez
Mitonic	Mitonic
Motozintla	Motozintla de Mendoza
Nicolás Ruiz	Nicolás Ruiz
Ocosingo	Ocosingo
Ocotepec	Ocotepec
Ocozocuautla de Espinosa	Ocozocuautla de Espinosa
Ostuacán	Ostuacán
Osumacinta	Osumacinta
Oxchuc	Oxchuc
Palenque	Palenque
Pantelhó	Pantelhó
Pantepec	Pantepec
Pichucalco	Pichucalco
Pijijiapan	Pijijiapan
Porvenir, El	El Porvenir
Pueblo Nuevo Comaltitlán	Pueblo Nuevo Comaltitlán

Pueblo Nuevo Solistahuacán	Pueblo Neuvo Solistahuacán
Rayón	Rayón
Reforma	Reforma
Rosas, Las	Las Rosas
Sabanilla	Sabanilla
Salto de Agua	Salto de Agua
San Cristóbal de las Casas	San Cristóbal de las Casas
San Fernando	San Fernando
Siltepec	Siltepec
Simojovel de Allende	Simojovel de Allende
Sitalá	Sitalá
Socoltenango	Socoltenango
Solusuchiapa	Solusuchiapa
Soyaló	Soyaló
Suchiapa	Suchiapa
Suchhate	Ciudad Hidalgo
Sunuapa	Sunuapa
Tapachula	Tapachula
Tapalapa	Tapalapa
Tapilula	Tapilula
Tecpatán	Tecpatán
Tenejapa	Tenejapa
Teopisca	Teopisca
Terán	Terán
Tila	Tila
Tonalá	Tonalá
Tololapa	Totolapa
Trinitaria, La	La Trinitaria
Tumbalá	Tumbalá
Tuxtla	Tuxtla Gutiérrez
Tuxtla Chico	Tuxtla Chico
Tuzantán	Tuzantán
Tzimol	Tzimol
Unión Juárez	Unión Juárez
Venustiano Carranza	Venustiano Carranza
Villa Corzo	Villa Corzo
Villa Flores	Villa Flores
Yajalón	Yajalón
Zapotal, El	El Zapotal
Zinacantán	Zinacantán

CHIHUAHUA

Capital: Chihuahua
Population: 1,933,856

State abbreviation: Chih
Area: 244,938 sq km

County	**County Seat**
Ahumada	Ahumada
Aldama	Aldama
Allende	Valle de Allende
Aquiles Serdán	Aquiles Serdán
Ascensión	Ascensión
Bachíniva	Bachíniva
Balleza	Balleza
Batopilas	Batopilas
Bocoyna	Bocoyna

Buenaventura
Camargo
Carichic
Casas Grandes
Coronado
Coyame
Cruz, La
Cuauhtémoc
Cusihuiriáchic
Chihuahua
Chínipas
Delicias
Dr. Belisario Domínguez
Galeana
General Trías
Gómez Farías
Gran Morelos
Guachochi
Guadalupe
Guadalupe y Calvo
Guazapares
Guerrero
Hidalgo del Parral
Huejotitán
Ignacio Zaragosa
Janos
Jiménez
Juárez
Julimes
López
Madera
Maguarichic
Manuel Benavides
Matachic
Matamoros
Meoqui
Morelos
Moris
Namiquipa
Nonoava
Nuevo Casas Grandes
Ocampo
Ojinaga
Praxedis G. Guerrero
Riva Palacio
Rosales
Rosario
San Francisco de Borja
San Francisco de Conchos
San Francisco del Oro
Santa Bárbara
Satevó
Saucillo
Temósachic
Tule, El
Urique
Uruáchic
Valle de Zaragoza

Buenaventura
Ciudad Camargo
Carichic
Casas Grandes
Coronado
Coyame
La Cruz
Cuauhtémoc
Cusihuiriáchic
Chihuahua
Chínipas
Delicias
Dr. Belisario Domínguez
Galeana
General Trías
Gómez Farías
Gran Morelos
Guachochi
Guadalupe
Guadalupe y Calvo
Témoris
Guerrero
Hidalgo del Parral
Huejotitán
Ignacio Zaragosa
Janos
Jiménez
Juárez
Julimes
López
Madera
Maguarichic
Manuel Benavides
Matachic
Matamoros
Meoqui
Morelos
Moris
Namiquipa
Nonoava
Nuevo Casas Grandes
Ocampo
Ojinaga
Praxedis G. Guerrero
Riva Palacio
Rosales
Rosario
San Francisco de Borja
San Francisco de Conchos
San Francisco del Oro
Santa Bárbara
Satevó
Saucillo
Temósachic
El Tule
Urique
Uruáchic
Valle de Zaragoza

DISTRITO FEDERAL o CUIDAD de MEXICO

Capital: Cuidad de México
Population: 9,373,353

District abbreviation: DF
Area: 1,479 sq km

Districts

Benito Juárez
Cuauhtémoc
Miguel Hidalgo
Venustiano Carranza
Azapotzalco
Coyoacán
Cuajimalpa
Gustavo A. Madero
Ixacalco
Ixtapalapa
Magdalena Contreras, La
Milpa Alta
Obregón
Tláhuac
Tlalpan
Xochimilco

District Seat

The first four districts are
divided within the City of
México.

Azapotzalco
Coyoacán
Cuajimalpa
Villa de Guadalupe Hidalgo
Ixacalco
Ixtapalapa
La Magdalena Contreras
Milpa Alta
Villa Obregón
Tláhuac
Tlalpan
Xochimilco

DURANGO

Capital: Victoria de Durango
Population: 1,160,196

State abbreviation: Dgo
Area: 123,181 sq km

County

Canatlán
Canelas
Coneto de Commonfort
Cuencamé
Durango
General Simón Bolívar
Gómez Palacio
Guadalupe Victoria
Guanaceví
Hidalgo
Indé
Lerdo
Mapimí
Mezquital
Nazas
Nombre de Dios
Ocampo
Oro, El
Otáez
Pánuco de Coronado
Peñón Blanco
Poanas
Pueblo Nuevo
Rodeo
San Bernardo

County Seat

Canatlán
Canelas
Coneto de Comonfort
Cuencamé de Ceniceros
Victoria de Durango
General Simón Bolívar
Gómez Palacio
Guadalupe Victoria
Guanaceví
Villa Hidalgo
Indé
Cuidad Lerdo
Mapimí
Mezquital
Nazas
Nombre de Dios
Villa Ocampo
Santa María del Oro
Santa María Otáez
Francisco I. Madero
Peñón Blanco
Villa Unión
El Salto
Rodeo
San Bernardo

San Dimas	Tayoltita
San Juan de Guadalupe	San Juan de Guadalupe
San Juan del Río	San Juan del Río, Cuna del Centauro del Norte
San Luis del Cordero	San Luis del Cordero
San Pedro del Gallo	San Pedro del Gallo
Santa Clara	Santa Clara
Santiago Papasquiaro	Santiago Papasquiaro
Súchil	Súchil
Tamazula	Tamazula
Tepehuanes	Santa Catarina de Tepehuanes
Tlahualilo de Zaragoza	Tlahualito de Zaragoza
Topia	Topia
Vicente Guerrero	Vicente Guerrero

GUANAJUATO

Capital:	Guanajuato	*State abbreviation:*	Gto
Population:	3,044,402	*Area:*	30,491 sq km

County	**County Seat**
Abasolo	Abasolo
Acámbaro	Acámbaro
Allende	San Miguel de Allende
Apaseo el Alto	Villa de Apaeso el Alto
Apaseo el Grande	Apaseo el Grande
Atarjea	Atarjea
Celaya	Celaya
Ciudad Manuel Doblado	Ciudad Manuel Doblado
Comonfort	Comonfort
Coroneo	Coroneo
Cortazar	Cortazar
Cuerámaro	Cuerámaro
Doctor Mora	Villa Doctor Mora
Dolores Hidalgo	Ciudad de Dolores Hidalgo, Cuna de la Independencia Nacional
Guanajuato	Guanajuato
Huanímaro	Huanímaro
Irapuato	Irapuato
Jaral del Progreso	Jaral del Progreso
Jarécuaro	Jarécuaro
León	León
Moroleón	Moroleón
Ocampo	Ocampo
Pénjamo	Pénjamo
Pueblo Nuevo	Pueblo Nuevo
Purísima del Rincón	Purísima del Rincón o de Bustos
Romita	Romita
Salamanca	Salamanca
Salvatierra	Salvatierra
San Diego de la Unión	San Diego de la Unión
San Felipe	San Felipe
San Francisco del Rincón	San Francisco del Rincón
San José Iturbide	San José Iturbide
San Luis de la Paz	San Luis de la Paz
Santa Catarina	Santa Catarina
Santa Cruz de Juventino Rosas	Santa Cruz de Juventino Rosas
Santiago Maravatío	Santiago Maravatío

Silao
Tarandacuao
Tarímoro
Tierrablanca
Uriangato
Valle de Santiago
Victoria
Villagrán
Xichú
Yuriria

Silao
Tarandacuao
Tarímoro
Tierrablanca
Uriangato
Valle de Santiago
Victoria
Villagrán
Xichú
Yuriria

GUERRERO

Capital: Chilpancingo de los Bravo
Population: 2,174,162

State abbreviation: Gro
Area: 64,281 sq km

County	County Seat
Acapulco	Acapulco de Juárez
Ahuacuotzingo	Ahuacuotzingo
Ajuchitlán	Ajuchitlán del Progresso
Alcozauca	Alcozauca de Guerrero
Alpoyeca	Alpoyeca
Apaxtla	Apaxtla de Castrejón
Arcelia	Arcelia
Atenango del Río	Atenango del Río
Atlamajalcingo del Monte	Atlamajalcingo del Monte
Atlixtac	Atlixtac
Atoyac de Alvarez	Atoyac de Alvarez
Ayutla	Ayutla de los Libres
Azoyú	Azoyú
Benito Juárez	San Jerónimo de Juárez
Buenavista de Cuéllar	Buenavista de Cuéllar
Coahuayutla	Coahuayutla de Guerrero
Cocula	Cocula
Copala	Copala
Copalillo	Copalillo
Copanatoyac	Copanatoyac
Coyuca de Benítez	Coyuca de Benítez
Coyuca de Catalán	Coyuca de Catalán
Cuajinicuilapa	Cuajinicuilapa
Cualac	Cualac
Cuautepec	Cuautepec
Cuetzala	Cuetzala del Progreso
Cutzamala de Pinzón	Cutzamala de Pinzón
Chilapa	Chilapa de Alvarez
Chilpancingo	Chilpancingo de los Bravo
Florencio Villarreal	Cruz Grande
General Canuto A. Neri	Acapetlahuaya
General Heliodoro Castillo	Tlacotepec
Huamuxtitlán	Huamuxtitlán
Huitzuco	Huitzuco de la Figueroa
Iguala	Iguala de la Independencia
Igualapa	Igualapa
Ixcateopan	Ixcateopan de Cuauhtémoc
José Azueta	Zihuatanejo
Juan R. Escudero	Tierra Colorada
Leonardo Bravo	Chichihualco

Malinaltepec	Malinaltepec
Mártir de Cuilapan	Apango
Metlatonoc	Metlatonoc
Mochitlán	Mochitlán
Olinalá	Olinalá
Ometepec	Ometepec
Pedro Ascencio Alquisiras	Ixcapuzalco
Petatlán	Petatlán
Pilcaya	Pilcaya
Pungarabato	Ciudad Altamirano
Quechultenango	Quechultenango
San Luis Acatlán	San Luis Acatlán
San Marcos	San Marcos
San Miguel Totolapan	San Miguel Totolapan
Taxco	Taxco de Alarcón
Tecoanapa	Tecoanapa
Tecpan	Tecpan de Galeana
Teloloapan	Teloloapan
Tepecoacuilco	Tepecoacuilco de Trujano
Tetipac	Tetipac
Tixtla	Tixtla de Guerrero
Tlacoachistlahuaca	Tlacoachistlahuaca
Tlacoapa	Tlacoapa
Tlalchapa	Tlalchapa
Tlalixtaquilla	Tlalixtaquilla
Tlapa	Tlapa de Comonfort
Tlapehuala	Tlapehuala
Unión, La	La Unión
Xalpatláhuac	Xalpatláhuac
Xochihuehuetlán	Xochihuehuetlán
Xochistlahuaca	Xochistlahuaca
Zapotitlán Tablas	Zapotitlán Tablas
Zirándaro	Zirándaro
Zitlala	Zitlala
Zumpango del Río	Zumpango del Río

HIDALGO

Capital: Pachuca de Soto   *State abbreviation:* Hgo
Population: 1,516,511   *Area:* 20,813 sq km

County	County Seat
Acatlán	Acatlán
Acaxochitlán	Acoxochitlán
Actopan	Actopan
Agua Blanca Iturbide	Agua Blanca Iturbide
Ajacuba	Ajacuba
Alfajayucán	Alfajayucán
Almoloya	Almoloya
Apan	Apan
Arenal, El	El Arenal
Atitalaquia	Atitalaquia
Atlapexco	Atlapexco
Atotonilco el Grande	Atotonilco el Grande
Atotonilco Tula	Atotonilco Tula
Calnali	Calnali
Cardonal	Cardonal

Cuautepec	Cuautepec de Hinojosa
Chapantongo	Chapantongo
Chapulhuacán	Chapulhuacán
Chilcuautla	Chilcuautla
Eloxochitlán	Eloxochitlán
Emiliano Zapata	Emiliano Zapata
Epazoyucan	Epazoyucan
Francisco I. Madero	Tepatepec
Huasca	Huasca de Ocampo
Huautla	Huautla
Huazalingo	Huazalingo
Huehuetla	Huehuetla
Huejutla de Reyes	Huejuetla de Reyes
Huichapan	Huichapan
Ixmiquilpan	Ixmiquilpan
Jacala	Jacala
Jaltocán	Jaltocán
Juárez Hidalgo	Juárez
Lolotla	Lolotla
Metepec	Metepec
Metzquititlán	Metzquititlán
Metztitlán	Metztitlán
Mineral de Chico	Mineral de Chico
Mineral del Monte	Mineral del Monte
Misión, La	La Misión
Mixquiahuala	Mixquiahuala
Molango	Molango
Nicolás Flores	Nicolás Flores
Nopala	Nopala
Omitlán de Juárez	Omitlán de Juárez
Orizatlán	Orizatalán
Pacula	Pacula
Pachuca	Pachuca de Soto
Pisaflores	Pisaflores
Progreso	Progreso de Obregón
Reforma, La	Pachuquilla
San Agustín Tlaxiaca	San Agustín Tlaxiaca
San Bartolo Tutotepec	San Bartolo Tutotepec
San Salvador	San Salvador
Santiago	Santiago de Anaya
Santiago Tulantepec	Santiago Tulantepec
Singuilucan	Singuilucan
Tasquillo	Tasquillo
Tecozautla	Teozautla
Tenango de Doria	Tenango de Doria
Tepeapulco	Tepeapulco
Tepehuacán de Guerrero	Tepehuacán de Guerrero
Tepeji del Río	Tepeji del Río
Tepetitlán	Tepetitlán
Tetepango	Tetepango
Tezontepec	Tezontepec
Tezontepec de Aldama	Tezontepec de Aldama
Tianguistengo	Tianguistengo
Tizayuca	Tizayuca
Tlahuelilpan	Tlahualilpan de Ocampo
Tlahuiltepa	Tlahuiltepa
Tlanalapan	Tlanalapan
Tlanchinol	Tlanchinol
Tlaxcoapan	Tlaxcoapan
Tolcayuca	Tolcayuca

Tula de Allende	Tula de Allende
Tulancingo	Tulancingo
Xochiatipan	Xochiatipan
Xochicoatlán	Xochicoatlán
Yahualica	Yahualica
Zacualtipán	Zacualtipán
Zapotlán de Juárez	Zapotlán de Juárez
Zempoala	Zempoala
Zimapán	Zimapán

JALISCO

Capital:	Guadalajara	*State abbreviation:*	Jal
Population:	4,293,549	*Area:*	80,836 sq km

County	**County Seat**
Acatic	Acatic
Acatlán de Juárez	Acatlán de Juárez
Ahualulco de Mercado	Ahualulco de Mercado
Amacueca	Amacueca
Amatitán	Amatitán
Ameca	Ameca
Antonio Escobedo	Antonio Escobedo
Arandas	Arandas
Arenal, El	El Arenal
Atemajac de Brizuela	Atemajac de Brizuela
Atengo	Atengo
Atenguillo	Atenguillo
Atotonilco	Atotonilco el Alto
Atoyac	Atoyac
Autlán	Autlán de Navarro
Ayo el Chico	Ayo el Chico
Ayutla	Ayutla
Barca, La	La Barca
Bolaños	Bolaños
Cabo Corrientes	El Tuito
Casimiro Castillo	Casimiro Castillo
Cihuatlán	Cihuatlán
Ciudad Guzmán	Ciudad Guzmán
Cocula	Cocula
Colotlán	Colotlán
Concepción de Buenos Aires	Concepción de Buenos Aires
Cuautitlán	Cuautitlán
Cuautla	Cuautla
Cuquío	Cuquío
Chapala	Chapala
Chimaltitán	Chimaltitán
Chiquilistlán	Chiquilistlán
Degollado	Degollado
Ejutla	Ejutla
Encarnación de Díaz	Encarnación de Díaz
Etzatlán	Etzatlán
Gómez Farias*	Gómez Farias
Grullo, El	El Grullo
Guachinango	Guachinango
Guadalajara	Guadalajara

Hostoipaquillo	Hostoipaquillo
Huejúcar	Huejúcar
Huejuquilla el Alto	Huejuquilla el Alto
Huerta, La	La Huerta
Ixtlahuacán de los Membrillos	Ixtlahuacán de los Membrillos
Ixtlahuacán del Río	Ixtlahuacán del Río
Jalostotitlán	Jalostotitlán
Jamay	Jamay
Jesús María	Jesús María
Jilotlán de los Dolores	Jilotlán de los Dolores
Jocotepec	Jocotepec
Juanacatlán	Juanacatlán
Juchitlán	Juchitlán
Lagos de Moreno	Lagos de Moreno
Limón, El	El Limón
Magdalena	Magdalena
Manuel M. Diéguez	Manuel M. Diéguez
Manzanilla de la Paz	La Manzanilla
Mascota	Mascota
Mazamitla	Mazamitla
Mexticacán	Mexticacán
Mezquitic	Mezquitic
Mixtlán	Mixtlán
Ocotlán	Ocotlán
Ojuelos de Jalisco	Ojuelos de Jalisco
Pihuamo	Pihuamo
Poncitlán	Poncitlán
Puerto Vallarta	Puerto Vallarta
Purificación	Purificación
Quitupan	Quitupan
Salto, El	El Salto
San Cristóbal de la Barranca	San Cristóbal de la Barranca
San Diego de Alejandría	San Diego de Alejandría
San Juan de los Lagos	San Juan de los Lagos
San Julián	San Julián
San Marcos	San Marcos
San Martín de Bolaños	San Martín de Bolaños
San Martín Hidalgo	San Martín Hidalgo
San Miguel el Alto	San Miguel el Alto
San Sebastián	San Sebastián
Santa María de los Angeles	Santa María de los Angeles
Sayula	Sayula
Tala	Tala
Talpa de Allende	Talpa de Allende
Tamazula de Gordiano	Tamazaula de Goriano
Tapalpa	Tapalpa
Tecalitlán	Tecalitlán
Tecolotlán	Tecolotlán
Tachaluta	Tachaluta
Tenamaxtlán	Tenamaxtlán
Teocaltiche	Teocaltiche
Teocuitatlán de Corona	Teocuitatlán de Corona
Tepatitlán de Morelos	Tepatitlán de Morelos
Tequila	Tequila
Teuchitlán	Teuchitlán
Tizapán el Alto	Tizapán el Alto
Tlajomulco	Tlajomulco de Zuniga
Tlaquepaque	Tlaquepaque
Tolimán	Tolimán

Tomatlán	Tomatlán
Tonalá	Tonalá
Totatiche	Totatiche
Tototlán	Tototlán
Tuxcacuesco	Tuxcacuesco
Tuxcueca	Tuxcueca
Tuxpan	Tuxpan
Unión de San Antonio	Unión de San Antonio
Unión de Tula	Unión de Tula
Valle de Guadalupe	Valle de Guadalupe
Valle de Juárez	Valle de Juárez
Venustiano Carranza	Venustiano Carranza
Villa Corona	Villa Corona
Villa Guerrero	Villa Guerrero
Villa Hidalgo	Villa Hidalgo
Villa Obregón	Villa Obregón
Yahualica de González Gallo	Yahualica de González Gallo
Zacoalco de Torres	Zacoalco de Torres
Zapopan	Zapopan
Zapotiltic	Zapotiltic
Zapotitlán de Vadillo	Zapotitlán
Zapotlán del Rey	Zapotlán del Rey
Zapotlanejo	Zapotlanejo

*This has also been listed as the County of San Sebastián.

MEXICO

Capital:	Toluca de Lerdo	*State abbreviation:*	Mex
Population:	7,545,692	*Area:*	21,355 sq km

County	County Seat
Acambay	Acambay
Acolman	Acolman de Netzahualcóyotl
Aculco	Aculco de Espinosa
Almoloya de Alquisiras	Almoloya de Alquisiras
Almoloya de Juárez	Almoloya de Juárez
Almoloya de Río	Almoloya de Río
Amanalco	Amanalco de Becerra
Amatepec	Amatepec
Amecameca	Amecameca de Juárez
Apaxco	Apaxco de Ocampo
Atenco	San Salvador Atenco
Atizapán	Atizapán
Atlacomulco	Atlacomulco de Fabela
Atlautla	San Miguel Atautla
Axapusco	Axapusco
Ayapango de Gabriel Ramos Millán	Ayapango
Calimaya	Calimaya de Díaz González
Capulhuac	Capulhuac de Mirafuentes
Coacalco	Coacalco de Berriozábal
Coatepec Harinas	Coatepec Harinas
Coctitlán	Cocotitlán
Coyotepec	Coyotepec
Cuautitlán	Cuautitlán
Cuautitlán Izcali	Cuautitlán Izcali

Chalco	Chalco de Díaz Covarrubias
Chapa de Mota	Chapa de Mota
Chapultepec	Chalpultepec
Chiautla	Chiautla
Chicolopan	Chicolopan de Juárez
Chiconcuac	Chiconcuac de Juárez
Chimalhuacán	Chimalhuacán
Donato Guerra	Asunción Donato Guerra
Ecatepec	Ecatepec de Morelos
Ecatzingo	Ecatzingo de Hidalgo
Huehuetoca	Huehuetoca
Hueypoxtla	Hueypoxtla
Huixquilucan	Huixquilucan de Degolado
Isidro Fabela	Tlazala de Fabela
Ixtapaluca	Ixtapaluca
Ixtapan de la Sal	Ixtapan de la Sal
Iztapan del Oro	Iztapan del Oro
Ixtlahuca	Ixtlahuaca de Rayón
Jalatlaco	Jalatlaco
Jaltenco	Jaltenco
Jilotepec	Jilotepec de Abasolo
Jilotzingo	Santa Ana Jilotzingo
Jiquipilco	Jiquipilco
Jocotitlán	Jocotitlán
Joquicingo	Joquicingo
Juchitepec	Juchitepec de Mariano Riva Palacio
Lerma	Lerma de Villada
Malinalco	Malinalco
Melchor Ocampo	Melchor Ocampo
Metepec	Metepec
Mexicalcingo	Mexicalcingo
Morelos	San Bertolo Morelos
Naucalpan	Naucalpan de Juárez
Netzahualcóyotl	Ciudad Netzahualcóyotl
Nextlapan	Nextlapan
Nicolás Romero	Nicolás Romero
Nopaltepec	Nopaltepec
Ocoyoacac	Ocoyoacac
Ocuilan	Ocuilan de Arteaga
Oro, El	El Oro de Hidalgo
Otumba	Otumba de Gómez Farías
Otzoloapan	Otzoloapan
Otzolotepec	Villa Cuauhtémoc
Ozumba	Ozumba de Alzate
Papalotla	Papalotla
Paz, La	Los Reyes
Polotitlán	Polotitlán de la Ilustración
Rayón	Rayón
San Antonio la Isla	San Antonio la Isla
San Felipe del Progreso	San Felipe del Progreso
San Martín de las Pirámides	San Martín de las Pirámides
San Mateo Atenco	San Mateo Atenco
San Simón de Guerrero	San Simón de Guerrero
Santo Tomás de los Plátanos	Nuevo Santo Tomás de los Plátanos
Soyaniquilpan	Soyaniquilpan
Sultepec	Sultepec de Pedro Ascencio de Alquisiras
Tecámac	Tecámac de Felipe Villanueva
Tejupilco	Tejupilco de Hidalgo
Temamatla	Temamatla

Temascalapa	Temascalapa
Temascalcingo	Temascalcingo de José María Velasco
Temascaltepec	Temascaltepec de Gonzalez
Temoaya	Temoaya
Tenancingo	Tenancingo de Degollado
Tenango del Aire	Tenango de Tepopula
Tenango del Valle	Tenango de Arista
Teoloyucan	Teoloyucan
Teotihuacan	Teotihuacan de Arista
Tepetlaoxtoc	Tepetlaoxtoc de Hidalgo
Tepetlixpa	Tepetlixpa
Tepotzotlán	Tepotzotlán
Tequixquiac	Tequixquiac
Texcaltitlán	Texcaltitlán
Texcalyacac	Texcalyacac
Texcoco	Texcoco de Mora
Tezoyuca	Tezoyuca
Tianguistenco	Tianguistenco de Galeana
Timilpan	Timilpan
Tlalmanalco	Tlalmanalco de Velázquez
Tlalnepantla	Tlalnepantla de Comonfort
Tlatlaya	Tlatlaya
Toluca	Toluca de Lerado
Tonatico	Tonatico
Tultepec	Tultepec
Tultilán	Tultilán de Mariano Escobedo
Valle de Bravo	Valle de Bravo
Villa de Allende	San José Allende
Villa del Carbón	Villa del Carbón
Villa Guerrero	Villa Guerrero
Villa Victoria	Villa Victoria
Xonacatlán	Xonacatlán
Zacazonapan	Zacazonapan
Zacualpan	Zacualpan
Zaragoza	Cuidad López Mateos
Zinacantepec	San Miguel Zinacantepec
Zumpahuacán	Zumpahuacán
Zumpango	Zumpango de Ocampo

MICHOACAN de OCAMPO

Capital: Morelia
Population: 3,048,704

State abbreviation: Mich
Area: 59,928 sq km

County	County Seat
Acuitzio	Acuitzio del Canje
Aguililla	Aguililla
Alvaro Obregón	Alvaro Obregón
Angamacutiro	Angamacutiro de la Unión
Angangueo	Mineral de Angangueo
Apatzingán	Apatzingán de la Constitución
Aporo	Aporo
Aquila	Aquila
Ario	Ario de Rosales

Arteaga	Arteaga
Briseñas de Matamoros	Briseñas
Buenavista	Buenavista Tomatlán
Carácuaro	Carácuaro de Morelos
Coahuayana	Coahuayana
Coalcomán	Coalcomán de Matamoros
Coeneo	Coeneco de la Libertad
Contepec	Contepec
Copándaro de Galeana	Copándaro de Galeana
Cotija	Cotija de la Paz
Cuitzeo	Cuitzeo del Porvenir
Charapán	Charapán
Charo	Charo
Chavinda	Chavinda
Cherán	Cherán
Chilchota	Chilchota
Chinicuila	Villa Victoria
Chucándiro	Chucándiro
Churintzio	Churintzio
Churumuco	Churubusco de Morelos
Ecuandureo	Ecuandureo
Epitacio Huerta	Epitacio Huerta
Erongarícuaro	Erongarícuaro
Gabriel Zamora	Gabriel Zamora
Hidalgo	Ciudad Hidalgo
Huacana, La	La Huacana
Huandacareo	Huandacareo
Huaniqueo	Huaniqueo de Morales
Huetamo	Huetamo de Núñez
Huiramba	Huiramba
Indapareo	Indapareo
Irimbo	Irimbo
Ixtlán	Ixtlán
Jacona	Jacona de Plancarte
Jiménez	Villa Jiménez
Jiquilpan	Jiquilpan de Juárez
José Sixto Verduzco	Pastor Ortiz
Juárez	Benito Juárez
Jungapeo	Jungapeo de Juárez
Lagunillas	Lagunillas
Lázaro Cárdenas	Lázaro Cárdenas
Madero	Villa Madero
Maravatío	Maravatío de Ocampo
Marcos Castellanos	Ornelas
Morelia	Morelia
Morelos	Villa Morelos
Múgica	Múgica
Nahuatzen	Nahuatzen
Nocupétaro	Nocupétaro
Nuevo Parangaricutiro	Pueblo Nuevo San Juan Parangaricutiro
Nuevo Urecho	Nuevo Urecho
Numarán	Numarán
Ocampo	Ocampo
Pajacuarán	Pajacuarán
Panindícuaro	Panindícuarto
Parácuaro	Parácuaro
Paracho	Paracho de Verduzco
Pátzcuaro	Pátzcuaro

Penjamillo	Penjamillo de Degollado
Peribán	Peribán de Ramos
Piedad, La	La Piedad Cabadas
Purépero	Purépero
Puruándiro	Puruándiro
Queréndaro	Queréndaro
Quiroga	Quiroga
Régules	Cojumatlán de Régules
Reyes, Los	Los Reyes de Salgado
Sahuayo	Sahuayo de José Ma. Morelos
San Lucas	San Lucas
Santa Ana Maya	Santa Ana Maya
Santa Clara	Santa Clara
Senguío	Senguío
Susupuato	Susupuato de Guerrero
Tacámbaro	Tacámbaro de Codallos
Tancítaro	Tancítaro
Tangamandapio	Tangamandapio
Tangancícuaro	Tangancícuaro de Arista
Tanhuato	Tanhuato de Guerrero
Taretan	Taretan
Tarímbaro	Tarímbaro
Tepalcatepec	Tepaltepec
Tingambato	Tingambato
Tingüindín	Tingüindín
Tiquicheo	Tiquicheo
Tlalpujahua	Tlalpujahua
Tlazazalca	Tlazazalca
Tocumbo	Tocumbo
Tumbiscatío de Ruiz	Tumbiscatío de Ruiz
Turicato	Turicato
Tuxpan	Tuxpan
Tuzantla	Tuzantla
Tzintzuntzan	Tzintzuntzan
Tzitzio	Tzitzio
Uruapan	Uruapan del Progreso
Venustiano Carranza	Venustiano Carranza
Villamar	Villamar
Vistahermosa	Vistahermosa de Negrete
Yurécuaro	Yurécuaro
Zacapu	Zacapu
Zamora	Zamora de Hidalgo
Zináparo	Zináparo
Zinapécuaro	Zinapécuaro de Figueroa
Ziracuaretiro	Ziracuaretiro
Zitácuaro	Heroica Zitácuaro

MORELOS

Capital:	Cuernavaca	*State abbreviation:*	Mor
Population:	931,675	*Area:*	4,950 sq km

County	County Seat
Amacuzac	Amacuzac
Atlatlahucan	Atlatlahucan
Axochiapan	Axochiapan

Ayala	Ayala
Coatlán del Río	Coatlán del Río
Cuautla	Cuautla Morelos
Cuernavaca	Cuernavaca
Emiliano Zapata	Emiliano Zapata
Huitzilac	Huitzilac
Jantetelco	Jantetelco
Jiutepec	Jiutepec
Jojutla	Jojutla de Juárez
Jonacatepec	Jonacatepec
Mazatepec	Mazatepec
Miacatlán	Miacatlán
Ocuituco	Ocuituco
Puente de Ixtla	Puente de Ixtla
Temixco	Temixco
Temoac	Cahuecán
Tepalcingo	Tepalcingo
Tepoztlán	Tepoztlán
Tetecala	Tetecala
Tetela del Volcán	Tetela del Volcán
Tlalnepantla	Tlalnepantla
Tlaltizapán	Tlaltizapán
Tlaquiltenango	Tlaquiltenango
Tlayacapan	Tlayacapan
Totolapan	Totolapan
Xochitepec	Xochitepec
Yautepec	Yautepec
Yecapixtla	Yecapixtla
Zacatepec	Zacatepec
Zacualpan	Zacualpan de Amilpas

NAYARIT

Capital:	Tepic	*State abbreviation:*	Nay
Population:	730,024	*Area:*	26,979 sq km

County	County Seat
Acaponeta	Acaponeta
Ahuacatlán	Ahuacatlán
Amatlán de Cañas	Amatlán de Cañas
Compostela	Compostela
Huajicori	Huajicori
Ixtlán	Ixtlán del Río
Jala	Jala
Jalisco	Jalisco
Nayar	Nayar
Rosamorada	Rosamorada
Ruiz	Ruiz
San Blas	San Blas
San Pedro Lagunillas	San Pedro Lagunillas
Santa María del Oro	Santa María del Oro
Santiago Ixcuintla	Santiago Ixcuintla
Tecuala	Tecuala
Tepic	Tepic
Tuxpan	Tuxpan
Yesca, La	La Yesca

NUEVO LEON

Capital: Monterrey
Population: 2,463,298

State abbreviation: NL
Area: 64,924 sq km

County	County Seat
Abasolo	Abasolo
Agualenguas	Agualenguas
Aldamas, Los	Los Aldamas
Allende	Allende
Anáhuac	Anáhuac
Apodaca	Apodaca
Aramberri	Aramberri
Bustamante	Bustamante
Cadereyta Jiménez	Cadereyta Jiménez
Carmen	Carmen
Cerralvo	Cerralvo
Ciénega de Flores	Ciénega de Flores
China	China
Doctor Arroyo	Doctor Arroyo
Doctor Coss	Doctor Coss
Doctor González	Doctor González
Galeana	Galeana
García	Villa de García
Garza García	Garza García
General Bravo	General Bravo
General Escobedo	General Escobedo
General Terán	General Terán
General Treviño	General Treviño
General Zaragoza	Zaragoza
General Zuazua	General Zuazua
Guadalupe	Guadalupe
Herreras, Los	Los Herreras
Higueras	Higueras
Hualahuises	Hualahuises
Iturbide	Iturbide
Juárez	Juárez
Lampazos de Naranjo	Lampazos de Naranjo
Linares	Linares
Marín	Marín
Melchor Ocampo	Melchor Ocampo
Mier y Noriega	Mier y Noriega
Mina	Mina
Montemorelos	Montemorelos
Monterrey	Monterrey
Parás	Parás
Pesquería	Pesquería
Ramones, Los	Los Ramones
Rayones	Rayones
Sabinas Hidalgo	Sabinas Hidalgo
Salinas Victoria	Salinas Victoria
San Nicolás de los Garza	San Nicolás de los Garza
San Nicolás Hidalgo	San Nicolás Hidalgo
Santa Catarina	Santa Catarina
Santiago	Santiago
Vallecillo	Vallecillo
Villaldama	Villaldama

OAXACA

Capital: Oaxaca de Juárez
Population: 2,518,157

State abbreviation: Oax
Area: 95,952 sq km

County	County Seat
Abejones	Abejones
Acatlán de Pérez Figueroa	Acatlán de Pérez Figueroa
Asunción Cacalotepec	Asunción Cacalotepec
Asunción Cuyotepejí	Asunción Cuyotepejí
Asunción Ixtaltepec	Asunción Ixtaltepec
Asunción Nochixtlán	Asunción Nochixtlán
Asunción Ocotlán	Asunción Ocotlán
Asunción Tlacolulita	Asunción Tlacolulita
Ayotzintepec	Ayotzintepec
Barrio, El	El Barrio
Calihualá	Calihualá
Candelaria Loxicha	Candelaria Loxicha
Ciénega, La	La Ciénega
Ciudad Ixtepec	Ixtepec
Coatecas Altas	Coatepec Altas
Coicoyán de las Flores	Coicoyán de las Flores
Compañía, La	La Compañía
Concepción Buenavista	Concepción Buenavista
Concepción Pápalo	Concepción Pápalo
Conatancia del Rosario	Constancia del Rosario
Cosolapa	Cosolapa
Cosoltepec	Cosoltepec
Cuilapan de Guerrero	Cuilapan de Guerrero
Cuyamecalco Villa de Zaragoza	Cuyamecalco Villa de Zaragoza
Chahuites	Chahuites
Chalcatongo de Hidalgo	Chalcatongo de Hidalgo
Chiquihuitlán de Benito Juárez	Chiquihuitlán de Benito Juárez
Díaz Ordaz	Villa Díaz Ordaz
Ejutla de Crespo	Ejutla de Crespo
Eloxochitlán de Flores Magón	San Antonio Eloxochitlán
Espinal, El	El Espinal
Espíritu Santo Tamazulapan	Espíritu Santo Tamazulapan
Fresnillo de Trujano	Fresnillo de Trujano
Guadalupe Etla	Guadalupe Etla
Guadalupe Ramírez	Guadalupe de Ramírez
Guelatao de Juárez	Guelatao de Juárez
Guevea de Humboldt	Guevea de Humboldt
Hidalgo	Hidalgo
Hidalgo Yalalag	Hidalgo Yalalag
Huajuapan de León	Huajuapan de León
Huautepec	Huautepec
Huautla de Jiménez	Huautla de Jiménez
Ixtlán de Juárez	Ixtlán de Juárez
Juchitán de Zaragoza	Juchitán de Zaragoza
Loma Bonita	Loma Bonita
Magdalena Apasco	Magdalena Apasco
Magdalena Jaltepec	Magdalena Jaltepec
Magdalena Jicotlán	Magdalena Jicotlán
Magdalena Mixtepec	Magdalena Mixtepec
Magdalena Ocotlán	Magdalena Ocotlán
Magdalena Peñasco	Magdalena Peñasco

Magdalena Teitipac

Magdalena Tequisistlán

Magdalena Tlacotepec

Magdalena Zahuatlán

Mariscala de Juárez

Mártires de Tacubaya

Matías Romero

Mazatlán de Flores

Miahuatlán de Porfirio Díaz

Monjas

Natividad

Nazareno Etla

Nejapa de Madero

Nieves Ixpantepec

Niltepec

Oaxaca de Juárez

Ocotlán de Morelos

Pe, La

Pinotepa de Don Luis

Pluma Hidalgo

Progreso

Putla de Guerrero

Quiquitani

Reforma de Pineda

Reforma, La

Reyes Etla

Rojas de Cuauhtémoc

Salina Cruz

San Agustín Amatengo

San Agustín Atenango

San Agustín Chayuco

San Agustín de las Juntas

San Agustín Etla

San Agustín Loxicha

San Agustín Tlacotepec

San Agustín Yatareni

San Andrés Cabecera Nueva

San Andrés Dinicuiti

San Andrés Huaxpaltepec

San Andrés Huayapan

San Andrés Ixlahuaca

San Andrés Lagunas

San Andrés Nuxiño

San Andrés Paxtlán

San Andrés Sinaxtla

San Andrés Solaga

San Andrés Teotilapan

San Andrés Tepetlapa

San Andrés Yaá

San Andrés Zabache

San Andrés Zautla

San Antonio Acutla

San Antonio el Alto

San Antonio de la Cal

San Antonio Castillo Velasco

San Antonio Huitepec

San Antonio Monteverde

San Antonio Nanahuatipan

San Antonio Sinicahua

Magdalena Teitipac

Magdalena Tequisistlán

Magdalena Tlacotepec

Magdalena Zahuatlán

Mariscala de Juárez

Mártires de Tacubaya

Matías Romero

Mazatlán de Flores

Miahuatlán de Porfirio Díaz

Monjas

Natividad

Nazareno Etla

Nejapa de Madero

Nieves Ixpantepec

Niltepec

Oaxaca de Juárez

Ocotlán de Morelos

La Pe

Pinotepa de Don Luis

Pluma Hidalgo

Progreso

Putla de Guerrero

Quiquitani

Reforma de Pineda

La Reforma

Reyes Etla

Rojas de Cuauhtémoc

Salina Cruz

San Agustín Amatengo

San Agustín Atenango

San Agustín Chayuco

San Agustín de las Juntas

San Agustín Etla

San Agustín Loxicha

San Agustín Tlacotepec

San Agustín Yatareni

San Andrés Cabecera Nueva

San Andrés Dinicuiti

San Andrés Huaxpaltepec

San Andrés Huayapan

San Andrés Ixtlahuaca

San Andrés Lagunas

San Andrés Nuxiño

San Andrés Paxtlán

San Andrés Sinaxtla

San Andrés Solaga

San Andrés Teotilapan

San Andrés Tepetlapan

San Andrés Yaá

San Andrés Zabache

San Andrés Zautla

San Antonio Acultla

San Antonio el Alto

San Antonio de la Cal

San Antonio Castillo Velasco

San Antonio Huitepec

San Antonio Monteverde

San Antonio Nanahuatipan

San Antonio Sinicahua

San Antonio Tepetlapa
San Baltasar Chichicapan
San Baltasar Loxicha
San Baltasar Yatzachi el Bajo
San Bartolomé Ayautla
San Bartolomé Loxicha
San Bartolomé Quialana
San Bartolomé Yucuañe
San Bartolomé Zoogocho
San Bartolo Coyotepec
San Bartolo Soyaltepec
San Bartolo Yautepec
San Bernardo Mixtepec
San Blas Atempa
San Carlos Yautepec
San Cristóbal Amatlán
San Cristóbal Amoltepec
San Cristóbal Lachirioag
San Cristóbal Suchixtlahuaca
San Dionisio del Mar
San Dionisio Ocotepec
San Dionisio Ocotlán
San Esteban Atatlahuca
San Felipe Jalapa de Díaz
San Felipe Tejalapan
San Felipe Usila
San Francisco Cahuacuá
San Francisco Cajonos
San Francisco Chapulapa
San Francisco Chindúa
San Francisco del Mar
San Francisco Huehuetlán
San Francisco Ixhuatán
San Francisco Jeltepetongo
San Francisco Lachigoló
San Francisco Logueche
San Francisco Nuxaño
San Francisco Ozolotepec
San Francisco Sola
San Francisco Telixtlahuaca
San Francisco Teopan
San Francisco Tlapancingo
San Gabriel Mixtepec
San Ildefonso Amatlán
San Ildefonso Sola
San Ildefonso Villa Alta
San Jacinto Amilpas
San Jacinto Tlacotepec
San Jerónimo Coatlán
San Jerónimo Silacayoapilla
San Jerónimo Sosola
San Jerónimo Taviche
San Jerónimo Tecoatl
San Jorge Nuchita
San José Ayuquila
San José Chiltepec
San José del Peñasco
San José Estancia Grande
San José Independencia

San Antonio Tepetlapa
San Baltasar Chichicapan
San Baltasar Loxicha
San Baltasar Yatzachi el Bajo
San Bartolomé Ayautla
San Bartolomé Loxicha
San Bartolomé Quialana
San Bartolomé Yucuañe
San Bartolomé Zoogocho
San Bartolo Coyotepec
San Bartolo Soyaltepec
San Bartolo Yautepec
San Bernardo Mixtepec
San Blas Atempa
San Carlos Yautepec
San Cristóbal Amatlán
San Cristóbal Amoltepec
San Cristóbal Lachirioag
San Cristóbal Suchixtlahuaca
San Dionisio del Mar
San Dionisio Ocotepec
San Dionisio Ocotlán
San Esteban Atatlahuca
San Felipe Jalapa de Díaz
San Felipe Tejalapan
San Felipe Usila
San Francisco Cahuacuá
San Francisco Cajonos
San Francisco Chapulapa
San Francisco Chindúa
San Francisco del Mar
San Francisco Huehuetlán
San Francisco Ixhuatán
San Francisco Jeltepetongo
San Francisco Lachigoló
San Francisco Logueche
San Francisco Nuxaño
San Francisco Ozolotepec
San Francisco Sola
San Francisco Telixtlahuaca
San Francisco Teopan
San Francisco Tlapancingo
San Gabriel Mixtepec
San Ildefonso Amatlán
San Ildefonso Sola
San Ildefonso Villa Alta
San Jacinto Amilpas
San Jacinto Tlacotepec
San Jerónimo Coatlán
San Jerónimo Silacayoapilla
San Jerónimo Sosola
San Jerónimo Taviche
San Jerónimo Tecoatl
San Jorge Nuchita
San José Ayuquila
San José Chiltepec
San José del Peñasco
San José Estancia Grande
San José Independencia

San José Lachiguirí	San José Lachiguirí
San José Tenango	San José Tenango
San Juan Achiutla	San Juan Achiutla
San Juan Atepec	San Juan Atepec
San Juan Bautista Animas Trujano	San Juan Bautista Animas Trujano
San Juan Bautista Atatlahuca	San Juan Bautista Atatlahuca
San Juan Bautista Coixtlahuaca	San Juan Bautista Coixtlahuaca
San Juan Bautista Cuicatlán	San Juan Bautista Cuicatlán
San Juan Bautista Guelache	San Juan Bautista Guelache
San Juan Bautista Jayacatlán	San Juan Bautista Jayacatlán
San Juan Bautista lo de Soto	San Juan Bautsita lo de Soto
San Juan Bautista Suchitepec	San Juan Bautista Suchitepec
San Juan Bautista Tlacoatzintepec	San Juan Bautista Tlacoatzintepec
San Juan Bautista Tlachichilco	San Juan Bautista Tlachichilco
San Juan Bautista Tuxtepec	San Juna Bautista Tuxtepec
San Juan Bautista Cachuatepec	San Juan Bautista Cacahuatepec
San Juan Cieneguilla	San Juan Cieneguilla
San Juan Coatzopan	San Juan Coatzopan
San Juan Colorado	San Juan Colorado
San Juan Comaltepec	San Juan Comaltepec
San Juan Cotzocón	San Juan Cotzocón
San Juan Chicomezúchil	San Juan Chicomezúchil
San Juan Chilateca	San Juan Chilateca
San Juan del Estado	San Juan del Estado
San Juan del Río	San Juan del Río
San Juan Diuxi	San Juan Diuxi
San Juan Evangelista Analco	San Juan Evangelista Analco
San Juan Guelavía	San Juan Guelavía
San Juan Guichicovi	San Juan Guichicovi
San Juan Igualtepec	San Juan Igualtepec
San Juan Juquila Mixes	San Juan Juquila Mixes
San Juan Juquila Vijanos	San Juan Juquila Vijanos
San Juan Lachao	San Juan Lachao
San Juan Lachigalla	San Juan Lachigalla
San Juan Lajarcia	San Juan Lajarcia
San Juan Lalana	San Juan Lalana
San Juan los Cures	San Juan los Cures
San Juan Mazatlán	San Juan Mazatlán
San Juan Mixtepec-Juxtlahuaca	San Juan Mixtepec-Juxtlahuaca
San Juan Mixtepec-Miahuatlán	San Juan Mixtepec-Miahuatlán
San Juan Numi	San Juan Numi
San Juan Ozolotepec	San Juan Ozolotepec
San Juan Petlapa	San Juan Petlapa
San Juan Quiahije	San Juan Quiahije
San Juan Quiotepec	San Juan Quiotepec
San Juan Sayultepec	San Juan Sayultepec
San Juan Tabaá	San Juan Tabaá
San Juan Tamazola	San Juan Tamazola
San Juan Teita	San Juan Teita
San Juan Tepeuxila	San Juan Tepeuxila
San Juan Teposcolula	San Juan Teposcolula
San Juan Yaé	San Juan Yaé
San Juan Yatzona	San Juan Yatzona
San Juan Yucuita	San Juan Yucuita
San Lorenzo	San Lorenzo
San Lorenzo Albarradas	San Lorenzo Albarradas
San Lorenzo Cacaotepec	San Lorenzo Cacaotepec
San Lorenzo Cuaunecuiltitla	San Lorenzo Cuaunecuiltitla
San Lorenzo Texmelucan	San Lorenzo Texmelucan

San Lorenzo Victoria
San Lucas Camotlán
San Lucas Ojitlán
San Lucas Quiaviní
San Lucas Zoquiapan
San Luis Amatlán
San Marcial Ozolotepec
San Marcos Arteaga
San Martín de los Canseco
San Martín Huamelulpan
San Martín Itunyoso
San Martín Lachilá
San Martín Peras
San Martín Tilcajete
San Martín Toxpalan
San Martín Zacatepec
San Mateo Cajonos
San Mateo Capulapan
San Mateo del Mar
San Mateo Eloxochitlán
San Mateo Etlatongo
San Mateo Nejapan
San Mateo Peñasco
San Mateo Piñas
San Mateo Río Hondo
San Mateo Sindihui
San Mateo Tlapiltepec
San Melchor Betaza
San Miguel Achiutla
San Miguel Ahuehuetitlán
San Miguel Aloápam
San Miguel Amatitlán
San Miguel Amatlán
San Miguel Coatlán
San Miguel Chicahua
San Miguel Chimalapa
San Miguel del Puerto
San Miguel del Río
San Miguel Ejutla
San Miguel el Grande
San Miguel Huautla-Nochixtlán
San Miguel Mixtepec
San Miguel Panixtlahuaca
San Miguel Peras
San Miguel Piedras
San Miguel Quetzaltepec
San Miguel Santa Flor
San Miguel Sola de Vega
San Miguel Soyaltepec-Temascal
San Miguel Suchixtepec
San Miguel Talea de Castro
San Miguel Tecomatlán
San Miguel Tenango
San Miguel Tequixtepec
San Miguel Tilquiapan
San Miguel Tlacamama
San Miguel Tlacotepec
San Miguel Tulancingo
San Miguel Yotao

San Lorenzo Victoria
San Lucas Camotlán
San Lucas Ojitlán
San Lucas Quiaviní
San Lucas Zoquiapan
San Luis Amatlán
San Marcial Ozolotepec
San Marcos Arteaga
San Martín de los Canseco
San Martín Huamelulpan
San Martín Itunyoso
San Martín Lachilá
San Martín Peras
San Martín Tilcajete
San Martín Toxpalan
San Martín Zacatepec
San Lateo Cajonos
Capulapan de Méndez
San Mateo del Mar
San Mateo Eloxochitlán
San Mateo Etlatongo
San Mateo Nejapan
San Mateo Peñasco
San Mateo Piñas
San Mateo Río Hondo
San Mateo Sindihui
San Mateo Tlapiltepec
San Melchor Betaza
San Miguel Achiutla
San Miguel Ahuehuetitlán
San Miguel Aloápam
San Miguel Amatitlán
San Miguel Amatlán
San Miguel Coatlán
San Miguel Chicahua
San Miguel Chimalapa
San Miguel del Puerto
San Miguel del Río
San Miguel Ejutla
San Miguel el Grande
San Miguel Huautla-Nochixtlán
San Miguel Mixtepec
San Miguel Panixtlahuaca
San Miguel Peras
San Miguel Piedras
San Miguel Quetzaltepec
San Miguel Santa Flor
San Miguel Sola de Vega
Temscal
San Miguel Suchixtepec
San Miguel Talea de Castro
San Miguel Tecomatlán
San Miguel Tenango
San Miguel Tequixtepec
San Miguel Tilquiapan
San Miguel Tlacamama
San Miguel Tlacotepec
San Miguel Tulancingo
San Miguel Yotao

San Nicolás
San Nicolás Hidalgo
San Pablo Coatlán
San Pablo Cuatro Venados
San Pablo Etla
San Pablo Huitzo
San Pablo Huixtepec
San Pablo Macuiltianguis
San Pablo Tijaltepec
San Pablo Villa de Mitla
San Pablo Yaganiza
San Pedro Amuzgos
San Pedro Apóstol
San Pedro Atoyac
San Pedro Cajonos
San Pedro Cántaros
San Pedro Comitancillo
San Pedro el Alto
San Pedro Huamelula
San Pedro Huilotepec
San Pedro Ixcatlán
San Pedro Ixtlahuaca
San Pedro Jaltepetongo
San Pedro Jicayán
San Pedro Jocotipac
San Pedro Juchatengo
San Pedro Mártir
San Pedro Mártir Quiechapa
San Pedro Mártir Yucuxaco
San Pedro Mixtepec-Juquila
San Pedro Mixtepec-Miahuatlán
San Pedro Molinos
San Pedro Nopala
San Pedro Ocopetatillo
San Pedro Ocotepec
San Pedro Pochutla
San Pedro Quiatoni
San Pedro Sochiapan
San Pedro Tapanatepec
San Pedro Taviche
San Pedro Teozacoalco
San Pedro Teutila
San Pedro Tidaá
San Pedro Topiltepec
San Pedro Totolapan
San Pedro Tututepec
San Pedro Yaneri
San Pedro Yólox
San Pedro Yucunama
San Pedro y San Pablo Ayutla
San Pedro y San Pablo Etla
San Pedro y San Pablo Teposcolula
San Pedro y San Pablo Tequixtepec
San Raymundo Jalpan
San Sebastián Abasolo
San Sebastián Coatlán
San Sebastián Ixcapa
San Sebastián Nicananduta
San Sebastián Río Hondo

San Nicolás
San Nicolás Hidalgo
San Pablo Coatlán
San Pablo Cuatro Venados
San Pablo Etla
San Pablo Huitzo
San Pablo Huixtepec
San Pablo Macuiltianguis
San Pablo Tijaltepec
San Pablo Villa de Mitla
San Pablo Yaganiza
San Pedro Amuzgos
San Pedro Apóstol
San Pedro Atoyac
San Pedro Cajonos
San Pedro Cántaros
San Pedro Comitancillo
San Pedro el Alto
San Pedro Huamelula
San Pedro Huilotepec
San Pedro Ixcatlán
San Pedro Ixtlahuaca
San Pedro Jaltepetongo
San Pedro Jicayán
San Pedro Jocotipac
San Pedro Juchatengo
San Pedro Mártir
San Pedro Mártir Quiechapa
San Pedro Mártir Yucuxaco
San Pedro Mixtepec-Juquila
San Pedro Mixtepec-Miahuatlán
San Pedro Molinos
San Pedro Nopala
San Pedro Ocopetatillo
San Pedro Ocotepec
San Pedro Pochutla
San Pedro Quiatoni
San Pedro Sochiapan
San Pedro Tapanatepec
San Pedro Taviche
San Pedro Teozarcoalco
San Pedro Teutila
San Pedro Tidaá
San Pedro Topiltepec
San Pedro Totolapan
San Pedro Tututepec
San Pedro Yaneri
San Pedro Yólox
San Pedro Yucunama
San Pedro y San Pablo Ayutla
San Perdo y San Pablo Etla
San Pedro y San Pablo Teposcolula
San Pedro y San Pablo Tequixtepec
San Raymundo Jalpan
San Sebastián Abasolo
San Sebastián Coatlán
San Sebastián Ixcapa
San Sebastián Nicananduta
San Sebastián Río Hondo

San Sebastián Tecomaxtlahuaca
San Sebastián Teitipac
San Sebastián Tutla
San Simón Almolongas
San Simón Zahuatlán
Santa Ana
Santa Ana Ateixtlahuaca
Santa Ana Cuauhtémoc
Santa Ana del Valle
Santa Ana Tavela
Santa Ana Tlapacoyan
Santa Ana Yareni
Santa Ana Zegache
Santa Catalina Quieri
Santa Catarina Cuixtla
Santa Catarina Ixtepeji
Santa Catarina Juquila
Santa Catarina Lachatao
Santa Catarina Loxicha
Santa Catarina Mechoacán
Santa Catarina Minas
Santa Catarina Quiané
Santa Catarina Tayata
Santa Catarina Ticuá
Santa Catarina Yosonotú
Santa Catarina Zapaquila
Santa Cruz Acatepec
Santa Cruz Amilpas
Santa Cruz de Bravo
Santa Cruz Itundujia
Santa Cruz Mixtepec
Santa Cruz Nundaco
Santa Cruz Papalutla
Santa Cruz Tacache de Mina
Santa Cruz Tacahua
Santa Cruz Tayata
Santa Cruz Xitla
Santa Cruz Xoxocotlán
Santa Cruz Zenzontepec
Santa Gertrudis
Santa Inés del Monte
Santa Inés Yatzeche
Santa Lucía del Camino
Santa Lucía Miahuatlán
Santa Lucía Monteverde
Santa Lucía Ocotlán
Santa María Alotepec
Santa María Apasco
Santa María Asunción
Santa María Asunción Tlaxiaco
Santa María Ayoquezco de Aldama
Santa María Azompa
Santa María Camotlán
Santa María Colotepec
Santa María Cortijos
Santa María Coyotepec
Santa María Chachoapan
Santa María Chilapa de Díaz
Santa María Chilchotla

San Sebastián Tecomaxtlahuaca
San Sebastián Teitipac
San Sebastián Tutla
San Simón Almolongas
San Simón Zahuatlán
Santa Ana
Santa Ana Ateixtlahuaca
Santa Ana Cuauhtémoc
Santa Ana del Valle
Santa Ana Tavela
Santa Ana Tlapacoyan
Santa Ana Yareni
Santa Ana Zegache
Santa Catalina Quieri
Santa Catarina Cuixtla
Santa Catarina Ixtepeji
Santa Catarina Juquila
Santa Catarina Lachatao
Santa Catarina Loxicha
Santa Catarina Mechoacán
Santa Catarina Minas
Santa Catarina Quiané
Santa Catarina Tayata
Santa Catarina Ticuá
Santa Catarina Yosonotú
Santa Catarina Zapaquila
Santa Cruz Acatepec
Santa Cruz Amilpas
Santa Cruz de Bravo
Santa Cruz Itundujia
Santa Cruz Mixtepec
Santa Cruz Nundaco
Santa Cruz Papalutla
Santa Cruz Tacache de Mina
Santa Cruz Tacahua
Santa Cruz Tayata
Santa Cruz Xitla
Santa Cruz Xoxocotlán
Santa Cruz Zenzontepec
Santa Gertrudis
Santa Inés del Monte
Santa Inés Yatzeche
Santa Lucía del Camino
Santa Lucía Miahuatlán
Santa Lucía Monteverde
Santa Lucía Ocotlán
Santa María Alotepec
Santa María Apasco
Santa María Asunción
Santa María Asunción Tlaxiaco
Santa María Ayoquezco de Aldama
Santa María Azompa
Santa María Camotlán
Santa María Colotepec
Santa María Cortijos
Santa María Coyotepec
Santa María Chachoapan
Santa María Chilapa de Díaz
Santa María Chilchotla

Santa María Chimalapa	Santa María Chimalapa
Santa María del Rosario	Santa María del Rosario
Santa María del Tule	Santa María del Tule
Santa María Ecatepec	Santa María Ecatepec
Santa María Guelaxé	Santa María ʻGuelaxé
Santa María Guienagati	Santa María Guienagati
Santa María Huatulco	Santa María Huatulco
Santa María Huazolotitlán	Santa María Huazolotitlán
Santa María Ipalapa	Santa María Ipalapa
Santa María Ixcatlán	Santa María Ixcatlán
Santa María Jacatepec	Santa María Jacatepec
Santa María Jalapa del Marqués	Santa María Jalapa del Marqués
Santa María Jaltianguis	Santa María Jaltianguis
Santa María Lachixío	Santa María Lachixío
Santa María Mixistlán	Mixistlán de la Reforma
Santa María Mixtequilla	Santa María Mixtequilla
Santa María Natívitas	Santa María Natívitas
Santa María Nduayaco	Santa María Nduayaco
Santa María Ozolotepec	Santa María Ozolotepec
Santa María Pápalo	Santa María Pápalo
Santa María Peñoles	Santa María Peñoles
Santa María Petapa	Santa María Petapa
Santa María Quiegolani	Santa María Quiegolani
Santa María Sola	Santa María Sola
Santa María Tataltepec	Santa María Tataltepec
Santa María Tecomavaca	Santa María Tecomavaca
Santa María Temaxcalapan	Santa María Temaxcalapan
Santa María Temaxcaltepec	Santa María Temaxcaltepec
Santa María Teopoxco	Santa María Teopoxco
Santa María Tepantlali	Santa María Tepantlali
Santa María Texcatitlán	Santa María Texcatitlán
Santa María Tlahuitoltepec	Santa María Tlahuitoltepec
Santa María Tlalixtac	Santa María Tlalixtac
Santa María Tonameca	Santa María Tonameca
Santa María Totolapilla	Santa María Totolapilla
Santa María Xadani	Santa María Xadani
Santa María Yalina	Santa María Yalina
Santa María Yavesía	Santa María Yavesía
Santa María Yolotepec	Santa María Yolotepec
Santa María Yosoyúa	Santa María Yosoyúa
Santa María Yucuhiti	Santa María Yucuhiti
Santa María Zacatepec	Santa María Zacatepec
Santa María Zaniza	Santa María Zaniza
Santa María Zoquitlán	Santa María Zoquitlán
Santiago Amoltepec	Santiago Amoltepec
Santiago Apoala	Santiago Apoala
Santiago Apóstol	Santiago Apóstol
Santiago Astata	Santiago Astata
Santiago Atitlán	Santiago Atitlán
Santiago Ayuquililla	Santiago Ayuquililla
Santiago Cacaloxtepec	Santiago Cacaloxtepec
Santiago Camotlán	Santiago Camotlán
Santiago Comaltepec	Santiago Comaltepec
Santiago Chazumba	Santiago Chazumba
Santiago Choapan	Santiago Choapan
Santiago del Río	Santiago del Río
Santiago Huajolotitlán	Santiago Huajolotitlán
Santiago Huauclilla	Santiago Huauclilla
Santiago Igüitlán Plumas	Santiago Igüitlán Plumas

Santiago Ixcuintepec
Santiago Ixtayutla
Santiago Jamiltepec
Santiago Jocotepec
Santiago Juxtlahuaca
Santiago Lachiguirí
Santiago Lalopa
Santiago Laollaga
Santiago Laxopa
Santiago Llano Grande
Santiago Matatlán
Santiago Miltepec
Santiago Minas
Santiago Nacaltepec
Santiago Nejapilla
Santiago Nundichi
Santiago Nuyoó
Santiago Pinotepa Nacional
Santiago Suchilquitongo
Santiago Tamazola
Santiago Tapextla
Santiago Tejupan
Santiago Tenango
Santiago Tepetlapa
Santiago Tetepec
Santiago Texcalcingo
Santiago Textitlán
Santiago Tilantongo
Santiago Tillo
Santiago Tlazoyaltepec
Santiago Xanica
Santiago Xiacuí
Santiago Yaitepec
Santiago Yaveo
Santiago Yolomécatl
Santiago Yosondúa
Santiago Yucuyachi
Santiago Zacatepec
Santiago Zoochila
Santiago Zoquiapan
Santo Domingo
Santo Domingo Albarradas
Santo Domingo Armenta
Santo Domingo Chihuitán
Santo Domingo de Morelos
Santo Domingo Ixcatlán
Santo Domingo Nuxaá
Santo Domingo Ozolotepec
Santo Domingo Petapa
Santo Domingo Roayaga
Santo Domingo Tehuantepec
Santo Domingo Teojomulco
Santo Domingo Tepuxtepec
Santo Domingo Tlatayapan
Santo Domingo Tomaltepec
Santo Domingo Tonalá
Santo Domingo Tonaltepec
Santo Domingo Xagacía
Santo Domingo Yanhuitlán

Santiago Ixcuintepec
Santiago Ixtayutla
Santiago Jamiltepec
Monte Negro
Santiago Juxtlahuaca
Santiago Lachiguirí
Santiago Lalopa
Santiago Laollaga
Santiago Laxopa
Santiago Llano Grande
Santiago Matatlán
Santiago Miltepec
Santiago Minas
Santiago Nacaltepec
Santiago Nejapilla
Santiago Nundichi
Santiago Nuyoó
Santiago Pinotepa Nacional
Santiago Suchilquitongo
Santiago Tamazola
Santiago Tepextla
Santiago Tejupan
Santiago Tenango
Santiago Tepetlapa
Santiago Tetepec
Santiago Texcalcingo
Santiago Textitlán
Santiago Titantongo
Santiago Tillo
Santiago Tlazoyaltepec
Santiago Xanica
Santiago Xiacuí
Santiago Yaitepec
Santiago Yaveo
Santiago Yolomécatl
Santiago Yosondúa
Santiago Yucuyachi
Santiago Zacatepec
Santiago Zoochila
Santiago Zoquiapan
Santo Domingo
Santo Domingo Albarradas
Santo Domingo Armenta
Santo Domingo Chihuitán
Santo Domingo de Morelos
Santo Domingo Ixcatlán
Santo Domingo Nuxaá
Santo Domingo Ozolotepec
Santo Domingo Petapa
Santo Domingo Roayaga
Santo Domingo Tehuantepec
Santo Domingo Teojomulco
Santo Domingo Tepuxtepec
Santo Domingo Tlatayapan
Santo Domingo Tomaltepec
Santo Domingo Tonalá
Santo Domingo Tonaltepec
Santo Domingo Xagacía
Santo Domingo Yanhuitlán

County	County Seat
Santo Domingo Zanatepec	Santo Domingo Zanatepec
Santo Tomás Jalieza	Santo Tomás Jalieza
Santo Tomás Mazaltepec	Santo Tomás Mazaltepec
Santo Tomás Ocotepec	Santo Tomás Ocotepec
Santo Tomás Tamazulapan	Santo Tomás Tamazulapan
Santos Reyes Nopala	Santos Reyes Nopala
Santos Reyes Nopala	Santos Reyes Nopala
Santos Reyes Pápalo	Santos Reyes Pápalo
Santos Reyes Tepejillo	Santos Reyes Tepejillo
Santos Reyes Yucuná	Santos Reyes Yucuná
San Vicente Coatlán	San Vicente Coatlán
San Vicente Lachixió	San Vicente Lachixió
San Vicente Nuñú	San Vicente Nuñú
Silacayoapan	Silacayoapan
Sitio de Xitlapehua	Sitio de Xitlapehua
Soledad Etla	Soledad Etla
Tamazulapan del Progreso	Tamazulapan del Progreso
Tanetze de Zaragoza	Tenetze de Zaragoza
Taniche	Taniche
Tataltepec de Valdés	Tataltepec de Valdés
Teococuilco de Marcos Pérez	Teococuilco de Marcos Pérez
Teotitlán del Camino	Teotitlán del Camino
Teotitlán del Valle	Teotitlán del Valle
Teotongo	Santiago Teotongo
Tepelmeme de Morelos	Tepelmeme de Morelos
Tezoatlán de Sagura y Luna	Tezoatlán de Segura y Luna
Tlacochahuaya de Morelos	Tlacochahuaya de Morelos
Tlacolula de Matamores	Tlacolula de Matamores
Tlacotepec Plumas	Tlacotepec Plumas
Tlalixtac de Cabrera	Tlalixtac de Cabrera
Totontepec Villa de Morelos	Totontepec Villa de Morelos
Trinidad de Zaachila	Trinidad de Zaachila
Trinidad Vista Hermosa, La	La Trinidad Vista Hermosa
Unión Hidalgo	Unión Hidalgo
Valerio Trujano	Valerio Trujano
Valle Nacional	Valle Nacional
Yaxe	Yaxe
Yocodono de Porfirio Díaz	Yocodono de Porfirio Díaz
Yogana	Yogana
Yutanduchi de Guerrero	Yutanduchi de Guerrero
Zaachila	Zaachila
Zapotitlán del Río	Zapotitlán del Río
Zapotitlán Lagunas	Zapotitlán Lagunas
Zapotitlán Palmas	Zapotitlán Palmas
Zaragoza	Zaragoza
Zimatlán de Alvarez Díaz Ordaz	Zimatlán de Alvarez Díaz Ordaz

PUEBLA

Capital:	Heroica Puebla de Zaragoza	*State abbreviation:*	Pue.
Population:	3,279,960	*Area:*	33,902 sq km

County	County Seat
Acajete	Acajete
Acateno	San José Acateno
Acatlán	Acatlán de Osorio
Acatzingo	Acatzingo de Hidalgo

Acteopan	Acteopan
Ahuacatlán	Ahuacatlán
Ahuatlán	Ahuatlán
Ahuazotepec	Ahuazotepec
Ahuehuetitla	Ahuehuetitla
Ajalpan	Ajalpan
Albino Zertuche	Acaxtlahuacán de Albino Zertuche
Aljojuca	Aljojuca
Altepexi	Altepexi
Amixtlán	Amixtlán
Amozoc	Amozoc de Mota
Aquixtla	Aquixtla
Atempan	Atempan
Atexcal	San Martín Atexcal
Atlixco	Atlixco
Atoyatempan	Atoyatempan
Atzala	Atzala
Atzitzihuacán	Santiago Atzitzihuacán
Atzitzintla	Atzitzintla
Axutla	Axutla
Ayotoxco	Ayotoxco de Guerrero
Calpan	San Andrés Calpan
Caltepec	Caltepec
Camocuautla	Camocuautla
Caxhuacan	Caxhuacan
Coatepec	Coatepec
Coatzingo	Coatzingo
Cohetzala	Santa María Cohetzala
Cohuecán	Cohuecán
Coronango	Santa María Coronango
Coxcatlán	Coxcatlán
Coyomeapan	Santa María Coyomeapan
Coyotepec	San Vicente Coyotepec
Cuapiaxtla de Madero	Cuapiaxtla de Madero
Cuautempan	San Estéban Cuautempan
Cuautinchán	Cuautinchán
Cuautlancingo	San Juan Cuautlancingo
Cuayuca	San Pedro Cuavuca
Cuetzalán del Progreso	Cuetzalán del Progreso
Cuyoaco	Cuyoaco
Chalchicomula de Sesma	Ciudad Serdán
Chapulco	Chapulco
Chiautla	Chiautla de Tapia
Chiautzingo	San Lorenzo Chiautzingo
Chiconcuautla	Chiconcuautla
Chichiquila	Chichiquila
Chietla	Chietla
Chigmecatitlán	Chigmecatitlán
Chignahuapan	Chignahuapan
Chignautla	Chignautla
Chila	Chila
Chila de la Sal	Chila de la Sal
Chila Honey	Honey
Chilchotla	Rafael J. García
Chinantla	Chinantla
Domingo Arenas	Domingo Arenas
Eloxochitlán	Eloxochitlán
Epatlán	San Juan Epatlán
Esperanza	Esperanza
Francisco Z. Mena	Metlaltoyuca

General Felipe Angeles	San Pedro de las Tunas
Guadalupe	Guadalupe
Guadalupe Victoria	Guadalupe Victoria
Hermenegildo Galeana	Bienvenido
Huaquechúla	Huaquechula
Huatlatlauca	Huatlatlauca
Huauchinango	Huauchinango
Huehuetla	Huehuetla
Huehuetlán el Chico	Huehuetlán el Chico
Huejotzingo	Huejotzingo
Hueyapan	Hueyapan
Hueytamalco	Hueytamalco
Hueytlalpan	Hueylalpan
Huitzilan de Serdán	Huitzilan
Huitziltepec	Santa Clara Huitziltepec
Ignacio Allende	Atlequizayan
Ixcamilpa de Guerrero	Ixcamilpa
Ixcaquixtla	San Juan Ixcaquixtla
Ixtacamaxtitlán	Ixtacamaxtitlán
Ixtepec	Ixtepec
Izúcar de Matamoros	Izúcar de Matamoros
Jalpan	Jalpan
Jolapan	Jolapan
Jonotla	Jonotla
Jopala	Jopala
Juan C. Bonilla	Cuanalá
Juan Galindo	Nuevo Necaxa
Juan N. Méndez	Atenayuca
Lafragua	Saltillo
Libres	Libres
Magdalena Tlalauquitepec, La	La Magdalena Tlalauquitepec
Mazapiltepec de Juárez	Mazapiltepec de Juárez
Mixtla	San Francisco Mixtla
Molcaxac	Molcaxac
Morelos Cañada	Morelos Cañada
Naupan	Naupan
Nauzontla	Nauzontla
Nealitcan	San Buenaventura Nealtican
Nicolás Bravo	Nicolás Bravo
Nopalucan	Nopalucan de la Granja
Ocotepec	Ocotepec
Ocoyucan	Santa Clara Ocoyucan
Olintla	Olintla
Oriental	Oriental
Pahuatlán	Pahuatlán del Valle
Palmar de Bravo	Palmar de Bravo
Pantepec	Pantepec
Petlalcingo	Petlalcingo
Piaxtla	Piaxtla
Puebla	Heroica Puebla de Zaragoza
Quecholac	Quecholac
Quimixtlán	Quimixtlán
Rafael Lara Grajales	Rafael Lara Grajales
Reyes de Juárez, Los	Los Reyes de Juárez
San Andrés Cholula	San Andrés Cholula
San Antonio Cañada	San Antonio Cañada
San Diego la Mesa Tochimiltzingo	Tochimiltzingo
San Felipe Teolalcingo	San Felipe Teolalcingo
San Felipe Tepatlán	San Felipe Tepatlán
San Gabriel Chilac	San Gabriel Chilac

San Gregorio Atzompa
San Jerónimo Tecuanipan
San Jerónimo Xayacatlán
San José Chiapa
San José Miahuatlán
San Juan Atenco
San Juan Atzompa
San Martín Texmelucan
San Martín Totoltepec
San Matías Tlalancaleca
San Miguel Ixitlán
San Miguel Xoxtla
San Nicolás de Buenos Aires
San Nicolás los Ranchos
San Pablo Amicano
San Pedro Cholula
San Pedro Yeloixtlahuacán
San Salvador el Seco
San Salvador el Verde
San Salvador Huixcolotla
San Sebastián Tlacotepec
Santa Catarina Tlaltempan
Santa Inés Ahuatempan
Santa Isabel Cholula
Santiago Miahuatlán
Santo Domingo Huehuetlán
Santo Tomás Hueyotlipan
Soltepec
Tecali de Herrera
Tecamachalco
Tecomatlán
Tehuacán
Tehuitzingo
Tenampulco
Teopantlán
Teotlalco
Tepanco de López
Tepango de Rodríguez
Tepatlaxco de Hidalgo
Tepeaca
Tepemaxalco
Tepeojuma
Tepetzintla
Tepexco
Tepexi de Rodríguez
Tepeyahualco
Tepeyahualco Cuauhtémoc
Tetela de Ocampo
Teteles de Avila Castillo
Teziutlán
Tianguismanalco
Tilapa
Tlacotepec de Benito Juárez
Tlacuilotepec
Tlachichuca
Tlahuapan
Tlaltenango
Tlanepantla
Tlaola
Tlapacoya

San Gregorio Atzompa
San Jerónimo Tecuanipan
San Jerónimo Xayacatlán
San José Chiapa
San José Miahuatlán
San Juan Atenco
San Juan Atzompa
San Martín Texmelucan de Labastida
San Martín Totoltepec
San Matías Tlalancaleca
San Miguel Ixitlán
San Miguel Xoxtla
San Nicolás de Buenos Aires
San Nicolás los Ranchos
San Pablo Amicano
Cholula de Rivadabia
San Pedro Yeloixtlahuacán
San Salvador el Seco
San Salvador el Verde
San Salvador Huixcolotla
Tlacotepec de Díaz
Santa Catarina Tlatempan
Santa Inés Ahuatempan
Santa Isabel Cholula
Santiago Miahuatlán
Santo Domingo Huehuetlán
Santo Tomás Hueyotlipan
Soltepec
Tecali de Herrera
Tecamachalco
Tecomatlán
Tehuacán
Tehuitzingo
Tenampulco
Teopantlán
Teotlalco
Tepanco de López
Tepango de Rodríguez
Tepatlaxco de Hidalgo
Tepeaca
San felipe Tepemaxalco
Tepeojuma
Tepetzintla
Tepexco
Tepexi de Rodríguez
Tepeyahualco
Tepeyahualco Cuauhtémoc
Tetela de Ocampo
Teteles de Avila Castillo
Teziutlán
Tianguismanalco
Tilapa
Tlacotepec de Benito Juárez
Tlacuilotepec
Tlachichuca
Santa Rita Tlahuapan
Tlaltenango
Tlanepantla
Tlaola
Tlapacoya

Tlapanalá	Tlapanalá
Tlatlauquitepec	Tlatlauquitepec
Tlaxco	Tlaxco
Tochimilco	Tochimilco
Tochtepec	Tochtepec
Totoltepec de Guerrero	Totoltepec de Guerrero
Tulcingo	Tulcingo de Valle
Tuzamapan de Galeana	Tuzamapan de Galeana
Tzicatlacoyan	Tzicatlacoyan
Venustiano Carranza	Venustiano Carranza
Vicente Guerrero	Santa María del Monte
Xayacatlán de Bravo	Xayacatlán de Bravo
Xicotepec	Xicotepec de Juárez
Xicotlán	Xicotlán
Xiutetelco	San Juan Xiutetelco
Xochiapulco	Xochiapulco Cinco de Mayo
Xochiltepec	Xochiltepec
Xochitlán	Xochitlán de Romero Rubio
Xochitlán Todas Santos	Xochitlán
Yaonahuac	Yaonahuac
Yehualtepec	Yehualtepec
Zacapala	Zacapala
Zacapoaxtla	Zacapoaxtla
Zacatlán	Zacatlán
Zapotitlán	Zapotitlán Salinas
Zapotitlán de Méndez	Zapotitlán de Méndez
Zaragoza	Zaragoza
Zautla	Santiago Zautla
Zihuateutla	Zihuateutla
Zinacatepec	San Sebastián Zinacatepec
Zongozotla	Zongozotla
Zoquiapan	Zoquiapan
Zoquitlán	Zoquitlán

QUERETARO de ARTEAGA

Capital: Querétaro
Population: 726,054

State abbreviation: Qto
Area: 11,449 sq km

County	County Seat
Amealco	Amealco
Amoles	Pinal de Amoles
Arroyo Seco	Arroyo Seco
Cadereyta	Cadereyta de Montes
Colón	Colón
Corregidora	Villa del Pueblito
Ezequiel Montes	Ezequiel Montes
Huimilpan	Huimilpan
Jalpan	Jalpan
Landa de Matamoros	Landa de Matamoros
Marqués, El	La Cañada
Pedro Escobedo	Pedro Escobedo
Peñamiller	Peñamiller
Querétaro	Querétaro
San Joaquín	San Joaquín
San Juan del Río	San Juan del Río
Tequisquiapan	Tequisquiapan
Tolimán	Tolimán

QUINTANA ROO

| *Capital:* | Chetumal |
| *Population:* | 209,858 |

| *State abbreviation:* | QR |
| *Area:* | 50,212 sq km |

County	**County Seat**
Cozumel	Cozumel
Felipe Carrillo Puerto	Felipe Carrillo Puerto
Isla Mujeres	Isle Mujeres
Othon P. Blanco	Chetumal
Benito Juárez	Cancún
José Ma. Morelos	José Ma. Morelos
Lázaro Cárdenas	Kantunil-Kin

SAN LUIS POTOSI

| *Capital:* | San Luis Potosi |
| *Population:* | 1,670,637 |

| *State abbreviation:* | SLP |
| *Area:* | 63,068 sq km |

County	**County Seat**
Ahualulco	Ahualulco
Alaquines	Alaquines
Aquismón	Aquismón
Armadillo de los Infante	Armadillo de los Infante
Cárdenas	Cárdenas
Catorce	Catorce
Cedral	Cedral
Cerritos	Cerritos
Cerro de San Pedro	Cerro de San Pedro
Ciudad del Maíz	Ciudad del Maíz
Ciudad Fernández	Ciudad Fernández
Ciudad Santos	General Pedro Antonio Santos
Ciudad Valles	Ciudad Valles
Coxcatlán	Coxcatlán
Charcas	Charcas
Ebano	Ebano
Guadalcázar	Guadalcázar
Huehuetlán	Huehuetlán
Legunillas	Legunillas
Metahuala	Metahuala
Mexquitic	Mexquitic
Moctezuma	Moctezuma
Rayón	Rayón
Ríoverde	Ríoverde
Salinas	Salinas de Hidalgo
San Antonio	San Antonio
San Ciro de Acosta	Pedro Montoya
San Luis Potosí	San Luis Potosí
San Martín Chalchicuautla	San Martín Chalchicuautla
San Nicolás Tolentino	San Nicolás Tolentino
Santa Catarina	Santa Catarina
Santa María del Río	Santa María del Río
Santo Domingo	Santo Domingo
San Vicente Tancuayalab	San Vicente Tancuayalab
Soledad Díaz Gutierrez	Soledad Díaz Gutiérrez
Tamascopo	Tamasopo
Tamazunchale	Tamazunchale

Tampacán	Tampacán
Tampamolón	Tampamolón Corona
Tamuín	Tamuín
Tanlajás	Tanlajás
Tanquián de Escobedo	Tanquián de Escobedo
Tierranueva	Tierranueva
Vanegas	Vanegas
Venado	Venado
Villa de Arista	Arista
Villa de Arriaga	Villa de Arriaga
Villa de Guadalupe	Villa de Guadalupe
Villa de la Paz	La Paz
Villa de Ramos	Ramos
Villa de Reyes	Villa de Reyes
Villa Hidalgo	Villa Hidalgo
Villa Juárez	Villa Juárez
Villa Terrazas	Alfredo M. Terrazas
Xilitla	Xilitla
Zaragoza	Zaragoza

SINALOA

Capital:	Culiacán Rosales	*State abbreviation:*	Sin
Population:	1,498,931	*Area:*	58,328 sq km

County	**County Seat**
Ahome	Los Mochis
Angostura	Angostura
Bediraguato	Bediraguato
Concordia	Concordia
Cosalá	Cosalá
Culiacán	Culiacán Rosales
Choix	Choix
Elota	La Cruz
Escuinapa	Escuinapa de Hidalgo
Fuerte, El	El Fuerte
Guasave	Guasave
Mazatlán	Mazatlán
Mocorito	Mocoroito
Rosario	Rosario
Salvador Alvarado	Guamúchil
San Ignacio	San Ignacio
Sinaloa	Sinaloa de Leyva

SONORA

Capital:	Hermosillo	*State abbreviation:*	Son
Population:	1,880,098	*Area:*	182,052 sq km

County	**County Seat**
Aconchi	Aconchi
Agua Prieta	Agua Prieta

Alamos
Altar
Arivechi
Arizpe
Atil
Bacdéhuachi
Bacanora
Bacerac
Bacoachi
Bácum
Banámichi
Baviácora
Bavispe
Benjamín Hill
Caborca
Cajeme
Cananea
Carbó
Colorada, La
Cucurpe
Cumpas
Divisaderos
Empalme
Etchojoa
Fronteras
Granados
Guaymas
Hermosillo
Huáchinera
Huásabas
Huatabampo
Huépac
Imuris
Magadelena
Mazatán
Moctezuma
Naco
Nácori Chico
Nacozari de García
Navojoa
Nogales
Onavas
Opodete
Oquitoa
Pitiquito
Puerto Peñasco
Quiriego
Rayón
Rosario
Sahuaripa
San Felipe
San Javier
San Luis Río Colorado
San Miguel de Horcasitas
San Pedro de la Cueva
Santa Ana
Santa Cruz
Sáric
Soyopa

Alamos
Altar
Arivechi
Arizpe
Atil
Bacedéhuachi
Bacanora
Bacerac
Bacoachi
Bácum
Banámichi
Baviácora
Bavispe
Benjamín Hill
Caborca
Ciudad Orbregón
Cananea
Carbó
La Colorada
Cucurpe
Cumpas
Divisaderos
Empalme
Etchojoa
Fronteras
Granados
Heroica Guaymas
Hermosillo
Huáchinera
Huásabas
Huatabampo
Huépac
Imuris
Magadelena de Kino
Mazatán
Moctezuma
Naco
Nácori Chico
Nacozari de García
Navojoa
Heroica Nogales
Onavas
Opodete
Oquitoa
Pitiquito
Puerto Peñasco
Quiriego
Rayón
Rosario
Sahuaripa
San Felipe de Jesús
San Javier
San Luis Río Colorado
San Miguel de Horcasitas
San Pedro de la Cueva
Santa Ana
Santa Cruz
Sáric
Soyopa

Suaqui Grande
Tepache
Trincheras
Tubutama
Ures
Villa Hidalgo
Villa Pesqueria
Yécora

Suaqui Grande
Tepache
Trincheras
Tubutama
Ures
Villa Hidalgo
Villa Pesqueira
Yécora

TABASCO

Capital: Villahermosaa
Population: 1,149,756

State abbreviation: Tab
Area: 25,267 sq km

County	County Seat
Balancán	Balancán de Domínguez
Cárdenas	Heroica Cárdenas
Centla	Frontera
Centro	Villahermosa
Comalcalco	Comalcalco
Cunduacán	Cunduacán
Emiliano Zapata	Emiliano Zapata
Huimanguillo	Huimanguillo
Jalapa	Jalapa
Jalpa	Jalpa de Méndez
Jonuta	Jonuta
Macuspana	Mascuspana
Nacajuca	Nacajuca
Paraíso	Paraíso
Tacotalpa	Tacotalpa
Teapa	Teapa
Tenosique	Tenoisque de Pino Suárez

TAMAULIPAS

Capital: Ciudad Victoria
Population: 1,924,934

State abbreviation: Tamps
Area: 79,384 sq km

County	County Seat
Abasolo	Abasolo
Aldama	Aldama
Altamira	Altamira
Antiguo Morelos	Antiguo Morelos
Burgos	Burgos
Bustamante	Bustamante
Camargo	Ciudad Camargo
Casas	Casas
Ciudad Madero	Ciudad Madero
Cruillas	Cruillas
Gómez Farías	Loma Alta de Gómez Farías
González	González

Güémez	Güémez
Guerrero	Nueva Ciudad Guerrero
Gustavo Díaz Ordaz	Gustavo Díaz Ordaz
Hidalgo	Hidalgo
Jaumave	Jaumave
Jiménez	Santander Jiménez
Llera	Llera de Canales
Mainero	Villa Mainero
Mante	Ciudad Mante
Matamoros	Heroica Matamoros
Méndez	Méndez
Mier	Mier
Miguel Alemán	Ciudad Miguel Alemán
Miquihuana	Miquihuana
Nuevo Laredo	Nuevo Laredo
Nuevo Morelos	Nuevo Morelos
Ocampo	Ocampo
Padilla	Padilla
Palmillas	Palmillas
Reynosa	Reynosa
Río Bravo	Ciudad Río Bravo
San Carlos	San Carlos
San Fernando	San Fernando
San Nicolás	San Nicolás
Soto La Marina	Soto La Marina
Tampico	Tampico
Tula	Tula
Valle Hermoso	Valle Hermoso
Victoria	Ciudad Victoria
Villagrán	Villagrán
Xicoténcatl	Xicoténcatl

TLAXCALA

Capital:	Tlaxcala de Xichténcatl	*State abbreviation:*	Tlax
Population:	547,261	*Area:*	4,016 sq km

County	County Seat
Amaxac de Guerrero	Amaxac de Guerrero
Apetatitlán	Apetatitlán
Atlangatepec	Atlangatepec
Atlzayanca	Atlzayanca
Barrón y Escandón	Apizaco
Calpulalpan	Calpulalpan
Carmen, El	Tequixquitla
Cuapiaxtla	Cuapiaxtla
Cuaxomulco	Cuaxomulco
Chiautempan	Chiautempan
Domingo Arenas	Munoz
Españita	Españita
Huamantla	Huamantla
Hueyotlipan	Heuyotlipan
Ixtacuixtla	Villa Mariano Matamoros
Ixtenco	Ixtenco
José María Morelos	Mazatecochco

Juan Cuamatzi	Contla
Lardizábal	Tepetitla
Lázaro Cárdenas	Sanctorum
Mariano Arista	Nanacamilpa
Miguel Hidalgo	Acuamanala
Natívitas	Natívitas
Panotla	Panotla
San Pablo del Monte	Villa Vicente Guerrero
Santa Cruz Tlaxcala	Santa Cruz Tlaxcala
Tenancingo	Tenancingo
Teolocholco	Teolocholco
Tepeyanco	Tepeyanco
Terrenate	Terrenate
Tetla	Tetla
Tetlatlahuca	Tetlatlahuca
Tlaxcala	Tlaxcala de Xicohtencatl
Tlaxco	Tlaxco
Tocatlán	Tocatlán
Totolac	Totolac
Trinidad Sánchez Santos	Zitlaltépetl
Tzompantepec	Tzompantepec
Xalostoc	Xalostoc
Xaltocan	Xaltocan
Xicohténcatl	Papalotla
Xicohzingo	Xicohzingo
Yauhquemehcan	Yauhquemehcan
Zacatelco	Zacatelco

VERACRUZ-LLAVE

Capital:	Jalapa Enriquez	*State abbreviation:*	Ver
Population:	5,264,611	*Area:*	71,699 sq km

County	**County Seat**
Acajete	Acajete
Acatlán	Acatlán
Acayucan	Acayucan
Actopan	Actopan
Acula	Acula
Acultzingo	Acultzingo
Adalberto Tejeda	Villa Tejeda
Alpatlahua	Alpatlahua
Alto Lucero	Alto Lucero
Altotonga	Altotonga
Alvarado	Heroica Alvarado
Amatlán	Naranjos
Amatlán de los Reyes	Amatlán de los Reyes
Angel R. Cabada	Angel R. Cabada
Antigua, La	Jose Cardel
Apazapan	Apazapan
Aquila	Aquila
Astacinga	Astacinga
Atlahuilco	Atlahuilco
Atoyac	Atoyac

Atzacan	Atzacan
Atzalan	Atzalan
Axocuapan	Tlaltetela
Ayahualulco	Ayahualulco
Banderilla	Banderilla
Benito Juárez	Benito Juárez
Boca del Río	Boca del Río
Calcahualco	Calcahualco
Camerino Z. Mendoza	Ciudad Mendoza
Carrillo Puerto	El Tamarindo
Catemaco	Catemaco
Cazones	Villa de Cazones de Herrera
Cerro Azul	Cerro Azul
Citlaltepec	Citlaltepec
Coacoatzintla	Coacoatzintla
Coahuitlán	Progreso de Zaragoza
Coatepec	Coatepec
Coatzacoalcos	Coatzacoalcos
Coatzintla	Coatzintla
Coetzala	Coetzala
Colipa	Colipa
Comapa	Comapa
Córdoba	Heroica Córdoba
Cosautlán de Carvajal	Cosautlán de Carvajal
Coscomatepec	Coscomatepec de Bravo
Cosoleacaque	Cosoleacaque
Cotaxtla	Cotaxtla
Coxquihui	Coxquihui
Coyutla	Coyutla
Cuichapa	Cuichapa
Cuitláhuac	Cuitláhuac
Chacaltianguis	Chacaltianguis
Chalma	Chalma
Chiconamel	Chiconamel
Chiconquiaco	Chiconquiaco
Chicontepec	Chicontepec de Tejeda
Chinameca	Chinameca
Chinampa de Gorostiza	Chinampa de Gorostiza
Choapas, Las	Las Choapas
Chocamán	Chocamán
Chontla	Chontla
Chumatlán	Chumatlán
Emiliano Zapata	Dos Ríos
Espinal	Espinal
Filomeno Mata	Filomeno Mata
Fortín	Fortín de las Flores
Gutiérrez Zamora	Gutiérrez Zamora
Hidalgotitlán	Hidalgotitlán
Huatusco	Huatusco de Chicuellar
Huayacocotla	Huayacocotla
Hueyapan de Ocampo	Hueyapan de Ocampo
Huiloapan	Huiloapan de Cuauhtémoc
Ignacio de la Llave	Ignacio de la Llave
Ilmatlán	Ilmatlán
Isla	Isla
Ixcatepec	Ixcatepec
Ixhuacán	Ixhuacán de los Reyes

Ixhuatlán	Ixhualtán del Café
Ixhuatlancillo	Ixhuatlancillo
Ixhustlán del Sureste	Nanchital
Ixhuatlán de Madero	Ixhuatlán
Ixmatlahuacan	Ixmatlahuacan
Ixtaczoquitlán	Ixtaczoquitlán
Jalacingo	Jalacingo
Jalapa	Jalapa-Enríquez
Jalcomulco	Jalcomulco
Jáltipan de Morelos	Jáltipan de Morelos
Jamapa	Jamapa
Jesús Carranza	Jesús Carranza
Jico	Jico
Jilotepec	Jilotepec
Juan Rodríquez Clara	Juan Rodríquez Clara
Juchique de Ferrer	Juchique de Ferrer
Landero y Cos	Landero y Cos
Lerdo de Trejada	Landero de Tejada
Magdalena	Magdalena
Maltrata	Maltrata
Manlio Fabio Altamirano	Manlio Fabio Altamirano
Mariano Escobedo	Mariano Escobedo
Martínez de la Torre	Martínez de la Torre
Mecatlán	Mecatlán
Mecayapan	Mecayapan
Medellín	Medellín de Bravo
Miahuatlán	Miahuatlán
Minas, Las	Las Minas
Minatitlán	Minatitlán
Misantla	Miscantla
Mixtla de Altamirano	Mixtla de Altamirano
Moloacán	Moloacán
Naolinco	Naolinco de Victoria
Naranjal	Naranjal
Nautla	Nautla
Nogales	Nogales
Oluta	Oluta
Omealca	Omealca
Orizaba	Orizaba
Otatitlán	Otatitlán
Oteapan	Oteapan
Ozuluama	Ozuluama
Pajapan	Pajapan
Pánuco	Pánuco
Papantla	Papantla de Olarte
Paso del Macho	Paso del Macho
Paso de Ovejas	Paso de Ovejas
Perla, La	La Perla
Perote	Perote
Platón Sánchez	Platón Sánchez
Playa Vicente	Playa Vicente
Poza Rica de Hidalgo	Poza Rica de Hidalgo
Profesor Rafael Ramírez	Profesor Rafael Ramírez
Pueblo Viejo	Villa Cuauhtémoc
Puente Nacional	Puente Nacional
Rafael Delgado	Rafael Delgado
Rafael Lucio	Rafael Lucio
Reyes	Reyes

Río Blanco	Tenango de Río Blanco
Saltabarranca	Saltabarranca
San Andrés Tenejapa	San Andés Tenejapa
San Andrés Tuxtla	San Andrés Tuxtla
San Juan Evangelista	San Juan Evangelista
Santiago Tuxtla	Santiago Tuxtla
Sayula	Sayula de Alemán
Soconusco	Soconusco
Sochiapa	Sochiapa
Soledad Atzompa	Soledad Atzompa
Soledad de Doblado	Soledad de Doblado
Soteapan	Soteapan
Tamalín	Tamalín
Tamiahua	Tamiahua
Tampico Alto	Tampico Alto
Tancoco	Tencoco
Tantima	Tantima
Tantoyuca	Tantoyuca
Tatatila	Tatatila
Teayo	Castillo de Teayo
Tecolutla	Tecolutla
Tehuipango	Tehuipango
Temapache	Alamo
Tempoal	Tempoal de Sánchez
Tanampa	Tenampa
Tenochtitlan	Tenochtitlan
Teocelo	Teocelo
Tepatlaxco	Tepatlaxco
Tepetlán	Tepetlán
Tepetzintla	Tepetzintla
Tequila	Tequila
Tesechoacán	Villa Azueta
Texhuacán	Texhuacán
Texistepec	Texistepec
Tezonapa	Tezonapa
Tierra Blanca	Tierra Blanca
Tihuatlán	Tihuatlán
Tlacojalpan	Tlacojalpan
Tlacolulan	Tlacolulan
Tlacotalpan	Tlacotalpan
Tlacotepec	Tlacotepec de Mejía
Tlachichilco	Tlachichilco
Tlalixocoyan	Tlalixcoyan
Tlalnelhuayocan	Tlalnelhuayocan
Tlapacoyan	Heroica Tlapacoyan
Tlaquilpa	Tlaquilpa
Tlilapan	Tlilapan
Tomatlán	Tomatlán
Tonayán	Tonayán
Totutla	Totutla
Tuxpan	Tuxpan de Rodríguez Cano
Tuxtilla	Tuxtilla
Ursulo Galván	Ursulo Galván
Vega de Alatorre	Vega de Alatorre
Veracruz	Heroica Veracruz
Villa Aldama	Villa Aldama
Xoxocotla	Xoxocotla
Yanga	Yanga

Yecuatla	Yecuatla
Zacualpan	Zacualpan
Zaragoza	Zaragoza
Zentla	Colonia Manuel González
Zongolica	Zongolica
Zontecomatlán	Zontecomatlán
Zozocolco de Hidalgo	Zozocolco de Hidalgo

YUCATAN

Capital: Mérida
Population: 1,034,648

State abbreviation: Yuc
Area: 38,402 sq km

County	County Seat
Abalá	Abalá
Acanceh	Acanceh
Akil	Akil
Baca	Baca
Bokobá	Bokobá
Buctzotz	Buctzotz
Cacalchén	Cacalchén
Calotmul	Calotmul
Cansahcab	Cansahcab
Cantamayec	Cantamayec
Celstún	Celstún
Cenotillo	Cenotillo
Conkal	Conkal
Cuncunul	Cuncunul
Cuzamá	Cuzamá
Chaczinkín	Chaczinkín
Chan-kom	Chan-kom
Chapab	Chapab
Chemax	Chemax
Chicxulub	Chicxulub
Chichimilá	Chichimilá
Chikindzonot	Chikindzonot
Chocholá	Chocholá
Chumayel	Chumayel
Dzan	Dzan
Dzemul	Dzemul
Dzidzantún	Dzidzantún
Dzilam de Bravo	Dzilam de Bravo
Dzilam González	Dzilam González
Dzitás	Dzitás
Dzoncauich	Dzoncauich
Espita	Espita
Halachó	Halachó
Hocabá	Hocabá
Hoctún	Hoctún
Homún	Homún
Huhí	Huhí
Hunucmá	Hunucmá
Ixil	Ixil
Izamal	Izamal
Kanasín	Kanasín
Kantunil	Kantunil
Káua	Káua
Kinchil	Kinchil

Kopomá	Kopomá
Mama	Mama
Maní	Maní
Maxcanú	Maxcanú
Mayapán	Mayapán
Mérida	Mérida
Mocochá	Mocochá
Motul	Motul de Felipe Carrillo Puerto
Muna	Muna
Muxupip	Muxupip
Opichén	Opichén
Oxkutzcab	Oxkutzcab
Panabá	Panabá
Peto	Peto
Progreso	Progreso
Quintana Roo	Quintana Roo
Río Lagartos	Río Lagartos
Sacalum	Sacalum
Samahil	Samahil
Sanahcat	Sanahcat
San Felipe	San Felipe
Santa Elena	Santa Elena
Seyé	Seyé
Sinanché	Sinanché
Sotuta	Sotuta
Sucilá	Sucilá
Sudzal	Sudzal
Suma	Suma
Tahdziú	Tahdziú
Tahmek	Tahmek
Teabo	Teabo
Tecoh	Tecoh
Tekal de Venegas	Tekal de Venegas
Tekantó	Tekantó
Tekax	Tekax de Alvaro Obregón
Tekit	Tekit
Tekom	Tekom
Telchac Pueblo	Telchac
Telchac Puerto	Telchac Puerto
Temax	Temax
Temozón	Temozón
Tepakán	Tepakán
Tetiz	Tetiz
Teya	Teya
Ticul	Ticul
Timucuy	Timucuy
Tinum	Tinum
Tixcacalcupul	Tixcacalcupul
Tixkokob	Tixkokob
Tixméuac	Tixméuac
Tixpéual	Tixpéual
Tizimín	Tizimín
Tunkás	Tunkás
Tzucacab	Tzucacab
Uayma	Uayma
Ucú	Ucú
Umán	Umán
Valladolid	Valladolid
Xocchel	Xocchel

Yaxcabá
Yaxkukul
Yobain

Yaxcabá
Yaxkukul
Yobain

ZACATECAS

Capital: Zacatecas
Population: 1,145,327

State abbreviation: Zac
Area: 73,252 sq km

County	**County Seat**
Apozol	Apozol
Apulco	San Pedro Apulco
Atolinga	Atolinga
Benito Juárez	Benito Juárez
Calera	Víctor Rosales
Cañitas de Felipe Pescador	Cañitas de Felipe Pescador
Concepción del Oro	Concepción del Oro
Cuauhtémoc	San Pedro Piedra Gorda
Chalchihuites	Chalchihuites
Fresnillo	Fresnillo de González Echeverría
García de la Cadena	García de la Cadena
Genaro Codina	Genaro Codina
General Enrique Estrada	General Enrique Estrada
General Francisco Murguía	Nieves
Gereral Joaquín Amaro	General Joaquín Amaro
General Pánfilo Natera	General Pánfilo Natera
Guadalupe	Guadalupe
Huanusco	Huanusco
Jalpa	Jalpa
Jerez	Jarez de García Salinas
Jiménez de Téul	Jiménez de Téul
Juan Aldama	Juan Aldama
Juchipila	Juchipila
Loreto	Loreto
Luis Moya	Luis Moya
Mazapil	Mazapil
Melchor Ocampo	Melchor Ocampo
Mezquital del Oro	Mezquital del Oro
Miguel Auza	Miguel Auza
Momax	Momax
Monte Escobedo	Monte Escobedo
Morelos	Morelos
Moyahua de Estrada	Moyahua
Nochistlán de Mejía	Nochistlán
Noria de Angeles	Noria de Angeles
Ojo Caliente	Ojo Caliente
Pánuco	Pánuco
Pinos	Pinos
Río Grande	Ciudad de Río Grande
Saín Alto	Saín Alto
Salvador, El	El Salvador
Sombrerete	Sombrerete
Susticacán	Susticacán
Tabasco	Tabasco
Tepechitlan	Tepechitlan
Tepetongo	Tepetongo
Téul de González Ortega	Téul de González Ortega

Tlaltenango de Sánchez Román	Tlaltenango de Sánchez Román
Valparaíso	Valparaíso
Vetagrande	Vetagrande
Villa de Cos	Villa de Cos
Villa García	Villa García
Villa González Ortega	Villa González Ortega
Villa Hidalgo	Villa Hidalgo
Villanueva	Villanueva
Zacatecas	Zacatecas

LIST OF ENTRIES

Note: An asterisk indicates minerals that were first discovered in México.